THE CITY

MEANING IN THE URBAN ENVIRONMENT

SOCIOLOGY OF THE CITY

MEANING IN THE URBAN ENVIRONMENT

M. KRAMPEN

Routledge
Taylor & Francis Group
LONDON AND NEW YORK

First published in 1979

This edition published in 2007
Routledge
2 Park Square, Milton Park, Abingdon, Oxon, OX14 4RN
605 Third Avenue, New York, NY 10017

Routledge is an imprint of Taylor & Francis Group, an informa business

First issued in paperback 2011

British Library Cataloguing in Publication Data
A CIP catalogue record for this book
is available from the British Library

Meaning in the Urban Environment

ISBN13: 978-0-415-41832-4 (hbk)
ISBN13: 978-0-415-51264-0 (pbk)
ISBN13: 978-0-415-41931-4 (subset)
ISBN13: 978-0-415-41318-3 (set)

Routledge Library Editions: The City

meaning in the urban environment

M Krampen

Research in Planning and Design

Series editor Allen J Scott

1 **Place and placelessness** E Relph
2 **Environmentalism** T O'Riordan
3 **The new urban economics** H W Richardson
4 **The automated architect** N Cross
5 **Meaning in the urban environment** M Krampen
6 **Feedback from tomorrow** A J Dakin
7 **Birds in egg/Eggs in bird** G Olsson

Pion Limited, 207 Brondesbury Park, London NW2 5JN

meaning in the urban environment

M Krampen

Pion Limited, 207 Brondesbury Park, London NW2 5JN

ISBN 0 85086 067 9

Printed in Great Britain

Preface

This book is concerned with some psychological effects of urban environments as wholes and architectural objects as their parts, studied from the point of view of semiotics or (in keeping with European tradition) of semiology. It contains an attempt to clarify the way in which architectural objects are processed cognitively, if not as signs of communication, at least as meaningful instruments, in order for the urban environment to become viable for human beings.

That there was a need for this book to be written stemmed from the fact that no monograph or coherent text on architectural or urban semiotics existed, with the exception of Gamberini's and Koenig's pioneering efforts in the fifties and sixties, unfortunately written only in Italian. When I started writing, there were many scattered articles, some chapters in books concerned basically with other issues, some conference reports, and one reader on this subject matter. Apart from Scalvini's 1975 monograph, likewise written in Italian, this situation has changed only this year, when two important books on the topic have appeared—both by Donald Preziosi: *The Semiotics of the Built Environment* (Indiana University Press) and *Architecture, Language and Meaning* (Mouton). Those interested in the problem of meaning in the urban environment will find that despite some differences in detail, there is an amazing amount of agreement between this book and these two by Preziosi, although they have been written completely independently of each other and in different continents—a circumstance which I find encouraging with respect to semiotics as a science. One common factor of the three books is, for instance, a front against the facile use of linguistic analogy in constructing a semiotics of the built environment.

There are, however, some issues in the present book, which are not within the scope of the tasks Preziosi set himself. For example, this book presents the first major attempt at a critical summary of the different approaches to, and of attempts at a theory of, the semiotics of objects generally and in architectural and urban semiotics in particular. These summaries have been developed from two reports given to the plenary session of the first Congress of the International Association of Semiotic Studies in Milan, 1974 (published this year in *A Semiotic Landscape*, Mouton). Another feature of this book is that it proceeds not only by theoretical analysis but also records experimental evidence to back up its theoretical claims. The experiments described have all been conducted especially for the research presented here, and the book can thus be considered to have been written in the true tradition of Psychological Monographs. In carrying out the research, the hypothetico-deductive method was generally followed (albeit based on a consideration of the historical background of the problem as a whole and of its various details). The monograph character of the book meant that careful consideration of methodological questions had to be included, but this does not mean that only specialists have been addressed. On the contrary,

I dare to hope that the experimental methods presented will be of benefit to students and practitioners alike in the fields of semiotics, architecture, urban design, environmental psychology, and aesthetics in the continuation and development of their own experiments. Certainly the text is directed at all such students and practitioners, although as an introduction it is not considered to require any particular skill but is intended to provide a stimulation for those who are curious to find out how meaning may be injected into or extracted from today's urban environment.

When Johann Sebastian Bach wrote "Soli Deo Gloria" above his cantatas, he probably meant that he had written the composition alone as an everyday job. I was tempted to use some similar motto for this book, since there were no grants and little institutional or personal help to finish it—on the contrary, there were obstacles of many kinds.

Consequently, where credits are due I have as a rule incorporated them into the text. However, apart from acknowledging occasional discussions with Geoffrey Broadbent, David Canter, Charles Jencks, and Luis Prieto (whose ideas I present from my own point of view and must assume the whole responsibility), and a continuous dialogue on semiotics with Thure von Uexküll and Hermann Kalkofen, I am very grateful to Donald Appleyard and Allen Scott for having worked critically through the first draft of the book. I am also indebted to Peter Fassheber, who tolerated me as a colleague during my appointment at the University of Goettingen Institute of Social Psychology, which he chairs, in a phase when I was trying to finish the final draft. Anne McClintock and Tim Goodwin corrected my English, and Johannes Eisenbarth prepared the graphics for the book.

Martin Krampen
Ulm-Donau, July 23rd, 1979

Contents

Part 1

Current ideas and theories about meaning in the urban environment

1

Introduction

1.1 Meaning in the urban environment—communication or signification?

This book presupposes, without specific discussion, that questions concerning meaning in the urban environment are of both theoretical and practical relevance. Such questions are theoretically relevant since the solution of practical problems will be made much easier if it is founded on a body of theoretical knowledge which has previously undergone empirical examination. The practical value of such questions lies in their relation to the physical and psychological effects of urban conditions on the human being.

The book is thus concerned less with the legitimacy of the search for meaning in the urban environment than with the potential mechanics by which meaning is injected into the urban environment and extracted from it. Since these mechanics are intrinsically connected to human activity, the science of their functioning cannot be simply a 'philosophy'. It must be a 'human science'. The human science by which meaning is studied was called 'semiology' by de Saussure (1971), who considered it a branch of social psychology. The research in this book should therefore be considered 'semiological' in this Saussurean sense.

From its beginning the task of semiology was the investigation of the mechanics of signs operating in human communication. It was no accident that for a long time linguistic signs played a dominant role in semiological studies. Later the investigation of meaning was generalized to include phenomena transcending the process of intentional communication. An animal may leave traces, a criminal fingerprints: these 'indices' carry meaning without being conceived as signs intended for communication. In the same way sultry weather can be seen as conveying information to the experienced organism (either human or animal) about an imminent storm. Semiology must thus account for meaning both in communication and beyond communication.

In applying this principle to the urban environment, the fundamental questions are the following. What kind of meaning is connected to the city by what kind of mechanisms? Is the city as a whole a compound of meaningful 'messages' transmitted more or less intentionally by 'senders'? If this is not the case, what other kinds of meaning can be perceived in the urban environment? If there is any meaning other than communicative, how does it relate to architectural objects in the city? What conditions must be fulfilled to enable one to attribute communicative functions to objects in general and to architectural objects in particular?

The mechanics of the transmission of meaning by processes of communication seem at first sight straightforward. This impression is justifiable as the functions of communication—at least those of verbal communication—have been sufficiently analyzed.

According to Prieto (1975b) the functioning of the communicative act can be defined most simply in the following way. In verbal communication a social rapport is established between an active and a passive partner. The active partner produces a vocal signal and furnishes by means of it an indication of the nature of this social rapport. The passive partner then 'interprets' this index. The social rapport itself is the 'sense' of the communicative act. The 'sense' is contained in the attempt of the sender to influence the receiver. Prieto considers two types of sense: either the sender tries to influence the receiver towards some action, in which case the influence is called an injunction, or the sender tries to influence the receiver by letting him know something, in which case the influence is called information. This distinction between 'injunction' and 'information' is, however, by no means novel, having already been made by Skinner (1957) in his behaviouristic analysis of language. Skinner called the injunctive sense 'mand': "The term 'mand' has a certain mnemonic value derived from 'command', 'demand', 'countermand' ...", and the informative sense 'tact': "The term carries a mnemonic suggestion of behaviour which 'makes contact with' the physical world ...".

One of the questions of this book is to what extent indices functioning as verbal signals carry meaning in the urban environment. One is of course immediately reminded of the enormous number of verbal messages imposed on the citizen by advertising, neon lights, and other parts of the 'verbal crust' of the buildings in a city.

Vocal or graphic signals in verbal behaviour constitute those indices which are at the centre of linguistic preoccupation with social rapport. They indicate to the listener or reader the nature of the influences that speakers or writers are trying to exert. But vocal or graphic signals are not the only kinds of indices in communication, the operation of indices being by no means confined to the domain of linguistics. Indices function in every act of communication, verbal or nonverbal, provided that they function as signals. Any signal may be an index of a 'sense', that is, of the influence intended by the sender to be exerted on the receiver. The study of indices must therefore be extended beyond the field of linguistics and be dealt with by a larger discipline called the semiology of communication (Prieto, 1975a), linguistics becoming merely a branch of that larger discipline.

As far as meaning in the urban environment is concerned, this entails the study of indices functioning as nonverbal signals. Pictorial advertising, traffic signs, and other nonverbal communicative devices clearly function as means of influencing people to do something, or letting them know something.

But indices are not to be seen as confined simply to the semiology of communication. Indices are operant not only in human verbal and communicative behaviour, but also in the whole sphere of human behaviour in general. By definition indices are facts which are capable of conveying information over and above the initial information conveyed by the facts themselves. For any behavioural element to be termed an index, it must fulfill more than its immediate function. In order for it to do so, a connection must be established between the basic function of the behaviour and an additional function. This connection is established by society. The connection established by society between behaviour functioning as an 'indicator' and an additional function indicated by that behaviour is called 'signification'. A particularized instance of this, where the behaviour entails signalling (a connection being likewise established by society between the signal and the 'sense' of the signal) is termed 'communication'. Consequently, the study of indices must be extended beyond the disciplines of linguistics and the semiology of communication and be incorporated into a still broader discipline called the 'semiology of signification' (Prieto, 1975a). If it is true that the city has meaning other than that conveyed by advertising or traffic signs, then 'signification' could well turn out to be the central issue in urban semiology.

A behavioural operation becomes an index only when contrasted with alternative operations fulfilling the same function. It is the selection by a society of a particular behaviour in preference to alternative behavioural modes which creates the index. Food, for example, may be eaten with one's hands or with eating utensils, the societal choice of one of these alternatives coming to bear a significance over and above the primary function of eating. A behaviour must therefore be 'stylized' in opposition to alternative behavioural modes in order to become an index. Out of a set of alternative behavioural modes, a society chooses a particular behaviour, which then becomes 'ritualized'.

Most of these behavioural 'rituals' involve the use of different tools for accomplishing the operations involved: different types of dress, eating utensils, seats, etc. These tools are a means of extending the physical behaviour. Certain alternative tools for executing the same operation may in turn be selected to fulfill functions beyond their basic ones. Such functions often have to do with the reinforcement of social distinctions. A suit and a dress both fulfill the same primary function of clothing the body. But by selecting one or the other, the operation of protection is 'transfunctionalized' into an operation of sex distinction.

The principle of 'transfunctionalization' of behavioural choice applies to the use and production of signals, as well as to the use and production of tools. Signals constitute a subcategory of the general category of tools, and serve the function of communication. The intentional selection or conception of alternative tools or signals for a given operation is the beginning of purposeful 'stylization'. As far as meaning in the urban environment is concerned this raises the question whether the city itself can be considered in terms of tools.

The moment a problemsolver conceives or chooses a particular tool in order to let others know about his particular 'style', the problem ceases to be one of the semiology of signification, but again becomes one of the semiology of communication. This seems to be the case in the 'Arts'. In some of the 'Arts' a particular method of selecting signals is used over and above the primary function of transmitting information (communicating information by narrative or visual devices). The selection of particular signals is used to communicate information about a particular way of communicating. The narrative or the visual art becomes a pretext for exhibiting a specific narrative or pictorial style. In other 'Arts' a particular way of selecting tools or objects is used over and above their primary utilitarian function (housing people in the case of a house or eating from it in the case of a plate). It is used to communicate information about a particular manner of fulfilling some practical goal. For the purpose of this book the question of artistic communication in the urban environment thus arises.

The function of indices has thus been examined by beginning with their role as verbal signals in linguistics, then broadening the scope of signals to cover the whole field of communication, and finally presenting indices in their most general form as conveyors of signification. The return from signification to communication was accomplished by considering the artistic transformation of tools into signals. This excursion was made to map out the context of research on meaning in the urban environment. Is it now possible to establish an urban semiology in the light of what has been said about behavioural operations and rituals in the semiology of signification, and about the 'Arts' in the semiology of communication? If it is possible to establish relationships of signification on the basis of behavioural operations involving the use of particular tools, the whole realm of human artefacts becomes involved in relationships of signification. Many of these ritualized uses of artefacts may be consciously or unconsciously 'transfunctionalized' to serve purposes other than their primary ones, that is, the real purpose of a painting may be the exhibition of its style rather than its literal content. If the whole realm of human artefacts is involved in relationships of signification, then the urban environment as a collection of artefacts supporting different behavioural rituals must be equally involved in such relationships.

To establish the existence of a relationship of signification between an urban environment and a set of 'indicated' facts, beyond the practical functioning of that environment, it has to be shown that out of alternative combinations of conceptions of urban environments a particular combination has been realized. Otherwise it would not have been possible to select an alternative which would contrast with others as a 'special' choice, thus enabling the connection between itself and a 'special' function to be made.

A semiological analysis of the urban environment may have a dual approach. On the one hand the analysis may concentrate on a specific urban environment together with its artefacts. In this case what will be investigated is the presence of signification of meaning in an existing urban environment. On the other hand the analysis may deal with different conceptions of urban design and planning as perceived specifically by architects and urban designers. In this case the process by which the urban environment acquires meaning or signification is the focus of analysis.

These two different starting points imply the investigation of two different points of view with regard to the city. In the one case, one studies the processes by which citizens classify the city within a system of possible urban environments of which they have knowledge. In the other case, one studies the urban environment as a projection of the conceptual schemata of its founders, planners, and designers. In the latter, one studies the process of choice made by urban designers between sets of solutions based on their specialized knowledge of systems of classification. The question whether citizens should participate in the planning of their environment is itself a question of style, that is, whether citizens are allowed to participate in urban planning is an indication of how and for whom urban planning is conceived in a given society. The different viewpoints of citizen and designer, rather than mutually excluding one another, should in fact be complementary. However, since a beginning has to be made, the point of view largely adopted in this book is that of the citizen.

2

Review of the literature

In order that the research presented in this book be placed in a broader context, it is first necessary to review current thought in those fields of semiology which are directly relevant to the problem at hand. There are two branches in the study of semiology which are closely connected to the question of meaning in the urban environment.

One is the semiology of objects, the other the semiology of architecture and the city. As has already been suggested, the urban environment may be considered simply a collection of objects or tools; for example, 'housing' may be seen as a system of tools for dwelling. If the city is seen thus as a collection of objects, then research into meaning in the urban environment entails scientific study into the process whereby objects acquire meaning.

The semiology of architecture and urban semiology are likewise fields of considerable importance, since both are directly concerned with the problem at hand.

Since the 1960s an interesting body of literature has accumulated which deals with the semiology of objects in general, with architectural objects in particular, and with urban semiology. As will become clear below, this literature separates into two major areas of concern: the definition of the meaning of objects, and the establishment of object semiology in relation to linguistics. The review of this literature constitutes the background against which the empirical studies that make up the core of this research are to be judged.

2.1 Work on the semiology of objects in general

2.1.1 The reduction of objects to signs

Although a great deal of specialized work has been published on architectural and urban semiology in the form of articles, publications on the more general theme of object semiology have more often taken the form of comprehensive books. Object semiology has been dealt with in depth since the 1960s, touched off largely by the seminal works of Lévi-Strauss (1958) and Barthes (1964). Barthes (1964, page 106) postulated the law of 'universal semantization of usage' whereby the use of an object is converted into a sign for that use. From the beginning of human society objects have acquired, in addition to their primary function, the secondary function of meaning. A raincoat is a protection against rain, but the use of the raincoat is directly connected to specific atmospheric conditions. It therefore becomes a sign for those conditions determining its use—a 'function sign' as Barthes calls it. This theory of the 'universal semantization of usage' was developed to its extreme by Baudrillard. In an early book *Le Système des Objets* (Baudrillard, 1968) he contended that objects lose their material and functional status by their integration into object systems. Within these systems they are simply signs analogous to the elements of a code which are themselves nothing but the building blocks of larger messages. According to Baudrillard this change in the status of the object is brought about by the particular nature of life in our 'consumer society'. Consuming—according to Baudrillard—is not a material practice, but the organization of material substance into signifying substance. "To become an object of consumption the object must become a sign" (Baudrillard, 1968, page 277).

In his later *Pour une Critique de l'Économie Politique du Signe* (1972) Baudrillard pushed the semantization of objects to an extreme. Objects do not only have use value and exchange value, but also 'sign value'. According to Baudrillard this additional value completes Marx's politico-economic analysis in which this aspect of value theory "was overlooked". Consumption of objects by a society does not simply involve their use or exchange. It involves 'conspicuous consumption' in Veblen's (1963) sense—the continuous 'potlatch' ceremony in which object signs of prestige are exchanged. In the study of the mechanics whereby objects acquire meaning, Baudrillard takes an extreme position: objects have no material existence of their own, but exist only through the symbolic activities of society. This point of view, when applied to the urban environment as a set of artefacts, overemphasizes the symbolic aspects of urban activity at the expense of the material aspects. In practice this would lead to neglect of the utilitarian functions of the city in favour of monumentalist and aestheticist preoccupations.

2.1.2 Objects and their use value

A counterpoint to this 'radical' point of view may be found in the technically oriented and sober *Théorie des Objets* of Moles (1972). Moles treats the object as a mediator between man and his material situation and between man and society. For Moles the object is related to communication, but is not in itself a message. Messages are associated with it. Unlike Baudrillard, Moles sees function as the meaning ('sens') of objects. Objects may be mapped into a semantic space according to distinctions in their functional meanings. A cup, for example, would belong in such a space in the same category as knives, forks, plates, and saucers. This group would be a certain distance from a group containing chairs, benches, taborets, etc. Like language, objects possess their internal syntax—the syntax according to which their parts and subfunctions of these parts are arranged.

By means of pertinent oppositions (not unlike those in a commutation test) different object styles may be objectively differentiated. Thus the presence of an old-fashioned rocking chair in a modern room gives the room a special stylistic meaning. Although Moles's book is a disciplined body of information and a methodological gold mine, it could be accused of failing to make a critical evaluation of the politico-economic problems of the dialectics of use value and exchange value. It gives the impression that the only value of objects is their use value. Their exchange value—

and with it the problems of production and appropriation of 'surplus value'—goes unmentioned. Applied to the urban environment this would amount to a concealment of the politico-economic conditions to which the artefacts of the city are subject. The meaning of the artefacts cannot be dissociated from these conditions.

2.1.3 The pragmatic meaning of objects

The idea of 'semantization of usage' was advanced long before Barthes by J von Uexküll (1940; 1957; von Uexküll and Kriszat, 1934; 1970). This German biologist first coined one of today's scientific key terms—the word 'Umwelt' (environment). For von Uexküll every organism is connected to a particular environment by the particular configuration of its sense organs. In this context he went on to account for meaning in the human environment (cf Sebeok, 1977). Reflecting on the processes whereby objects acquire meaning, he wondered what it is that enables one to place correctly in the category of chairs a variety of three-legged and four-legged objects made of different materials and variously coloured, styled, and constructed. He concluded that in order to know what a chair is one must have sat on one, that is, one must have used one.

To illustrate his point von Uexküll tells of an African tribesman he saw once staring uncomprehendingly at a ladder. To the tribesman the ladder was something built of pieces of wood with spaces between them. Only when he was shown how to climb it, could the pieces of wood and spaces together become 'something you climb up with'—a ladder (von Uexküll, 1935).

The moment one can categorize a chair as 'something you sit on', a cup as 'something you drink with', one has recognized the 'sense' of the object. According to von Uexküll this 'sense' is neither material nor of the nature of energy, but he calls it the essence ('Wesen') of the object.

Objects can be divided into two classes according to their 'sense'. To the one class belong those things you work with (Werkzeug): chairs, ladders, locomotives, airplanes, etc, and to the other those things you perceive with (Merkzeug): glasses, telescopes, telephones, loudspeakers, etc. In order for these objects to have a special 'sense' their material parts must be coordinated in such a way that they are capable of common action. This coordination, or internal 'plan', encompasses space and time and, while not visible as such, could perhaps be made visible graphically. The relationship between plan and sense is exemplified by the mechanics of a musical clock. The 'sense' of the clock is the melody it should make audible. The musical clock consists of a cylinder which, by means of small pegs, puts into vibration tiny metal tongues which in turn produce sound waves which strike the ears in a specific manner which is interpreted by the brain as melody. The musical clock is structured round the idea of a special melody, according to a plan which will exhibit the sense of the object.

T von Uexküll (1973) continued the work of his father by ordering objects into what he calls pragmatic systems. Sense does not appear except in pragmatic situations, and the sense of a specific object will therefore change from situation to situation. What J von Uexküll called 'sense' became in T von Uexküll's terminology the 'answer' of the object to a 'meaning hypothesis' which the potential user had advanced. This answer contains information about 'directions for use'. The 'dialogue' between object and user is part of a system of interacting pragmatic programmes in which men and objects function together. The pragmatic system of tea drinking, for example, is made up of several subprogrammes. The cycle is triggered off by a need for liquid transmitted by the dry tissue of the body. The drinker responds to the optical sign of the handle of the cup by gripping it. The tactile sign resulting from this contact evokes the lifting of the cup to the mouth. At the mouth a tactile sign is registered by the lips, to which the mouth responds by opening. At this point the role of the cup in the system is over. The last programmes are reserved for the tongue, throat, and stomach in the process of swallowing. In the pragmatic system of tea drinking the cup means 'something to drink with'. In the pragmatic system of commerce it means 'something to sell' or 'an object of value'. In the pragmatic system of aesthetics it becomes an 'object of art', etc. In each of these pragmatic systems men participate with other 'actors' in scenes which are preprogrammed. The meaning changes from one system to the other. As meaning does not exist outside pragmatic systems, objects which serve no purpose have no meaning.

The idea of pragmatic meaning propounded by T von Uexküll must be understood in relation to Morris's (1955) tripartition of semiotics into syntactics, the study of signs in relation to themselves; semantics, the study of signs in relation to the objects they stand for; and pragmatics, the study of signs in relation to their users. It will be shown in greater detail below that these three relationships stand in a further logical relationship of successive inclusions in each other: pragmatics including semantics, and semantics including syntactics. Morris's semiotics then becomes a closed system, although the ideological content of the pragmatic relationships is not explicitly spelled out. It must, however, be interpreted in terms of pragmatic philosophy and ethics as basically utilitarian and relativistic. The principle of sign use is its utility, which may change from situation to situation. This implies that signs do not indicate 'true' or 'objective' relationships. The application of this idea to the urban environment would lead to interpreting it as only having abstract utilitarian value. It is true that in connection with semiology the question of the sign user's purpose has to be answered. But this is an extrasemiological problem, since semiology deals only with the functioning of signs.

2.1.4 Objects and design

There are some theories about objects which are particularly design-oriented. In some of these (Menna, 1967; 1971) the current desemantization of objects is criticized. This loss of meaning is said to be a result of the uniformity and standardization of object design due to industrial production. The more uniform the appearance of objects with the same function becomes, the less variety there will be in the world of objects. But uniformity is now even creeping into the design of objects with different functions. In urban design 'modern' buildings with different functions are being constructed according to a uniform 'international' style. Buildings all over the world are beginning to resemble each other. According to Menna, design has to restore diversity and richness of meaning to the urban environment.

A similarly design-oriented theory is that of Bense (Walther, 1974). It is based on the semiotics of Peirce (1960), who is widely regarded as the founder of that science. Bense divides all object into four categories—natural objects, technical objects, objects of art, and design objects—by assigning to them different degrees of functional determination. Design objects, like all other 'artificial' objects, are constructed and, like all technical objects, are planned, but are not as fully determined in their functions as the latter. Objects in the category of art are the least functionally determined.

In the development of his semiotic classification of design objects, Bense (1971c) adopts Peirce's (1960) philosophy and his triadic conception of the sign. As will be shown in greater detail in a later chapter of this book, this theory defines each sign in relation to its material dimension (syntactics), in relation to the object it stands for (semantics), and in relation to its interpretation and use (pragmatics). According to Bense the design object is a special type of sign, in that it realizes a combination of particular characteristics from all three dimensions. It has a material aspect, which Bense calls hyletic. As an object it possesses a semantic dimension, which is called morphetic. Finally it exhibits a functional aspect, which is called synthetic. In less complex terminology a design object has the aspects of material, form, and function. Its pragmatic dimension (its use), is present in hyletics as well as in morphetics and synthetics: The application of the functional aspects (synthetics) to the material aspects (hyletics) creates its form (morphetics), which is then available for consumption and use. According to the theory, three phases in the creative design process result. In the first phase, the planning phase, the functional dimension is adapted to the material. In the production phase the object is realized in form. Finally, in the phase of utilization the object satisfies the human needs which gave rise to it (Walther, 1974).

What contributions can this terminological apparatus make to problems concerning meaning in the urban environment? So far only a rudimentary sketch of Peircean semiotics has been given. But this much can already be said: Bense's conversion of objects into signs by referring to Peirce's triadic theory of signs seems to be a nonsequitur. The logic of his argument is as follows: all signs are triadic, all objects are triadic, therefore all objects are signs. Moreover, the division of the design process into the phases of planning, production, and use does not need the backing of this complex type of semiotics as it is already commonly recognized.

2.1.5 Objects and human needs

In the object theories summarized so far there has been a tendency to overemphasize either the 'signifying' aspect of objects (their function as signs) at the expense of their functional aspects, or the functional aspects at the expense of their politico-economic aspects. In order to avoid an isolated view of either aspect the historical and dialectical method must be employed. Such a method would ensure that functional considerations are placed within the framework of the concrete historical conditions which determine the production of objects.

According to Marx (Marx and Engels, 1970, volume 23, page 192) men produce the conditions necessary for their survival in a metabolic relationship with nature. They appropriate natural materials and impose on them the new forms useful for their own survival. As men transform nature, their own nature is transformed. If one accepts this point of view, one cannot separate questions of object production from problems of the natural environment. The question of environmental pollution, for example, cannot be treated lightly, as it involves this fundamental relationship between man and nature. As a result environmental problems have kindled a spirited debate during the last few years within the camp of Marxist philosophers. The key terms of this debate, as reviewed by Maldonado (1973), are 'object', 'merchandise', and 'human needs'.

Marx's writings before 1844 and *Die deutsche Ideologie* contain much material on the theme of human needs. His pre-1844 writings are frequently referred to as 'Hegelian' and there is indeed a connection between Marx's early conception of human needs and Hegel's theory of human needs as presented in *Rechtsphilosophie*.

In Hegel's writings, too, the concept of 'proliferation of human needs' is already present. As far as the 'proliferation of human needs' and the corresponding proliferation of objects is concerned, Marx always favoured the enrichment of human conditions. This does not imply, however, that Marxism advocates a carte blanche policy of liberal expansion of all kinds of human needs. It is rather a question of practical experience determining which human needs are to be developed. This question cannot be answered by debating the legitimacy of isolated human needs but only by considering the entire system of needs in a given society at a specific moment.

This renewed interest in the question of human needs stirred up considerable controversy in Marxist circles. The structuralist wing of the Marxists opposed the opening up of such anthropological questions by pointing to Marx's break with philosophical humanism in his post-1844 writings. On the other hand, other Marxists elaborated theories of 'human needs' and theories of 'everyday life' in which objects play an important role. The work of some of Lukács's disciples and Lefebvre's early, but unresolved, work (Lefebvre, 1968) on 'everyday life' tended in this direction. According to Maldonado, theories about objects have been developed, particularly in France, which take the early work of Lefebvre as a starting point and give it either a 'leftist' (for example, Baudrillard) or a 'technocratic' (for example Moles) slant. In the Federal Republic of Germany Haug (1972) would be a representative of the 'leftist' current.

According to Maldonado, Baudrillard's underestimation of the utilitarian value of objects is an indication of the dangers of idealistic semiology: the material conditions of life are forgotten and everything is dissolved into 'meaning'. The process in which an object is used is reduced to the conceptual knowledge necessary for the use of the object. Certainly no object can be used without the attribution of meaning to the object and to the process involved. Each process of use includes a process of signification. But this 'semiosis' should not conceal the material base on which it rests. This material base is antecedent to every semantic act and could not therefore be meaningfully associated with properties it does not possess.

The relevance of the preceding discussion to the problem of meaning in the urban environment is in its dialectics. On the one hand the meaning of a city should not be considered merely a reflex of cultural and symbolic practices. These play an important role but have their foundation in the material and economic processes of human production. On the other hand, the restriction of meaning in the urban environment to its material functions would neglect the human conditions of its production. These conditions vary according to different social structures in the course of history. Any attempt to formulate an urban semiology would therefore have to come to grips with the problems of excessive symbolism or technocracy, idealistic humanism, or mere economism. Each one of these excesses has produced and continues to produce negative consequences for the maintenance and growth of meaning in the urban environment. The leading question in the field of urban semiology must be formulated from a philosophical standpoint which is capable of dealing with these pitfalls.

2.1.6 Object syntax

There have been some attempts to create a semiology of objects by taking linguistics as a promising methodological starting point. One fundamental linguistic concept is syntax. The notion of object syntax present in Moles's book (1972) was also developed by Boudon (1969). He differentiates between an 'internal' assembly of parts into a complete object (for example, cup + saucer) and an 'external' assembly of objects into object systems (for example, tea service: cups + teapot + teaspoons, etc). Since object systems cannot be assembled at will, they must be based on fundamental syntactic structures. These are hypothesized as analogous to the structures of Chomsky's transformational grammar, to which transformational operations can be applied by a limited set of appropriate rules. The idea of a transformational grammar applicable to object structures in general and to architectural or urban structures in particular is certainly fascinating. However, the disadvantages of approaching the problem of meaning in the urban environment in analogy to linguistic structures will become clearer in the course of this review. Here I should mention just one of the questions which the application of transformational grammar to all kinds of object structures would raise. Chomsky and many of his followers regard linguistic structures as genetically based. Would the syntax of object systems such as a city be likewise genetically determined? This is hardly conceivable.

2.1.7 Objects and language

Although an attempt to draw direct analogies between language and objects may be problematic, there may nevertheless be an anthropological relationship between them. Such a relationship has been the recurrent theme of the work of Rossi-Landi (1968; 1972; 1975), who maintains that a comparative study of the production of language and the production of objects brings out their 'homology'. Reasoning by homology, in contrast to reasoning by analogy, involves the investigation of the common logical roots of different entities. The common root of language and objects is that they are both human 'artefacts'. Artefacts are not simply traces left by human activity (like footprints) but are the results of transformations brought about in raw material by means of human labour. Human labour is in turn planned and intentional. The labourer must learn how to differentiate between and to assemble his materials. Human language sounds, too, are not simply cries, but are first learned in society and are intentionally formed and produced by men. Recently Eco (1976) has proposed a whole typology of signs based on what he calls 'sign production'.

According to Rossi-Landi one of the reasons why the homology between language and objects has only recently been discovered, is the fact that the recording of speech production has only been possible since the invention of the phonograph. The fortuitous circumstance that the sound of the spoken word 'disappears' immediately and, unless graphically represented, leaves no trace has conferred an aura of immateriality on language products.

The homology between material artefacts (like wooden planks, hammers, and cars) and language artefacts (like words, sentences, and discourses) lies in their common constitutive root which is human production.

This homology can be deduced by the 'genetic method' (including phylogenetic and ontogenetic reasoning). There are four specific arguments for a common origin of language production and material production.

1 Men have never produced linguistic artefacts without at the same time producing material artefacts (and vice versa).

2 The production of both kinds of artefacts coincides largely with the separation of hominids from the rest of primates, thus inaugurating human society.
3 It is unthinkable that material artefacts can be produced without the concomitant production of linguistic artefacts, and vice versa. Men would not have been able to work on any object without communicating with other workers. Conversely, verbal communication presupposes a world of real objects to which language refers, and thus the advent of the capacity to distinguish and manipulate such objects.
4 The social conditions which dominate both kinds of production are largely identical. The child enters the world of language artefacts by much the same process by which he enters the world of material artefacts: he enters both as ready-made systems by expending working energy in order to learn how to use them. He could not learn to speak without learning to differentiate and manipulate objects, and he could not learn to differentiate and manipulate objects according to the manner of his society without learning to speak.

Rossi-Landi devised a model summarizing the homology between production of material artefacts and production of language. In this model artefacts are classified on eleven levels according to varying degrees of complexity, and these eleven levels are as a whole divided into three major classes.

The zero level consists of intact, unworked natural elements, the nonlinguistic component being 'material nonsound substance', the linguistic component being sound. The first level consists of 'presignificant' artefacts: in material production, extracted raw materials; in language production phonemes in the secondary articulation of language. The second level consists of irreducibly significant units of material and linguistic production, namely, the characteristic transformations of raw material such as the sharp or obtuse edge on the head of a hammer ('objectemes'), or monemes in the primary articulation of language. The third level consists of completed components such as the head or handle of the hammer, or words, syntagms, expressions, parts of speech, phrases. These four levels constitute the first major group. The second group comprises artefacts of the degree of complexity of material utensils and sentences. The fourth level consists of complete utensils, simple tools such as the hammer, and simple sentences capable of transmitting self-contained messages. The fifth level consists of aggregated artefacts, that is, object compounds such as a tea service, and compound sentences or messages comprising a combination of sentences, etc. These two levels constitute the second major group.

The third major group consists of artefacts of a greater degree of complexity than complete utensils and sentences. The sixth level, the level of mechanisms, consists of simple machines such as a loom, and meaning-generating mechanisms such as syllogisms. The seventh level comprises aggregated and self-sufficient mechanisms such as machines capable of executing several working phases, and lectures, speeches, essays, and books. The eighth level accommodates total mechanisms or automation, the material artefacts of this level being automated machines, the language component being 'postlinguistic' artefacts such as subcodes and lexicons. The ninth level consists of nonrepeatable production of complex artefacts such as unique prototypes or special constructions like luxury cars, and 'original' literary or scientific production. The tenth level comprises 'global' production—the entire production of an individual, group, or culture in material or linguistic terms.

On each of these different levels the material or linguistic products flow from the production process and become 'frozen' as 'means of production'. These are adopted as 'natural' wholes by other workers, who in turn learn to produce with them. Just as individual material production would not be possible without collective work, individual speech is dependent on previous collective speech. Every individual who learns to speak starts the movement of collective speaking anew. By engaging further in the 'labour of language' he uses the accumulated wealth of 'language capital', just as raw materials or even money are used in the productive process.

Does Rossi-Landi's model imply the existence of a homology between the urban environment and any of the higher levels of linguistic production? It could well be considered an aggregate of artefacts, but would then be located at a level inferior to mechanisms such as the loom. Barthes (1967; 1971) considered the city as analogous to a text. Would it be reasonable to situate the urban environment on the level of aggregated mechanisms such as machines with several working phases, and their postlinguistic components, essays and books? It seems more probable that the city is on the level of 'global production', as comprising the entire production of a group or culture. The urban environment would then consist of a multiplicity of structures lying between artefacts and their functional and transfunctional meaning.

A study of such structures would fall under the 'semiology of culture' (Lotman and Uspenskij, 1975). The utility of Rossi-Landi's model for the study of urban semiology is therefore that it indicates the level of complexity on which the urban phenomenon is to be located. This does not, however, exclude the possibility of breaking it up into its component parts and analyzing it at lower levels of complexity.

2.1.8 The reduction of signs to objects

The methodological advantage of reducing signs to objects.

Barthes (1964) declared objects to be function signs and Baudrillard propounded the 'political economy' of signs. Prieto (1966; 1973; 1975a; 1975b) did the reverse in maintaining that in order to understand the functioning of signs in the communicative act one must first understand, in more general terms, the functioning of instruments in the instrumental act.

Signs are special types of instruments in a special type of instrumental act, called 'communication' (cf Krampen, 1973a). Considering speech and communication in general to be a special case of the instrumental act has a considerable methodological advantage. Since the beginning of the century methodologies for explaining the functioning of verbal communication have become highly developed. This advantage in development has encouraged the application of linguistic theories, by analogy, to other semiological questions, particularly those in object semiology. Analogies can be productive but unfortunately often foster logical confusion. Regarding linguistics as simply a branch of semiology would stimulate a search for a general core of linguistic methodology which would be applicable to semiological problems not by analogy, but by logical pertinence. For this reason Prieto's (1973; 1975a; 1975b) model of the instrumental act will be reviewed here in some detail.

2.1.9 The vocabulary of Prieto's model

The vocabulary of the model must first be defined. The execution of a particular operation presupposes two elements: the *operation* and the *tool.* With a given tool at least one, but usually more than one, operation may be executed. All those operations which may be executed by means of the same tool are members of a system of abstract sets of operations. This system of sets is called the '*utility*' of the given tool. All tools do not possess the same utility (tool 1 versus tool 2, etc). But all variants of a specific tool connected to the same utility (for example, all variants of a corkscrew) are members of a second system of abstract sets, that of tool variants. This system of sets is called the '*operant*' of a given tool. Each tool thus belongs at the same time to two systems of sets: its utility and its operant. These systems of sets may thus be considered as being coordinated by means of the tool, forming a 'bifacial' unit called the 'instrument'.

It is obvious that the linguistic model is a special case of the instrumental model: the 'signified' (le signifié) in the linguistic model corresponding to 'utility', the 'signifier' (le signifiant) to 'operant', 'sign' to 'instrument', and 'signal' to 'tool'. The concrete transmission of a message by means of a signal is therefore nothing but a particular way of executing a concrete operation by means of a tool.

2.1.10 The asymmetry of the coordination between operant and utility

Once the vocabulary of the model has been understood, the coordination established by the instrument between the operant and the utility may be further analyzed. A strict application of set theory to the problem reveals the asymmetrical nature of this coordination.

This can be shown as follows: one operation may conceivably be executed by means of two or more different tools. In this case each of these would be a member of its own operant and utility. The operation being now a member of two or more different utilities may then be considered the element of intersection between these utilities: it belongs to both (or all) of them. A family of four can be housed in an apartment or a bungalow, both of which constitute tools for housing. These tools have different utilities, but these intersect at at least one point: the operation of housing the family can be done by both.

On the other hand, all tools that belong to the same system of tool variants forming an operant can only serve to carry out those operations belonging to the one coordinated utility. For example, variants of five-roomed bungalows can serve to carry out only one system of sets of operations, namely, that group of operations which can be carried out by means of five-roomed bungalows. The asymmetry of the instrumental coordination between operant and utility is thus the following: a certain operation can be a member of various utilities which are coordinated with various operants; but a tool variant of a particular operant can carry out only those operations which are members of its one coordinated utility.

2.1.11 The logical consequence of asymmetrical coordination

The logical consequence of this asymmetry is that the utility of any given tool variant is always the smallest system of sets of operations which can be coordinated with that tool's operant. On the other hand, the tool's operant is always the largest system of sets of tool variants which can be coordinated with the above utility. It has been pointed out that the same operation may be carried out by different tools, each of which involves different utilities. It will be remembered that these utilities must be considered as being in a relationship of intersection. But this is not the only possible relationship between the utilities of different instruments. The following list includes all logically possible relationships between the utilities of two instruments:

a relationship of logical exclusion,
utility $I_1 \neq$ utility I_2;
a relationship of logical inclusion,
utility $I_1 \subset$ utility I_2;
a relationship of logical intersection or union,
utility $I_1 \cap$ utility I_2 or utility $I_1 \cup$ utility I_2.

The utilities of two instruments which would have a relationship of logical identity (utility $I_1 =$ utility I_2) actually belong to one and the same instrument. The asymmetry of the coordination between operant and utility can also be seen in the fact that these various relationships do not hold for the different operants of different instruments. Since a given operant is coordinated with only one utility the different operants of two instruments can only be in a relationship of logical exclusion.

This analysis of the asymmetry of coordination between operant and utility of an instrument again makes it clear that the linguistic model is a special case of the instrumental model. The coordination between signifier and signified is equally asymmetrical. The same meaning can be transmitted by many different signs. But a sign usually has only one particular meaning coordinated with it.

2.1.12 The two stages of conceptualization in an instrumental act

In psychological terms the carrying out of an instrumental act amounts to the distribution of operations and tools by a cognitive, conceptual process into abstract systems of sets of operations and abstract systems of sets of tool variants. This process of conceptualization has two consecutive stages. In order to carry out an instrumental act the agent must first know what he is going to do. He must thus have a basic preconception of a utility. This utility is by definition a system of sets containing those operations which can be carried out by a given tool.

Having thus first conceived of the utility in relation to his purpose, he must next reconceptualize the utility as that system of sets of operations which can be carried out by a particular tool, which tool must in turn be conceptualized as belonging to a system of sets containing all those tool variants which are coordinated with the utility in question. In other words the utility is, in the second stage, always conceptually coordinated with an operant.

By regarding this two-phase conceptual process as a sequence of cognitive acts any operation would thus be distributed into two systems of sets of operations consecutively;

the system of sets of those operations which the agent has conceived of carrying out (that is, in relation to his purpose);

the system of sets of operations which can be carried out by the same tool variants.

The system of sets of operations forms the connection between these two conceptual phases. In the first phase it is coordinated with the purpose the agent has in mind, this purpose being in turn one set of a system of sets containing different purposes. In the second phase it is coordinated with a system of sets containing tool variants. Consequently the system of sets of operations has a double coordination.

The set which determines an operation (the purpose of the agent) is necessarily part of a system of such sets (everything an agent could conceive of doing within a given instrumental culture)—called here for the sake of brevity System I. Between the utility of a given instrument and the sets of System I there exist two possible relationships. Either the utility bears a one-to-one relationship with one of the sets of System I, or it is a logical sum of several of these sets.

2.1.13 Denotation and connotation

The two successive cognitive acts constituting the conceptualization of an operation are called denotation and connotation. First the agent 'denotes' the operation, that is, conceives of it in relation to his purpose. Then he 'connotes' it, that is, conceives of it in relation to the appropriate tools.

The connotative process is functionally neutral. The aim of an operation remains the same no matter how the utility is connoted by its coordination with an operant. Connotation is simply the manner of conceptualizing the coordination between a utility and an operant in order to carry out an operation. However, although the connotative process is functionally neutral, it may be subject to evaluation. The preference of one connotation over another (or others)—the preference of one way of carrying out an operation over another (or others)—may result in the connotation (or connotations) acquiring an additional function over and above its immediate function of fulfilling the purpose of the agent. In other words it may be 'transfunctionalized'. A typical example of an additional function acquired by connotation through the process of transfunctionalization is differentiation. Different ways of carrying out a specific operation may serve as a means for distinction. A luxury villa is one possible way of conceptualizing an operant for carrying out the operation of housing. This may, however, be transfunctionalized, becoming a 'tool' for distinguishing the rich from the poor.

2.1.14 The relevance of Prieto's model to urban semiology

The question must now be answered, what contribution this model makes to the clarification of the problem of meaning in the urban environment. The answer has been partly given by the review of the model itself. The advantage of hammering out the methodology and terminology of a more general semiology of objects, rather than using the ready-made linguistic apparatus, has already been spelled out. The terminology of this model can serve as a basis for the reconstruction of a linguistic vocabulary as well as the logical deduction of a vocabulary of urban semiology.

The principle of the asymmetrical coordination between utility and operant constitutes the logical basis for the principle of the different 'styles' of the instrumental act. A specific operation may be carried out by means of several different tools, whereas any tool is coordinated with only one set of operations. This concept guarantees, in the context of functional meaning in general as well as in the context of meaning in the urban environment in particular, the necessary openness, pluralism, and creative aspect of any problemsolving.

The two-phase conceptualization of the instrumental act links the question of style, the 'how' of the operation, to the 'what' of the agent's purpose. This irreversible sequence, beginning with the preconception of 'what' the operation should accomplish and ending with a solution to the question of 'how' it should be accomplished, precludes mere formalism. The basic meaning of the instrumental act is not intrinsic to the act itself, but derives from the purpose of the agent. Although it provides formal closure in its organization, Prieto's model has the advantage of remaining ideologically open. This should certainly be a prerequisite for any concept of urban semiology. The model also clarifies the terms denotation and connotation, generally applied in a very loose way in semiology. In contrast to Barthes (1964) connotation is for Prieto not a secondary meaning acquired by an object through the semantization of usage. There is no denotation (conceptualization of an operation in relation to a purpose) without connotation (conceptualization of the manner of execution). Prieto's principle of the transfunctionalization of style renders his model capable of absorbing the various loose notions of connotation and 'secondary' or 'indirect' function present in the literature on semiology. It is here that Prieto and Barthes, while using different terminologies, touch common ground. Semantization of usage is simply the transfunctionalization of a particular style in order that it may acquire another function over and above its primary one, that is, that it may function as more than just a particular style. Eco's (1968) notion of 'primary' and 'secondary' functions—developed in his semiology of architecture—can be interpreted in the same way. The distinction between 'direct' and 'indirect' ('symbolic') functions introduced by Mukařovský (1970), and that between the 'material' and 'semiotic' functions of objects advanced in the semiotics of culture by some Soviet authors (see Faccani and Eco, 1969) seems to be dealing with the same question. Prieto's model of the instrumental act thus reveals itself as a promising starting point for the deduction of a model of urban semiology.

2.1.15 Summary of current thought in the field of object semiology

This review of the literature on the semiology of objects has thus revealed two major groups. One is concerned with the definition of the meaning of objects. The other is concerned with establishing object semiology in relation to other semiological disciplines, notably linguistics. There is a tendency in the first group to define the meaning of objects by transforming them into signs, thereby neglecting their material and functional aspects (Barthes, 1964; Baudrillard, 1968; 1972). A similar problem arose in design-oriented object theory, where the design object was reduced to a sign by introducing it as a special case of Peirce's triadic sign concept (Bense, 1971c; Walther, 1974). The other group tends to derive meaning from the function of the object, thereby prejudging the extrasemiotic analysis of the historical and humanistic aspects of the problem.

From a more synthetic point of view, the interest in both these attempts is in their careful elaboration of particular aspects of object semiology (though at the expense of others). Only an extrasemiotic consideration of the historical development of the means and methods of production can conceivably synthesize these divergences (Maldonado, 1973).

Those attempting to establish a semiology of objects within the framework of general semiology are mainly concerned with the relationship between a specifically linguistic methodology and a more general one. The development of a semiology of objects in analogy to linguistics presents major difficulties, for example, seeing object relationships in terms of a set of transformational rules. If, however, a common semiology of linguistic and nonlinguistic objects is developed by considering the homology of their production, interesting insights into the varying levels of complexity of human artefacts would result (Rossi-Landi, 1968; 1972; 1975).

The most promising approach to the problem of the relationship between linguistics and the semiology of objects seems to be the reduction of signs to objects. By this reversal the highly developed methods and concepts of linguistics become special cases of a more general semiology of objects. Characteristics common to both the instrumental and the communicative act, such as sequence of purpose and style, of denotation and connotation, and phenomena such as the transfunctionalization of style for the purpose of communication can now be considered the core of semiological theory (Prieto, 1966; 1973; 1975a; 1975b).

What contributions can current thought on the semiology of objects make to the establishment of a semiology of the urban environment? From it one can learn the disadvantages of considering meaning in the urban environment from an isolated ideological position such as pansemiotics or functionalism. The historical and dialectical method helps to avoid this pitfall. It also becomes apparent that linguistic methods and concepts have a general semiological core which results from the homology of production of human artefacts (including the city). The instrumental act as a basis for the development of an urban semiology appears thus to be a fruitful starting point.

2.2 Work on the semiology of architecture and on urban semiology

2.2.1 Linguistic analogy in architectural semiology—obstacle or advantage?

In search of the constitutive elements of architectural 'language'

It is not surprising that semiology began taking an interest in objects in general, since nonlinguistic signs impress the beholder as having manifest object characteristics that are apparently lacking in verbal signs. What is surprising is that a special branch of semiology should have developed to deal specifically with architectural objects, and the city as a collection of architectural objects. Why has there been this particular interest in the semiological status of architectural objects and the city when there has been no corresponding development in, for example, 'rural semiology'? The reason seems to lie in recent cultural history.

The intellectual climate in Italy in the early 1950s was particularly vigorous and stimulating. A new generation of painters had emerged, the question of social realism versus abstract art was hotly debated, and a new brand of art critics had begun to publish their ideas. In the field of architecture one began to witness the first negative consequences of urban sprawl unleashed by the building boom and reckless speculation. It was in this atmosphere that the first concepts of architectural semiology began to be developed. The recurrent theme was the lack of meaning in the urban environment and the uniformity of architectural design (Bettini, 1958;

Dorfles, 1959; 1962; Brandi, 1960a, 1960b; 1967; della Volpe, 1964; Garroni, 1964). Out of this semantic criticism and the need for a unified terminology in the face of increasing chaos rose a view of architecture as a complex system of meaningful signs composed in accordance with a set of rules.

That there was a possible analogy between this and language became obvious. The Florence School of Architecture became the centre of this new development and in 1953 Gamberini published *Per una Analisi Degli Elementi dell' Architettura.* According to Gamberini (1953; 1959; 1961) the constitutive elements of architecture analogous to constitutive linguistic elements (words) were found to be:

1 floors, either directly posed on the ground or elevated;
2 vertical connections between floors, like ramps and stairs;
3 walls (either portant or nonportant, fixed or movable);
4 connections in walls from outside to inside or from inside to inside (like doors and windows);
5 roofs (either self-supporting or supported);
6 pillars and beams to support floors, walls, and roofs;
7 objects to accentuate the meaning of spaces (as a bed accentuates the meaning of a bedroom).

From the beginning, as in Gamberini, functional objects were considered part of the architectural language and thus devoid of any autonomous status. This is all the more surprising when one considers that the reverse is equally conceivable: architectural objects may just as logically be regarded as a special case of a general object system.

A semiotic slant to the original linguistic analogy was given by Draghetti (1958/1959; 1961) and more particularly by Koenig (1964), who introduced into architectural semiology Morris's notion that language is a system of signs organized in accordance with a set of rules. Signs, according to Morris, function as stimuli evoking a specific behaviour which would normally be evoked by an object which in this case is absent and which is represented by the sign. These signs carry meaning which is common to a group of interpreters and which remains the same in different situations. Koenig held that this would have equal application in architecture if it were seen as a language. Here an architectural object such as a house evokes and supports specific behaviour patterns which (in the case of the house) are normally associated with family life. The house is then a sign of a specific sphere of special relationships. Architectural sign complexes together with nonvoluminous objects (such as bridges) constitute the sign repertory of a city, where objects of use (as in Gamberini, 1953) serve to qualify the function of architectural spaces. In *Architettura e Communicazione* Koenig (1970) postulated the existence of a communication process in architecture, thus simply taking the original hypothesis that architecture is a language to its final conclusion. It must therefore then fulfill the functions of communication (cf also Broadbent, 1973).

The well-known Shannon model (below) of the communication process (Shannon and Weaver, 1949) is consistently applicable to this process of architectural communication:

1 the sender: the architect or the architectural team;
2 the codes and lexicons: the functional, legal, structural, and economic rules in accordance with which a building is designed;
3 the signal: the sum of drawings, models, and written specifications of an architectural design;
4 the channel: the construction-site;
5 the physical signal: the material building resulting from the design signal (3);
6 noise: environmental obstacles impeding perception and the physical disintegration of the architectural sign complex;
7 the receiver: the human being, by means of his sense organs;
8 the significant aspect of the message: architectural space including the specifying objects within it;
9 the codes and lexicons of the receiver: functional, legal, structural, and economic expectations;
10 semantic noise: the prejudice of the receiver;
11 the receiver as a collective: the city as a system of communications;
12 the meaning ('signified') aspect of the message: the original function which the architectural object denotes and the second function which it connotes.

Koenig drew the linguistic consequences of the application of Shannon's model. The physical signal (5) is articulated on the level of primary articulation by 'choremes' and on the level of secondary articulation by 'archemes' (Koenig's term 'archeme' here is a modification of the term 'architem' first proposed by Cappabianca, 1967). Koenig argued that the real stimulus in the architectural sign process is not the architectural object but a human need. The architectural object denotes by means of 'chorematic chains' those human needs which it is intended to satisfy. Thus in archaeology we are able by examination of architectural remains to reconstruct the functions and needs of past societies which are not present when their signs are encountered.

The first synthesis of architectural semiology was attempted by Eco (1968) in his classic *La Struttura Assente.* In this he criticized Koenig's behaviouristically oriented theory that the architectural sign is a preparatory stimulus standing for some behavioural pattern. Eco held that the architectural object is itself a significant stimulus whose meaning ('signified') is the function it fulfills. Architectural signs are thus 'signifiers' which are open to description and enumeration and whose relationship with the 'signified' is established and maintained by specific cultural codes. The architectural object denotes a particular cultural manner of using it.

Like Barthes Eco differentiated between architectural denotation and connotation, defining connotation as the acquisition by an architectural sign of a secondary (more or less 'symbolic') function. The original bifacial architectural unit (sign)—the 'signifier' together with its corresponding functional 'signified'—becomes as a whole a signifier which is coordinated through the mediation of a cultural code to a secondary function (secondary 'signified') which is called connotative.

Eco proposed the term 'choremes'—analogous to monemes in the primary articulation of language—for the most irreducibly significant elements in architecture. In Gamberini's classification of the constitutive elements of architecture Eco saw the possibility of a secondary articulation analogous to phonemes in the secondary articulation of language. But whereas the basic units of architectural language had thus been defined, its grammar and syntax remained to be established.

At this point Eco hypothesized that the existing architectural codes probably rely for their articulation on nonarchitectural codes, and that it is by reference to these that the behaviour-regulating messages of architectural objects are decoded. An example of a nonarchitectural code of anthropological provenance is the proxemic code of distance regulation (Hall, 1966; Segaud, 1973) whereby social situations are stylized differently in different cultures by the regulation of specific distances between its members.

In the introduction to his famous "Componential analysis of a column", first presented at a Symposium of Audiovisual Semiotics in Urbino, 1971, Eco concluded his analysis of the constitutive elements of architecture. To solve the problem of architectural semiotics it is convenient, he said, to refer to Hjelmslev's (1968) distinction between the level of content and the level of expression of any semiotic structure. Both level of content and level of expression are further subdivided into sublevels of form and substance. To study architecture as the communication of a special conception of space amounts to studying the syntactic, namely formal, aspects of an expression corresponding to the content organized into pertinent semantic units.

An architectural sign is then a bifacial unit consisting of two levels. The level of expression consists of an architectural morpheme M which realizes a particular form of the substances particular to architecture. The level of content consists of an architectural sememe A which realizes a particular function, itself part of a system of functions, into which the substance of all possible functions of a cultural system may be subdivided.

Morphemes and sememes may be further subdivided. Thus the architectural morpheme M_i realizing a desk may be subdivided into morphological features such as the table legs and the table top, which in turn have certain prearchitectural characteristics such as dimension, etc. The corresponding architectural sememe A_i = 'desk', representing a particular function belonging to a system of functions, may be subdivided into sememes representing such subfunctions as 'sustaining the table top', 'putting a distance between two people', etc.

Semiology of space

The semiology of architecture has recently come to be regarded as part of a general semiology of space (Hammad and others, 1973b). The members of the Paris 'Group 107' (Hammad and others, 1973b), basing their work on the structural semantics of Greimas (1966; 1974) placed specific emphasis on the advantages of broadening architectural semiology to include analyses of human behaviour and its spatial support. Space is thus seen as a stage upon which humans enter into relationships both with each other and with objects. These relationships can be described by what Greimas calls 'analyse actantielle', that is, the analysis of the roles and functions operating within a specific cultural 'tale'. Proceeding, like Eco, on the basis of Hjelmslev's model but giving it a different slant, this group also distinguished between two levels: the level of expression, which consists of sets containing objects and individuals, and the level of content, which consists of sets of activities. The constitutive unit of the level of expression is called 'topos', which may be analyzed in terms of its internal relationships (objects, individuals) or in terms of its relationship with other 'topoi'.

There is a noteworthy difference between Group 107's and Eco's use of the Hjelmslev model. Eco saw the spatial characteristics of architectural objects as only one set among a number of different characteristics—the spatial characteristics having no more special importance than these others. Eco saw the functions of architectural objects as constituting their level of content. Group 107 on the other hand placed specific emphasis on the dynamic spatial relationships existing between people and objects. They saw the level of content of these relationships as consisting of the activities by which they are sustained. They moreover defined architecture as a language on the basis of Hjelmslev's model simply by indicating that it does indeed exhibit a level of content coordinated with a level of expression, neglecting the distinction between the hypothesis that any language will contain these two levels and between the hypothesis that anything which contains these levels is therefore a language. In contrast to Eco, however, the latter's general thesis in taking human behaviour into consideration does nevertheless seem to have wider relevance and could perhaps prove (as will be suggested in greater detail below) to be more productive for architectural semiology as a whole.

As has been pointed out, spatial relationships did not play a predominant role in Eco's analysis. He saw space as neither the signified nor the signifier of an architectural object but simply as one of its morphological features. This view gave rise to an extensive controversy with the 'Neapolitan school' of de Fusco (de Fusco et al, 1966) and Scalvini (1968; 1972), who held that the signifier of an architectural object is the 'building' while the signified is its 'internal space'.

Architecture—mass medium, rhetoric, and style

The next step after regarding the architectural object as a sign complex was to regard it as a 'mass medium' (de Fusco, 1967; Menna, 1967; Eco, 1968). The content of the architectural message is then 'consolatory', persuasive and commercial. De Fusco (1967) observed that, in addition to this, functionalist architecture, although not industrially produced, portrays as a message the image of the tool of the Industrial Revolution—the machine.

Eisenbeis (1974) also took up the theme of architecture as mass communication. He held that architectural creation is a process of different types of communication being carried on by dialogues, sketches, scale models, etc. The moment architecture is determined by industrial production the citizen is forced into the same anonymous role when confronted with architecture as he is when confronted with the mass media. He becomes just a 'statistical entity'.

This view that architecture is a language leads to further consequences. Once architecture had been declared a language, it had to be established not only that it communicates but also that the stylistic characteristics of this communication can be analyzed. Ostrowetsky et al (1973) proceeded to analyze the stylistic differences between 'old-authentic' housing in the French Provence and its modern imitations, registering the permanence or change of certain elements of the old style as exhibited in the modern. Jencks (1969; 1974; 1972) drew an analogy between stylistic changes in architecture and different rhetorical styles. This idea of architectural 'rhetoric' is still in the very early stages, but research into those styles which were openly acknowledged to be 'rhetorical' (such as Fascist architecture) may prove to be of considerable value in the development of architectural semiology. The fundamental question is whether such rhetorical styles as Fascist architecture do in fact exert a psychological influence, as is assumed to be the case in linguistic rhetoric. Bentmann and Müller (1970) explored this possibility in their analysis of certain architectural styles (such as the 'villa' from the time of Palladio or the modern New York City penthouse) as 'architecture of domination'. This preoccupation with architectural style gave rise to two ways of looking at architecture; either from the point of view of the constructive aspects of buildings or from their semantic or expressive aspects. The Soviet architectural critic Markuzon (1972) and Scalvini (1971; 1975) quite independently of each other called the first aspect 'tectonics' and the second 'architecture' proper. Eco (1971) warned against the 'aesthetic fallacy' explicit in this approach, which would tend to exclude from semiological consideration the great bulk of architectural constructions which have no stylistic quality.

Summary

This review of the early developments of architectural semiology has revealed two principle themes. The first is that, if architecture is considered a language, it must be structured like a language, that is, one of its features must for example be double articulation. The second is that if architecture is a language it must communicate. Koenig and Eco both dealt with the first theme, with the difference that Koenig followed Morris's semiotic school, which is behaviouristically oriented and based on a threefold analysis of the sign, whereas Eco belonged to the Saussurean semiological school, which is structurally oriented and based on a dichotomous conception (signifier–signified). Despite these differences, both arrived at the conclusion that since architecture is a sign system it must communicate. Koenig then used the Shannon information theory model to explain the communication process whereas Eco explained it in terms of the paradigm of mass communication.

Thus from the very beginning of architectural semiology two different conceptions of the architectural sign evolved. The differences between them must be kept in mind if an attempt is to be made to develop a more advanced theory of architectural semiology. As far as architectural objects as 'messages' are concerned, two different models of the communication process developed. Koenig's, which originated in electrical engineering, seems mechanical and limited in its sphere of relevance, whereas Eco's simple declaration of architecture as a mass medium appears superficial and in need of a more detailed body of empirical support. It seems then that neither of these is sufficient for establishing the existence of meaning in the urban environment. The later part of this book will attempt to propose an alternative which may avoid these shortcomings.

2.2.2 'Deep structures' in architecture and the problem of 'generative grammar'

Chomsky's work (cf Chomsky, 1965; 1971), like other linguistic theories, has also had some influence on architectural semiology. A whole section of the 1972 Symposium on Architecture, History and Theory of Signs in Castelldefels/Barcelona was devoted to the hypothesis that there may be 'deep structures' underlying the production of architecture, similar to those which Chomsky postulated as underlying the production of language. Indications that this new type of linguistic analogy might hold were given by Eisenman (1974; see also Gandelsonas, 1972) and, with some caution, by Broadbent (1971; 1974), whereas Llorens (1974) took a negative view of their positions. At the 1972 International Semiotics Workshop (Ulm, FRG) and at the 1973 Colloque de Sémiotique de l'Architecture at the Centro Internazionale di Semiotica e Linguistica of Urbino some of the contributions (Castex et al, 1973; Hammad and others, 1973a) also aimed at postulating or describing architectural 'grammars' or 'syntaxes' which could be understood as generative devices.

However, at the same Colloque, Marcus (1973) gave a more rigorous exposition (based on his work on genetic grammar) of the problem of an architectural grammar. He pointed out that grammars are not simply sets of rules according to which 'lexical units' are permutated, but are sets of recursive rules permitting the generation of any 'well-formed string'.

Whenever a finite set (of rules) is capable of generating an infinite set (of strings) we have what is called a grammar. Thus an architectural grammar should be capable of generating every building past, present, and future.

Marcus (1974) described the Polish mathematician Pawlak's first attempt to define a genetic grammar as a set of (three) rules for the concatenation of triangles (the triangles corresponding to chains of amino acids in the RNA substratum of the 'genetic syntax'). This set of rules imposes certain restrictions on the formation of possible patterns out of a number of triangles. The patterns are formed by connecting the base of the second triangle from a finite collection of twenty triangles (figure 2.1) representing 'well-formed string' to one side of the initial triangle. One of the restrictions is that the side of the second triangle which is connected to the first must have the same number as the side it is being connected to, the same rule being valid for the addition of a third triangle, etc, as in figure 2.2(a). Marcus went on to show that the concatenation of triangles may be 'linearized' in graph form. Thus the triangle in figure 2.2(b) may be linearized as in figure 2.2(c) and figure 2.2(a) may be linearized as in figure 2.2(d). These graphs are called 'rooted trees'.

According to Marcus (1974) Vauquois demonstrated that Pawlak's rules could be transformed into a 'Chomskian context-free grammar' containing fifty rules for 'permitted' concatenations of triangles (belonging to the initial set of twenty triangles representing well-formed strings of amino acid molecules).

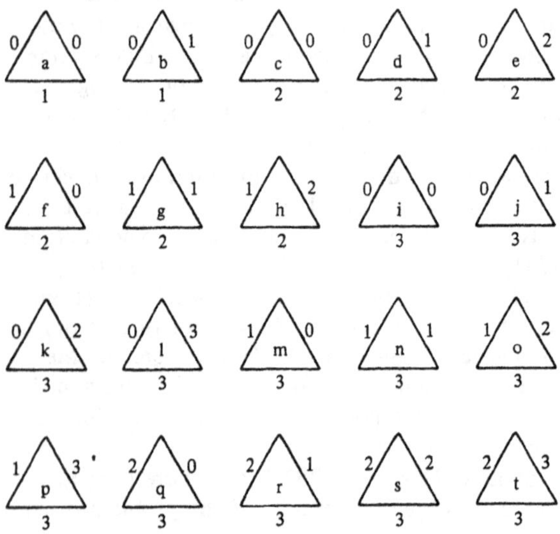

Figure 2.1. Pawlak's twenty triangles representing 'well-formed strings' of amino acids in the RNA substratum. The 'genetic grammar' consists of a set of three rules for the concatenation of these triangles (Marcus, 1974).

After weighing the strengths and weaknesses of both the Pawlak and the Vauquois methods, Marcus proposed a third method based on so-called Lindenmayer systems. The value of Marcus's analyses for architectural semiology is obviously the principle of linearization. As he himself pointed out, the linearization of two-dimensional structures is simply a special case of the linearization of polydimensional structures. Would this then mean that a given polyhedral figure in three-dimensional Euclidean space is also based on a specific set of grammars? Marcus pointed out that, as far as the two-dimensional problem is concerned, successful efforts have been made to construct grammars capable of generating pictures ('picture grammars'). The generative grammar of an architectural plan would simply fall under the category of 'picture grammars', whereas an architectural object would be based on grammars capable of generating polyhedral figures in space.

If it were possible to construct picture-generating grammars, these would be of considerable value in composing architectural layouts. By taking functional constraints such as sunlight, the nature of the site, its accessibility, etc into consideration in the concatenation of housing units, it might then be possible to obtain solutions automatically. An attempt to construct a computerized grammar for the layout of housing developments was recently made by Z'Graggen (1974) at the Ecole polytechnique fédérale de Lausanne (Switzerland).

The value of Marcus's approach to semiology of architecture appears to be that his analogy is not simply formal. The graphic elements of a two-dimensional architectural plan and the elements of a three-dimensional

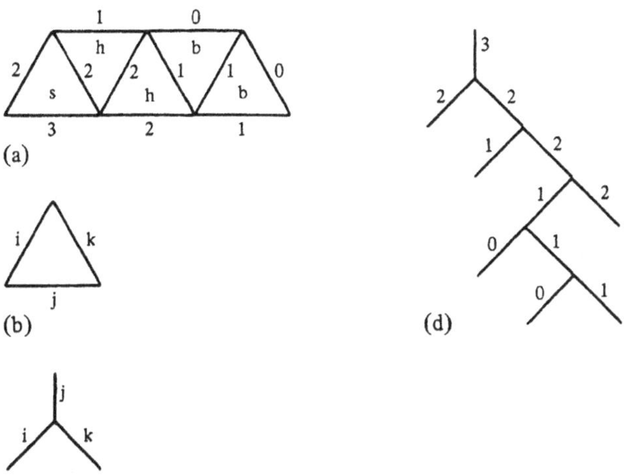

Figure 2.2. (a) Concatenation of five triangles permitted by the 'genetic grammar'. One of the rules is that the side of any second triangle connected to a first must have the same designated number as the side it is being connected to. (b) A triangle which may be 'linearized'. (c) Linearization of the triangle shown in (b). (d). Linearization of the concatenation of five triangles shown in (a).

building cannot be composed in an unrestricted way if the results are to be 'well-formed'. The old saying that you cannot build a house by beginning with its roof points to the irreversible sequence in which the parts of a building have to be put together. Architectural plans must also follow a certain architectural logic in the concatenation of their (geometrical) elements. Unfortunately nobody has as yet spelled out the restrictions governing the concatenation of architectural elements into 'well-formed strings'.

Chomsky's influence can also be seen in Boudon's theory of the 'sémiologie des lieux' (Boudon, 1973; 1974). Boudon, who had already attempted to apply transformational grammar to objects (Boudon, 1969), pointed out that 'language' in his term 'language des lieux' was intended to be used in the same sense as it is used in 'language of mathematics', the justification being that the methods required to describe and explain phenomena of 'lieux' are parallel to those required to describe and explain verbal language.

Boudon regarded 'lieux' in its broadest sense. When an ethnologist studies a village its mythology, the topology of its layout, and in our culture the plans for its construction are as relevant as the material aspects of the buildings. He warned against the dangers of 'substantialism', that is, taking only the material aspects into consideration, and also against the reduction of 'lieux' to such sociocultural determinants as exchange, that is, regarding it only as the expression of its economic structure.

He proposed instead a structuralist method whereby 'lieu' would be analyzed on different levels (mythological, topological, architectural, etc) as a system structured according to syntactical rules. 'Lieu' could then be explained in terms of the structural relationships operating between the different levels. The syntactical organization of the different elements would be established by reference to basic oppositional criteria. Every 'lieu' exhibits a series of distinctive features: internal closure versus external closure; accessibility versus nonaccessibility in different directions; hierarchization (the way in which it is connected with other 'lieux'); orientation (north, south, east, west, or above, below, in front of, behind, to the left and right); scale (the proportional relationship between one element and another); concentration (the measure of the density of the organization of elements); and different patterns of expansion (linear, circular, branching, repetitive, etc). These basic criteria allow for an elementary definition of 'lieu'.

Boudon went on to construct a grammar of 'lieux', a set of rules according to which well-formed strings of elements are organized to form a 'lieu'. These rules describe the permissible syntagmatic relationships between the immediate constituents of 'lieux', for example a sanctuary, a statue, people, a ceremonial avenue, and their articulation in terms of their access to each other. This method would, Boudon states, result in a taxonomy of 'lieux' according to 'syntagmatic types'. This taxonomy would constitute the basis for an explanation of space. According to Boudon such an explanation has not yet been produced either for phonology or for the phenomenon of 'lieux'.

The intuitive and heuristic advantages of linguistic analogy must apparently be paid for by serious shortcomings as to completeness of theory. This becomes evident in Boudon's (1973; 1974) 'theorie des lieux'. The description of the elements of 'lieu' is easy to follow. Whether the inventory is complete perhaps remains open to question. But how these elements must be lawfully connected into coherent wholes can only be answered by case studies of a restricted number of concrete situations. Whether this will ever permit the formulation of a complete theory is doubtful.

Hillier and Leaman (1975; Hillier et al, 1976) did, however, recently attempt to do precisely this. Their 'theory of space syntax' is based on a set of concepts which do not in themselves constitute the syntax but rather its elementary building blocks.

> "In order to proceed to the syntax proper, we must show how all these concepts are related to each other by being constructed out of an even more elementary set of concepts no more than four. In showing this 'deep structure' of concepts... we can also establish the basis of the ideography and the notation that make the space syntax model. The four elementary terms are: continuity, discontinuity, enclosure and permeability. The first pair are objects, the second pair relations ... . Our hypothesis is that all the elementary combinations of an object with a relation have a 'natural', intuitively convincing meaning, and that the set of these elementary combinations has, as its set of intuitive meanings, the foundation of the concepts of space we have in everyday language."

The authors sketched their theory as follows:

> "There exist only a limited number of basic pattern types for space. We learn about them as part of our unconsciously acquired knowledge of the world. To a large extent, they are embedded in language, in the set of spatial, or prepositional, terms that appear in most languages—words like 'between', 'inside', 'among', 'around', 'through', and so on. Built forms and space arrangement, however large and complex they become, are essentially elaborations on some selection of these basic themes. Moreover, human beings in all cultures creatively explore the combinatorial possibilities of one or some of these themes to turn physical space into social space. But however far this goes, it is in the nature of things that intelligibility is retained: our unconscious knowledge of the set of syntactic principles means that we can always decode space patterns, even though it may take time, exploration and mental work. Syntax can, therefore, at once be the key to understanding built form and space arrangement itself, and to understanding how we understand it."

Another attempt to apply generative grammar to architecture (Glassie, 1973) will be discussed later in this book (page 256).

2.2.3 Urban semiology—semantization or desemantization of the environment

A logical consequence of the application of linguistic analogy to urban semiology is that if architecture is a language then the city must be a text. The concept of urban semiology is mainly associated with Barthes (1965; 1967; 1971) and Choay (1967; 1970/1971; 1971). Barthes first advanced his concept of urban semiology at a conference in Naples (May 1967). He held that the city is in general a language consisting of relationships and oppositions of elements, becoming in effect a 'text' inscribed by man onto the soil. The 'meaning' of the city as a 'text' is to be found in the perception of its 'readers'. The discrete units of the city, comparable to semantemes, were recognized earlier by Lynch (1960) as its paths (streets), edges (linear elements forming a boundary), districts (those areas with a distinct internal unity of characteristics), nodes (junctions of paths) and landmarks (visual focal points). The meaning of a city is derived not only from its functions, but also from the structural configuration of the relationships between its elements. According to Barthes, the centre of the city as the place of human encounter and exchange is especially charged with meaning.

Barthes's urban semiology is based on his general thesis of the universal semantization of usage. In contrast Choay presented a theory of increasing desemantization. Her concept of urban semiology entails three semiological systems which have developed successively in history. If a city has a long historical past, all three systems may at present coexist in it. These systems are the Medieval, the Renaissance, and the Industrial. The medieval city was a 'global semiological system' in which the meaning of space was to promote human contact. The Renaissance city was an 'iconic semiological system', the meaning of its space being based on perceptual qualities, with space fulfilling primarily an aesthetic function. With the Industrial Revolution, space in the city lost its meaning and became simply a functional instrument of circulation. The modern city is subject to two 'chronologies', which Choay (after Braudel) called 'slow' and 'rapid'. Rapid chronology involves accelerated technological development, whereas slow chronology involves relatively permanent variables such as territorial, psychological, and cultural behaviour. Choay's semiological analysis aimed to show that a modern city, which is only functionally conceived and therefore 'monosemic', is not in fact monosemically used because of the territorial, psychological, and cultural behaviour of its inhabitants, corresponding to its long-term history. But as a product of a class society the infusion of meaning into the city is left to the dominating class and its merely functional interests. As a result Choay concluded that a new conception of the city as an open polysemic system has to be evolved.

Choay's theory of the desemantization of the modern city was criticized by Ledrut (1973a; 1973b). He described the city as a 'pseudo-text', at present revealing the different life-styles of the different social classes of our society. In this sense it 'communicates' as well as did the cities of the past. Today the content of its message is the alienation and contradictions of class society. The city, however, is not only a reflection of a steady state of alienation, but is also the place where historical action takes place. How such action can change the meaning of the city was analyzed by Castels (1972) in his description of different new types of symbolic usage of the city, which evolved for example during the 1968 Paris riots. Castels maintained that the city will be embued with new meaning when historical action finally overcomes this class society.

Greimas (1974) suggested moreover that desemantization is not only an urban problem but a general semiotic phenomenon based on the substitution of meaningful human behaviour by automated programmes (for example, the substitution of central heating for the open hearth fire). However, this process of desemantization may be compensated for by a process of resemantization (for example, the introduction of fireplaces into centrally heated apartments). What Greimas seems to say here is that semantization and desemantization are dialectically interdependent processes and not irreversible trends in human history.

Basing his work on Greimas's structural semantics (Greimas, 1966; 1974) Fauque (1973) developed an urban semiology based on the ways in which people perceive the city. According to Fauque the only possible starting point is the semantic oppositions [that is, of emplacement (central versus peripheral, close versus far) or of quality (dense versus clear)] which determine people's perception of the city and which may be established by analyzing their spoken views of the city. The preoccupation with the basic elements of the city (analogous to those of language) is also evident in Fauque. Every specific 'urbeme' consists of a set of specific characteristics, namely, its 'semes', which are related hierarchically or in separate constellations. Thus a given architectural structure (for example, a tower) will have specific qualities (for example, density), a specific emplacement (for example, central), and will acquire meaning only as a result of the particular arrangement of these characteristics; the meaning of the tower would change if its particular qualities or emplacement were to change. The question of the ways in which 'urbemes' are related to form the city as a whole involves thus the question of the ways in which people perceive spatial juxtapositions (above, below, at the side of, in front of, behind, etc), and which juxtapositions evoke different emotional responses (that is, which juxtapositions are seen as natural, pleasant, unpleasant, unusual, etc).

The current ideas on urban semiology are not very much different from those on architectural semiology. The conceptualization of the urban environment as a text parallels that of architectural objects as words of a language. If the city becomes a text, the citizen becomes a reader. Practice is thus reduced to the handling of symbolic aspects or to the perception of pertinent oppositions in the urban environment. A similar phenomenon was already criticized as the reduction of objects to signs at the expense of their material, historical, and economic aspects.

The other theme of urban semiology, the semantization hypothesis versus the desemantization hypothesis, could already be recognized as a primary motive in architectural semiology and in semiology of objects. If on the one hand the idea of semantization by use (Barthes) has been applied to the city, it is not surprising that the de facto desemantization (uniformization) has been brought into the discussion. This had been already the case as regards objects (Menna) and as a general trend at the beginning of architectural semiology in Italy.

Of the numerous psychological studies on the perception of the city, on 'cognitive maps', etc (cf Barbey and Gelber, 1973; Bunt et al, 1974; Canter, 1970; Downs and Stea, 1974; Graumann, 1975; Proshansky et al, 1970; Honikman, 1971) I have concentrated here largely on those written by self-declared semiologists. This choice is admittedly arbitrary but was imposed by the limited scope of the survey.

Studies on what I would like to call 'drafted' and 'written' architecture have not been treated in this book. Concerning 'drafted' architecture there are interesting attempts to propose a semiology of the architectural plan (Castex and Panerai, 1974; Castex et al, 1973; Hammad and others, 1973a; 1973c, 1976). The generative function of the architectural sketch has been stressed by Bonta (1972; 1975) and Ph Boudon (1974); the general problem of notation has been fully worked out by Goodman (1968), and the special problem of space notation by Appleyard et al (1964), Ceccato (1971), Halprin (1965), and Thiel (1961a; 1961b; 1962). In my opinion these treatments of 'drafted' architecture are not a part of the semiology of architecture but of 'graphic semiology' (cf Bertin, 1967). Likewise, semiologies of 'written architecture' as manifested in the works of Bachelard (1957), Choay (1973a; 1973b, 1974a, 1974b), Marin (1974) and Ostrowetsky (1974) belong in a semiology of literature or of 'texts'.

2.2.4 Is the urban environment a language?

The main discovery of this survey has been that most authors in the field of architectural semiology adhere in one form or another to linguistic analogy. This persistent analogy between the architectural 'sign' and such linguistic units as the word, moneme, morpheme, etc derived from the early Italian writings of Gamberini, Koenig, and others. Equally obvious is the linguistic analogy in the search for basic architectural units comparable to phonemes. There seems to be a strong tendency to think that since all languages are made up of words and all words are signs, all things made up of signs are languages.

Adding to this confusion are the terms 'grammar' and 'syntax', which played a central role in the Chomskian line of linguistic analogy. As Choay (1973b; 1974a; 1974b) has shown, the roots of architectural preoccupation with 'generative devices' go back to Alberti. But again does the fact that all languages are produced on the basis of grammars, and all grammars are generative devices mean that all things produced by means of generative devices are languages? One may agree with Hjelmslev that all languages exhibit a level of content and a level of expression, each of them further subdivided in form and substance, but if all languages are so subdivided, does this mean that everything which can be divided into a level of content and a level of expression must be a language?

Further complications arise in the notions of denotation and connotation. Is the subdivision of architectural functions (as signifieds of architecture) into two classes, primary (denotative) functions and secondary (connotative) functions (Eco, 1968) actually necessary? If it is, would we have to come to the conclusion that some buildings fulfill only 'symbolic' functions? Would churches thus always be connotative architectural signs, but without a denotative basis, while buildings such as factories which fulfill 'nonsymbolic' functions be only denotative architectural signs? Looking at the multiplicity of functions and motivations in human life, could we not dispense with compensating the narrow functionalist view by the complement of a symbolic function and come to the conclusion that all activities are by definition polyfunctional?

As Mukařovský (1970) wrote in 1942:

> "A building, especially a building for the purpose of housing, is not exhausted in one function only. It is the stage of human life and human life is polymorphous. Also, the function of a house and of each of its rooms is not multiple because the building serves a series of different functions (although this case is conceivable). Its function is multiple, since, even in serving only one purpose, the building must correspond to all those human needs, which may not be explicitly contained in the definition of that purpose, but whose fulfillment is indispensable for everybody using the building just because human existence is multiple and many sided. Thus, architects come increasingly to the conclusion, that a functional conception of a building does not mean a straightforward logical deduction from the building's purpose but a complex reflection, by which the concrete and multiple needs of the user are inductively taken into consideration. What is valid for functions in architecture is valid for functions in general: They must not be projected onesidedly into the object; in the first place one must take into consideration the subjects as the living source of all functions."

One does a disservice to de Saussure (who placed his semiology on a much broader basis than linguistics) if one insists on linguistic analogy in architectural semiology. Linguistic analogy may on some occasions have been heuristically and didactically useful [as for example in Summerson's (1963) *The Classical Language of Architecture*], but for the sake of logical clarification one should rather, like Prieto (1971), consider architectural objects a particular set of instruments belonging to the general class of instruments, of which signs serving the purpose of communication (and linguistic signs in particular) are nothing but a particular subset (Krampen, 1973b).

2.3 The semiotic approach

2.3.1 Different approaches to the study of meaning

Thus far the subject of this book has been defined as the investigation of the function and mechanics of meaning in the urban environment. I have also suggested that the study of indices falls within the sphere of semiology, whether the indices function in verbal communication, in communication in general (as in the arts), in meaningful behaviour, or as behaviour-regulating objects. I have in addition presented a review of the literature which is pertinent to this problem, dealing thus with literature on the semiology of objects in general and on the semiology of architectural and urban objects in particular.

It has become apparent in the course of the review that much of object semiology and architectural and urban semiology has been based on the ill-founded linguistic analogy. It has become equally apparent that at the root of research into the meaning of objects and meaning in the urban environment lie some fundamental theoretical differences. There have been, for example, the approach that takes as its starting point the reduction of objects to signs, and the reverse approach that begins with the reduction of signs to meaningful objects.

These differences in approach can be recognized on the surface by differences in the terminology. The most obvious of these is differences in designation of the field of enquiry as a whole, which has been called 'semiotics' by some and 'semiology' by others. [Sebeok (1972) traced the use of the term 'semiotics' from its origin in ancient Greek philosophy through Locke to Peirce, the founder of modern semiotics, whereas the term 'semiology' originated with de Saussure.] If, however, it was only a matter of differences in designation, one could easily be convinced to adopt one or the other. But there are far more profound differences underlying the various theories about the nature and structure of signs themselves.

So far the review has dealt largely with developments within the de Saussurean tradition. Therefore before any decision can be made as regards the theoretical framework of the material in this book, it is first necessary to balance the review by an examination of the development of the second major trend which began with Peirce's (1960) seminal work *Collected Papers*.

2.3.2 Semiotic classifications

The triadic metaphysics of Peirce

At the base of Peircean semiotics lies the conception of the sign as a relational element. This conception is in the last analysis not geared to communication (that is, the encoding or decoding either of the message of the sender or of processes of signification in behavioural sign systems) but is rooted rather in philosophical, epistemological, and logical considerations. It must therefore be seen in relation to fundamental metaphysical and ontological questions (Walther, 1969; 1974). Nevertheless Peircean metaphysics has an extremely simple basic structure. It contains only three categories:

the being existing without relationship to any other being,
the being existing in relationship to a second being,
the being establishing a relationship between a second and a third being.

On the basis of this metaphysics, the 'Stuttgart School' developed out of Peirce's scattered writings on signs a 'basis theory' of semiotics (for example Bense and Walther, 1973). According to this triadic 'basis theory' this metaphysics is particularly applicable to signs. Any sign may be placed in one of three categories:

in a one-membered relationship with itself (that is, the nature of its material structure, colour, etc),
in a two-membered relationship with the object for which it stands as a sign,
in a three-membered relationship established between itself, a user-interpretant, and the object for which the sign stands (the object being in this case not present to the interpretant).

It was independently of but similarly to this triadic conception that Morris (1955) formulated his three distinct branches of semiotic analysis:

syntactics, which deals with the way in which signs are organized in relation to each other;
semantics, which deals with the meaning of signs;
pragmatics, which deals with the origin, use, and effects of signs.

On each of the three Peircean relational levels signs may be subject to further trichotomous analysis, based again on the three-level metaphysics. If the sign is regarded in its one-membered relationship with itself, then the analysis focuses on its formal structure and its relation to the medium in which it is realized. In this case three aspects of the sign assume specific importance:

the quality of the sign's material elements (substance, form, colour, etc);
the particular relationship realized between these elements which results in the single sign's size, situation, and intensity of its qualities as an object or thing;
the typical formal characteristics of the sign within a more general system (that is, its being a type from which tokens may derive).

Peirce termed these three aspects qualisign, sinsign, and legisign, respectively.

If the sign is regarded in its two-membered relationship with the object for which it stands, three basic types of signs result:

iconic signs, which are constructed as images of the object they stand for and therefore have certain characteristics (at least one) in common with that object (as a building realizes in its physical construction the features contained in the plan);

indexical signs, which exist in a direct or causal relationship with the object they stand for (as the form of the building is causally related to a functional programme underlying it);

symbolic signs, which represent an object regardless of its formal or functional characteristics on the basis of pure convention. (The form of a traditional cathedral, for example, can be assigned by convention and not on any direct or causal relationship to the function of a university building.)

If the sign is regarded in the three-membered relationship which is established between itself, the object for which it stands, and the interpretant, there are three possible ways of interpreting it:

The interpretation can remain open (as in interpreting the economic value of a building which is dependent as a potential object of exchange on the capricious behaviour of the market).

The interpretation can be based on information (as in the objective interpretation of a building according to its use value, which can be tested).

The interpretation may be complete (as in a complex ideological interpretation of a building within the framework of cultural, scientific, historical, political, etc considerations).

Peirce called these three interpretive modes rhema, dicent, and argument, respectively.

Hierarchical nesting of the sign triad

According to Peirce and his followers a given sign must always realize the three sign relationships: sign–medium, sign–object, and sign–interpretant. These three relationships are, however, interdependent. The sign–object relationship presupposes the sign in its material form, that is, a sign–medium relationship. The sign–interpretant relationship presupposes a sign–object relationship, as without this there would be nothing to be interpreted. A hierarchy is thus established of which the base is the sign–medium relationship, the intermediate level the sign–object relationship, and the 'highest' level the sign–interpretant relationship, which presupposes and includes both 'lower' levels (figure 2.3).

A similar hierarchy exists between the qualisign, sinsign, and legisign of the sign–medium relationship, the qualisign constituting the lowest level, the sinsign the medium level, and the legisign the highest level. Similarly in the sign–object relationship the icon constitutes the lowest level, the index the medium and the symbol the highest. Finally in the sign–

interpretant relationship the rhema constitutes the lowest level, the dicent the medium level and the argument the highest level.

Walther (1969) summarized these hierarchical relationships in a tabular matrix (figure 2.4). Since any sign must involve all three relationships, only those combinations of relationships are possible which respect the hierarchical nature of the matrix. This means that the basic sign types are not simply a result of permutation of the nine sign aspects, but are a result of 'hierarchical degradation', which begins at the highest level of each sign and progresses through the medium level, which is included in the first, to the lowest level, which is included in both preceding levels. This in turn means that a sign related to the interpretant on the highest trichotomous level, the argument, cannot enter into object and medium relationships on any lower level, the result thus being an argumentic–symbolic–legisign (ccc). A sign which realizes its interpretant relationship on the intermediate level, the dicent, can realize its object and medium relationships only on its own level or on the higher level, thus the following combinations are allowed: dicentic–symbolic–legisign (bcc), dicentic–indexical–legisign (bbc), or dicentic–indexical–sinsign (bbb). Finally the sign which realizes its interpretant relationship on the lowest level, the rhema, can realize its object and medium relationships on all three levels, thus these combinations are allowed: rhematic–symbolic–legisigns (acc), rhematic–indexical–legisigns (abc), rhematic–indexical–sinsigns (abb), rhematic–iconic–legisigns (aac), rhematic–iconic–sinsigns (aab), and rhematic–iconic–qualisigns (aaa).

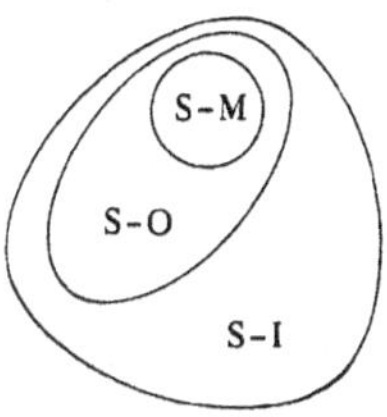

Figure 2.3. Hierarchy of logical inclusions of Peircean sign relationships according to Walther (1969). The sign-object (S-O) relationship presupposes the relationship of the sign to itself as a material form (S-M). The sign-interpretant (S-I) relationship in turn presupposes the sign-object relationship that it is interpreting.

	3	2	1
	interpretant	object	medium
c	argument ⟶	symbol ⟶	legisign
	↓	↓	↓
b	dicent ⟶	index ⟶	sinsign
	↓	↓	↓
a	rhema ⟶	icon ⟶	qualisign

Figure 2.4. Matrix of hierarchically nested sign trichotomies according to Walther (1969).

Ten basic types of signs

The following ten basic types of signs result (beginning at the lowest levels of the hierarchy and progressing upwards).

aaa: rhematic–iconic–qualisigns. Here a material quality such as colour functions as a sign. The relational rules at this level of the hierarchy dictate that a quality functioning as a sign must stand in an iconic relationship with the object for which it stands (that is, the sign must have some characteristics in common with the object). This would hold, for example, in the relationship between the colour orange and the fruit orange. On the other hand, a quality functioning as a sign remains open to interpretation and is therefore 'rhematic'.

aab: rhematic–iconic–sinsigns. In this case a single object or thing (for example the curve of the cost of housing in England in 1976, or the map of London in 1870) functions as a sign and depicts certain qualities of the object for which it stands. Again the sign–object in question must exhibit some characteristics which it has in common with the object it stands for (that is, fluctuations in the cost of housing, or the shape of London). On the other hand, there is no fixed interpretation of the shape of the curve or the map; it is therefore a rhematic sign.

aac: rhematic–iconic–legisigns. This is a general type of sign which depicts a particular quality of the object it stands for. Instead of taking the curve of the housing cost in London in its particular relationship to the specific situation of which it is made a sign, one can also regard it as a 'diagram' capable of depicting iconically any relationship between any two variables x and y and therefore as being open to any interpretation. Other examples of iconic signs are city maps as a class of signs capable of depicting iconically any city at any time no matter what the interpretation. Here another peculiarity of the Peircean sign typology must be noted. The same thing (like a diagram or a map) may be regarded by its interpreter at one time as being iconically related to a single object (London), thus being a sinsign, and at another time as belonging to a general type of relationships, thus being a legisign.

abb: rhematic–indexical–sinsigns. In this case a single object or thing again functions as a sign, but here, in contrast to the preceding example, a causal relationship exists between it and the object it stands for. A wet stain on the ceiling can be taken as sign of a leak in the roof. But the interpretation of the relationship remains open (rhematic): one interpreter may simply think its shape looks interesting, the other might see it as an occasion to call the repairmen, etc.

abc: rhematic–indexical–legisigns. This is a general type of sign which is directly or even causally connected to the object it stands for but which remains open to any interpretation. Directional pointers, no matter whether they are on instruments or on roadsigns, point in the same direction as the object they stand for. The arrow functions as a directional indicator because its directional disposition is that of the road, etc that it indicates. At the same time it is of course also iconic in that it has certain specific characteristics, for example straightness or bends, in common with the road, etc.

acc: rhematic-symbolic-legisigns. This is a general type of sign associated by convention to the object it stands for, while remaining open to interpretation. Triangular, circular, or square traffic signs have come by international law to be associated with different contents (warnings in the case of the triangle, injunctions in the case of the circle, and information in the case of the square), although the shapes are not directly or iconically related to their content. However, as such they remain open as to the interpretation of the specific nature of the danger, injunction, or information they convey.

bbb: dicentic-indexical-sinsigns. Again a single object or thing functions as a sign and stands in a causal relationship with the object for which it stands, but in this case it conveys concrete information about that object: and therefore it is open to precise interpretation. A weather vane is a single object (it may take a variety of forms), but as a sign it conveys information as to the direction of the wind (its object), which causally determines its position. At this point it is important to understand that the sign type in question must by logical inclusion also be iconic: although the direction of the weather vane is caused by the wind, the element, direction, is common to both.

bbc: dicentic-indexical-legisigns. This is again a general type of sign which is directly or causally related to the object it stands for, but in such a way that it furnishes concrete information about it. The words 'city hall' or 'hospital' on a building provide sufficient information (by virtue of their direct relationship with the buildings) for precise interpretations of the specific functions of the buildings.

bcc: dicentic-symbolic-legisigns. This is a general type of sign associated by convention to the object it stands for, but in such a way that it furnishes concrete information about it. This would be the case in a single compound sentence, which conveys information by the conventional association of its information to the object(s) it refers to. Again take the example of traffic signs: a circular shape with a red frame around the figure 50 specifies the negative injunction 'speed limit 50 miles', whereas a blue circular shape inscribed with a white 50 specifies the positive injunction 'minimum speed 50 miles' purely by conventional association.

ccc: argumentic-symbolic-legisigns. This is a general system of signs which conveys exhaustive information on a subject matter which is associated with it by convention. This would be the case,for example, in a system of traffic signs which could convey exhaustive information as to all potential dangers, and provide all the injunctions and information necessary in all possible road situations. Strictly speaking, existing traffic sign systems do not belong to this type as they are not exhaustive and contain indexical and iconic elements (for example, arrows and pictograms) which do not simply function by conventional association (although these could be regarded as similar to deictic or onomatopoeic elements in conventional language). The types of signs on this highest level of the hierarchy are governed by constraints similar to those on the lowest level. An argument must according to the rules of hierarchical inclusion be symbolic (conventional in its sign-object relationship) and typical (must belong in a general and complete system of material composition).

2.3.3 Peircean sign typology and the classification of buildings

Sign typology and the architectural object

The question that arises at this point is what relevance this sign typology has to the semiology of architecture and the urban environment. Can architectural or urban objects be assigned to any of these sign types? Walther (1974) proposed that architectural objects in general be classified as dicentic-indexical-sinsigns (bbb). The architectural object may be taken as a sign, as it is a single realized combination of elements (such as pillars, rooms, windows, doors, etc) existing in a direct or causal relationship with (as an index of) space, time, and its architect-constructor, and furnishing specific information ('dicent') as to the nature of its function and, more particularly, as to the style of its architect. This classification would be too general and of little help if architectural objects were not, as sinsigns, single combinations of elements (such as pillars, rooms, windows, doors, etc) which are themselves sinsigns with specific object relationships. These object relationships may be iconic in that the different elements of a building that function as signs have characteristics in common with what they stand for. This is the case with 'frame systems' such as those in limited spaces, rooms, courts, etc, where the walls, as iconic signs, have the same qualities of length, width, height, and depth and are subject to the same limits as the spaces they delimit. The object relationships of the building parts considered as signs may also be directly or causally determined by what they stand for. This is the case with 'directional systems' such as windows (which are situated in relation to the direction of the sunlight), entrances (which are situated in relation to the roads, etc by which access is given to the building), staircases and elevators (which depend on the number of storeys), etc. Finally, the object relationships of the building parts considered as signs may have a conventional relationship to what they stand for. This is particularly the case with 'symbolic systems' which concern the number, size, and proportion of the iconic and indexical systems. The interpretation of these three types of building parts that function as signs, however, remains open. As single elements (sinsigns) they are seen only in relation to themselves and they cannot therefore be interpreted. It is only when the individual elements are seen as together comprising the building as a whole that they convey specific information.

Sign typology and the city
The same semiotic analysis can be applied to the city as a supersign composed of signs which are in turn classified as sinsigns that have specific object relationships. There are in a particular city (being in this case analogous to a particular building) 'frame systems' consisting of houses, squares, parks, etc for which city parts could be taken as signs. There are 'directional systems' (of streets, rivers, canals, etc) that have indexical sign characteristics. Finally there are 'symbolic systems' which involve the proportional relationships that exist between the city parts and phenomena such as street names, traffic signs, etc which are based on conventional association.

Bense (1968b; 1969b; 1971b) repeatedly expounded this notion of the city as an urban sign system, sometimes under the designation of 'tertiary architecture' (Bense 1971a; 1973). Moreover a predominant feature of the city was seen to be its increasing crustation with what one of Bense's former students (Kiefer, 1970) called 'secondary architecture', namely, the layers of behaviour-regulating signs such as graphic signs, advertising, etc. Bense (not unlike Barthes) talked about the increasing 'semiotization' of the city and drew attention to the predominantly verbal character of this process, defining this aspect of the city as 'textual'. Whereas the city as a complete sign system was called 'tertiary architecture' and behaviour-regulating signs 'secondary architecture', the realized buildings of the city were regarded as 'primary architecture'.

Another application of sign typology
An alternative application of Peircean semiotics to architecture was recently presented by Blomeyer and Helmholtz (1976), who suggested that semiotics as a metamethod can only be used in connection with those specific architectural theories which incorporate semiotic principles. Accordingly, an 'architectural semiotic' as a self-contained theory does not appear to be a viable concept. Furthermore, semiotic analysis must always refer to concrete objects in single situations, the semiotic aspects and relationships being thus extracted from a given object rather than being impressed upon it. Questions as to the nature of the transformation process whereby an object becomes a sign (a semiotic object) were answered by demonstrating the object's relational character with respect to the material repertory from which its constitutive elements were drawn, with respect to the repertory of all architectural objects, and with respect to the repertory of contexts in which architectural objects are placed (that is, interpreted). Having thus defined the three basic repertories according to which the relational status of a specific concrete architectural object may be determined, the authors then defined the object further in terms of Peirce's trichotomous-sign schema. The relationship existing between the object and its medium was seen in the sensory and perceptual substance of its constitutive elements (qualisign), in its particular form (sinsign), and in the

physical, constructive laws and other conventional aspects concerning it (legisign). The relationship existing between the object as a sign and the object it stands for was seen in its frame systems (icon), its directional systems (index), and its metric and other conventional systems (symbol). Finally the interpretive relationship existing between the object as a sign, the object it stands for, and the interpretant was seen in its interpretation as an element in an open context (rhema), as a unit in a closed context (dicent), or as a necessary part in a complete context (argument) (table 2.1).

The relationship between the object as a sign and the object it stands for was regarded by Blomeyer and Helmholtz as being of particular importance in the semiotic analysis. Since architectural objects are always a combination of various elements, they can be seen as systems and, as relational objects, as sign systems. As Bense demonstrated, a general systems classification can be applied to the three different object relationships of Peirce's schema, regarding frame systems as the iconic sign—object relationship, directional systems as the indexical relationship, and selective (conventional) systems as the symbolic relationships. Because architectural frame systems subdivide space by separating exterior from interior space they provide a system of habitable units. Architectural directional systems connect these iconic habitation systems by providing access to them by transportation (of energy, information, human beings, etc). They can thus be called access systems. On the symbolic level the architectural object is subject to metric specifications, and is thus numerically determined in its proportion. This then would be called its metric system.

Blomeyer and Helmholtz held that the actual creative design work of an architect is not determined by semiotic considerations. Classification according to semiotic relationships allows rather for the ordering of data and the detection of problems within a general process of interpretation which precedes the design process.

Table 2.1. Matrix of sign trichotomies applied to architecture according to Blomeyer and Helmholtz (1976).

Medium relation, M	Qualisign sensory (haptic, visual) perceptible substance	Sinsign realization in an individual, singular form and its aesthetic states	Legisign physical, constructive and static laws, conventions
Object relation, O	Icon frame systems habitation systems	Index directional systems, access systems	Symbol selective systems metric systems
Interpretant relation, I	Rhema object interpreted as an element of an open connex (repertoire)	Dicent object interpreted as an element (part) in a closed connex (unit)	Argument object interpreted as a necessary part for a complete connex

The architect is an interpretant in the sense that he suggests new buildings to fulfill a specific function, or new functions for (old) buildings. In this interpretive role he is especially interested in the pragmatic dimension of the architectural object. It will be remembered that according to Bense's object theory (1971c) there are three different aspects of this dimension: the material state of an object (hyletic), its form (morphetic), and its technical functioning (synthetic). The object can be interpreted or used (pragmatics) only if the materials are applied to the form of an object in such a way that it functions (page 10). The effective organization of this process on the pragmatic level—called the design process—is according to Blomeyer and Helmholtz of special importance to the architect and designer.

Classification of buildings according to their predominant sign features
Dreyer (1974) proposed yet another approach to the classification of architectural objects on the lines of Peircean sign typology. He defined the architectural object as a sign by stating its relationship to its material realization (medium), to its underlying building programme, and to its evaluative interpretation.

The material dimension of the architectural sign involves the qualities of its elements (qualisign), the relation of the elements as constituting a single architectural unit to the building as a whole (sinsign), and the general system in which these elements belong (legisign).

The object (programme)-related aspect of the architectural sign involves the underlying construction plan (icon), the underlying functional programme (index), and the underlying formal ideas (symbol). The interpretation may be based on economic (rhematic), utilitarian (dicentic), or ideological (argument) criteria. Dreyer summarized the resulting trichotomies in his 'matrix of architectural semiotics' (table 2.2).

Any building may thus be classified by a combined consideration of the relationships. Here are three examples. In a typical Mies van der Rohe building the dominant feature of the relationship between the sign and its material realization is the impression given of a particular order repeated

Table 2.2. Matrix of sign trichotomies applied to architecture according to Dreyer (1974).

Relationship of the architectural sign			
To its medium (material realization)	element (qualisign)	relationships between elements (sinsign)	general order (legisign)
To its object (= programme)	construction type (icon)	functional type (index)	formal type (symbol)
To its value (interpretation)	economical value (rhema)	use value (dicent)	ideological value (argument)

systematically (legisign) over the building (in the same way that a rectangular gridwork represents a systematic repetition of a specific ordering of elements). The dominant feature of the relationship with the underlying programme is the importance placed on constructive considerations (in preference, for example, to functional or aesthetic considerations). The striking impression on the evaluative level is that of economy (rhematic). The typical Mies van der Rohe building would therefore be classified as a rhematic-iconic-legisign.

In a typical Gropius house of the Bauhaus period, on the other hand, the predominant feature of the relationship of the sign with its material realization is the appearance of the building as a single combination (sinsign) of building elements (unlike the van der Rohe building, which gives the appearance of a repeated combination of a particular pattern or order). In the Gropius house the underlying programme appears to have been determined by functional considerations (over and above aesthetic and constructive considerations) and the evaluative level by utilitarian considerations. The typical Gropius house would therefore be classified as a dicentic-indexical-sinsign.

If one compares these two types of architecture with Le Corbusier's "Notre-Dame-du-Haut" in Ronchamp, the dominant feature there on the level of material realization is the appearance of a systematic opposition of materials and forms (concrete versus wood, etc), which yields its own order. The underlying programme seems to have been determined by the particular form assigned to the building by its religious function (symbolic) —this form clearly deriving simply from convention and not from specific constructional or utilitarian considerations. The predominant interpretive value of the building would be ideological. It would therefore be classified as an argumentic-symbolic-legisign.

2.3.4 Critical appreciation of Peircean sign typology and its application

The utility of the Peircean sign typology

The immediate question is whether the complexity of the triadic semiotic apparatus actually provides a worthwhile alternative to the predominantly dyadic semiological approach. That there are difficulties inherent in semiotic terminology is clear, especially since some of the key terms used are at variance with their conventional use in everyday language or their use in semiology. The term icon, for example, which in semiotics refers to the medium-object relationship characterized by elements common to both medium and object, is very seldom used in semiology. In semiology the term symbol is frequently used to characterize the so-called 'motivated' class of signs, namely those signifiers which by their very nature bring the signified to mind. This use adheres to the Greek etymological origin of the word 'symbol', designating the 'falling together', namely the intersection, of two sets of characteristics. In semiotics the index is by definition distinct from the icon or symbol. A sign always falls in one of these three categories, although a particular object may at one time be interpreted as an icon (a weather vane indicating wind direction) and at another time as an index (the weather vane whose position is causally determined by the direction of the wind). On the other hand semiologists like Prieto hold that all signs, whether motivated or not, are indices since they are indicative of that with which they are associated as a sign. The use adheres to the Latinate origin of the word.

Nevertheless a methodology obviously cannot be rejected solely on the ground of difficulties in its terminology. The next step therefore is to establish the actual utility of the semiotic apparatus. Blomeyer and Helmholtz (1976) gave one answer in maintaining that semiotic terminology does not concern the whole design process [characterized by Bense (1971c) and Walther (1974) as consisting of a planning phase, a realization phase, and the phase where the object is put into use], but involves only the "determination of the various levels of planning (e.g. structuring of data about physical, social, political environments and individual experiences)... In using semiotics one cannot resolve controversial issues or innovatively show new aims or solutions (these being political matters or a question of design); one can, however, illuminate the various aspects of problems, indicate the need for more information and identify potential areas of difficulties."

One could agree that an analytic tool cannot fulfill synthetic functions and that the structuring of the levels and the data of the planning process is an important phase of the entire problemsolving process. Bense (1971c), on the other hand, held that the act of 'invention' as the transition from an existing 'inventory' to the 'invented' (by a recombination of the elements of the inventory) is comprehensible only as a 'semiosis', since everything regarding consciousness pertains to the category of signs which can only be explained by signs. According to Bense any recombination of an inventory will involve two characteristic operations: the construction of supersigns out of elementary signs and the transformation of signs (the coding of a sign from one repertory into a sign from another), the schema of creativity being thus reducible to a sign schema.

The contrast between Blomeyer and Helmholtz's unpretentious interpretation of semiotics and Bense's claim that creativity is reducible to sign operations casts some doubt on the value of adopting the Peircean terminology while this contradiction remains unresolved.

A more detailed criticism, however, is that, although the architectural and urban design process was regarded by the 'Stuttgart School' as an interpretive, that is pragmatic, activity involving the moulding of material aspects (hyletics) into form (morphetics) which is synthesized in such a way that the object functions (synthetics), designers do in any case adhere to this view without necessarily being aware of Peircean sign typology or of Bense's applications of it to design. So unless one specifically wants to 'extract' trichotomous relations from the semiotic process after the application of this knowledge, it seems that this knowledge is actually nonsemiotic.

Apart from these utilitarian questions, there do moreover seem to be contradictions inherent in the various applications of the Peircean schema to architecture and the urban environment.

Walther (1974), for example, held that all buildings fall into the category of dicentic-indexical-sinsigns. Blomeyer and Helmholtz (1976), on the other hand, held that each individual building must be exposed to the whole semiotic classification process, each building being therefore analyzed anew in terms of medium, object and interpretive relationships. If, as Walther says, all architectural objects fall within the category of dicentic-indexical-sinsigns, the repetitive process advocated by Blomeyer and Helmholtz would become redundant.

Dreyer (1974) classified the typical Gropius building as a dicentic-indexical-sinsign. Is Gropius then an architect's architect, realizing in his buildings the architectural prototype? What then is the status of Mies von der Rohe and Le Corbusier, who do not?

These are just some of the practical questions which have to be answered about the applications of Peircean semiotics. There are, however, more theoretical questions which have risen from the triadic metaphysics.

The ideology of the Peircean sign typology

George Klaus (1963) criticized the fundamental idealism of Morris's semiotic schema on the grounds that the whole system makes no provision for the epistemological origin of meaning, meaning being simply reduced to a mentalistic interpretive process. In contrast, Klaus held that meaning reflects in a more or less direct way the material reality which precedes it. In order to correct the mentalistic bias which he saw in the triadic schema, he then added to the triad (of syntactics, semantics, and pragmatics) a fourth relational dimension called sigmatics.

This sigmatic dimension involves the relationship between the meaning of the sign and material reality, that is, the relationship between the sign and the material object it stands for. The necessity for this additional dimension lies in the fact that the object in semiotics, in accordance with the hierarchical nature of the system, becomes included in the pragmatic relationship and must therefore be regarded as 'mental', since it is dependent on interpretation. This sigmatic relationship is not without general interest if one wants to keep the semiotic system open to criticism. It is certainly true that a sign stands for the mental process which is its meaning, and the question whether the mental process interprets the circumstances of material reality correctly or incorrectly may well be an extrasemiotic question. However, if semiotics wants to retain an epistemological basis the question of how men know reality cannot remain unanswered.

Resnikow (1968) acknowledged the importance of Klaus's criticism but argued that the problem need not be treated by a special discipline but could be handled within the semantic context if semantics were conceived dialectically as a relationship between meaning and sign meaning, both of which cannot exist independently of the two-way relationship involved in the subjective reflection of objective material reality. This may well be true, but it could be shown that the tetradic concept would in this case be more useful than the triadic. For example, in a semiotic analysis of the architectural plan this relationship would be of great value in studying the difference between the planning of the architectural object and its execution. An analysis of the architectural plan as a compound of signs would proceed as follows:

On the syntactic level the relationship between different graphic elements of the architectural plan would be the object of study. On the semantic level one would study the relationship between the mental conception of a building (which would be based on prior conceptions of existing buildings) and its graphic portrayal in the plan.

The pragmatic level would deal with the relationship between the graphic expression of the plan and those who put it to use, that is, those who produce the plan or 'read' it (taking into consideration the fact that the average person would be unable to read the architectural plan).

The sigmatic level would deal with the relationship between important architectural objects, materially executed and standing in the environment (for example, the works of the 'great architects'), and imitations of these or the influences they have had on the plans of epigones.

Whether one agrees with Klaus's or Resnikow's attempts to counteract the metaphysical idealism of triadic semiotics, their criticism of it is what is important. It points to the same error which was apparent in the reduction of signs to objects (discussed above), where the material status of the object was sacrificed in favour of its symbolic status. This symbolic status must by definition be a result of interpretation, that is, must fall under pragmatics, becoming thus in the process a mentalistic phenomenon. This is precisely what Klaus pointed out. The so-called syntactic or medium relationship of a sign with itself is mentalistic as it also involves the interpretive relationship. The medium-object relationship is likewise mentalistic, as both the medium and object are products of the interpretive relationship. The object in this case is not a real object but is merely the perception of an object in the mind of the interpreter, endowed with those selected properties which serve the interpreter's purposes (leaving aside such fictitious objects as the unicorn which has generated such interest in semiotic discussion).

If this were applied to the city, it would mean that the city would become reduced either to its 'cognitive map' or to a utopia (the urban equivalent of the unicorn). I can then conclude this review by regarding those of the Peircean school who applied the Peircean triad to architecture and the urban environment, as falling in the same category as those who reduced objects to signs. This in fact is what they say themselves in speaking of 'introducing' or 'establishing' an object as a triadic sign. I will head in the opposite direction by regarding the sign as a special kind of object and by basing the semiological status of architectural and urban objects on operations other than on their reduction to signs.

On the other hand, there could be some merit in using the three relational sign dimensions as one option among others for classifying planning data or for generating research questions. With regard to the latter possibility I would suggest an entire research programme based on semiotics.

The one-membered relationship of the sign to its medium suggests the following investigations:

a study of the various qualities of building materials in relation to human perception (qualisign),

a study of the individual properties of the relations between building materials or forms in order to optimize contrast and distinction in architecture (sinsign),

a study of the potentials for standardization of architectural components and the structural organization of these components with a view to minimizing uniformity (legisign).

The two-membered relationship of the sign to its object (for example to the underlying programme) suggests the following investigations.

a study of the scope and limits of the iconic manifestation of the idea of the plan in the realized construction (icon),

a study of the aesthetic effects of 'transinformation' (that is, of the visual information regarding the causal relationship between a buildings's external form and its internal functions) (index),

a study of architectural form and its traditional and potential symbolic capacity (symbol).

The three-membered relationship of the sign as relational element between object and interpretation suggests the following investigations:

a study of situations in which an architecture open to multiple use and interpretation would be warranted (rhema),

a study and classification of situations in which clear and unequivocal interpretation of architectural form is necessary (dicent),

a study of the scope and limits of comprehensive architectural prefabrication systems or of the feasibility of general architectural systems within a historical, political, and economic context (argument).

3

State of the theory

3.1 The need for a model of the cognitive act

Up to this point the question of the way in which meaning is conferred upon the urban environment has been discussed in terms of the literature available on the topic. Of the two recurrent possibilities, namely the reduction of objects to signs or of signs to objects, the latter appeared to be the more promising starting point. This became particularly clear in Prieto's (1973; 1975b) theory of the instrumental act, where the instrument was revealed to be a bifacial unit consisting of operant and utility, parallel to the sign as a bifacial unit consisting of signifier and signified. But can the city or a building be considered an instrument? Is there such a thing as an instrumental act which is applicable to an individual building or the urban environment as a whole? What constitutes an instrumental act or—for that matter—what constitutes any act?

To answer these questions one must first establish a very general model of how human action takes place. The point of view taken here is that activity and cognition are interdependent. As regards the specific content of this book this interdependence may be illustrated by Marx's comparison between the building activity of a bee, beaver, or ant and that of men (Marx and Engels, 1970). In the latter case the conscious preconception of a building by the human constructor precedes its material realization.

The act of cognition itself must in turn be regarded as a dialectical unit consisting of a sensory and a cognitive phase, the former serving as the perceptual input to the latter, the latter in turn correcting and improving the sensory and perceptual input.

What does this have to do with semiology? In the following pages the place of the cognitive act in semiology will first be established in a very general way and then will be applied to the specific case of the architect, the urban designer, and the consumer-citizen. However, some preliminary remarks should first be made about the historical development of semiology and its relation to linguistics.

3.2 Phonology and linguistics

Peircean semiotics is based on a triadic relational theory and the discipline of ordinal categorization. Semiology, on the other hand, developed out of linguistics, What had to be explained by linguistics was essentially the mechanics by which crucial phonetic differences are alternately produced and recognized.

As a result of advances in instrumental phonetics which made possible the direct recording of speech sounds, it was discovered that although there may be distinct differences in individual articulations of a particular speech sound such as /a/, these differences are for all practical purposes not recognized as significant by the average person. On the other hand, strict borderlines are drawn between the different articulations of /b/ and /p/. The conclusion was drawn that people recognize speech sounds by distributing them into abstract classes called phonemes. Phonemes are thus not 'real sounds' but are simply classes into which sounds are distributed. The perception of a speech sound is therefore an act of classification.

These discoveries gave rise to a new discipline called phonology. Its representatives are Trubetzkoy and Jakobson of the Linguistic Circle of Prague, who not only worked out the relationships operating between the phonemes of a particular language and constituting a structure of opposition, but showed by a procedure called commutation that changes in the phonemic structure (for example the replacement of one phoneme by another) result in changes in the semantic structure (as for example with /bark/ and /dark/). Similar thoughts had led de Saussure to proceed from the conception of the sign as a bifacial unit consisting of signifier and signified to the assertion that no sequence of sounds can be regarded as linguistic unless they are the medium of an idea, and that, in turn, concepts such as 'house', 'white', or 'look' become linguistic entities only by association with acoustic images (de Saussure, 1971). He then proceeded to the recognition of fundamental similarities between language as a sign system expressing ideas, and phenomena such as graphic notation, the deaf-and-dumb alphabet, symbolic rites, social codes, and military signals. It then became possible to conceive of a science the specific function of which would be the analysis of the particular role of signs in society. The science would be a branch of social psychology and therefore part of general psychology. For this science de Saussure proposed the name 'semiology' (de Saussure, 1971).

3.3 Linguistics and semiology

I have already drawn attention to the tendency to construct, in analogy to linguistics, 'semiologies' of various human activities (for example, object production, architecture and urban design). What is remarkable about these attempts is that although they are essentially based on linguistic analogy by the adoption of linguistic terminology and methodology, they have nevertheless disregarded the fundamental innovation which was made in phonology—that recognition (of speech sounds) is an act of classification.

Questions about the classificatory procedures whereby people distinguish between different speech sounds, about the precise nature of oppositional structures in linguistics, and about the processes whereby commutation engenders changes in meaning were at first asked by phonologists in reference to language. This does not exclude the possibility of asking similar questions about sign systems whose fundamental substances are not sound, but are other materials from which signifiers can be produced. If there is to be an urban, architectural, or object semiology, the questions that have to be asked are not linguistic, but will nevertheless stem in part from the scientific and methodological options which have been taken in linguistics since the beginning of phonology, these options being typical of any human science which has as its subject matter human behaviour and different activities based on conceptualizations of material reality.

We can ask how people recognize different buildings as belonging to the same or different classes (just as we ask how they recognize this phenomenon in different speech sounds) without having to resort to linguistic analogy. We can also ask which pertinent oppositional features distinguish one building or object from another, and how commutation of these features affects their perception, in the same way that we ask what constitutes oppositional structures in linguistics. In order to answer these questions, certain other fundamental questions about the perception and classification of objects (whether registered as acoustic or visual, or as 'images' of other sensory modalities) have to be dealt with. In addition to this some basic epistemological questions also need to be answered. The most general and coherent attempt to do this is Prieto's (1975b) model of the cognitive act, which he based on set theory. The following pages constitute a brief summary of his position (which was mentioned above in a less general form in the context of object semiology).

3.4 Basic problems of a semiological model of cognition

3.4.1 Set theory as a tool for constructing a model of cognition

As cognition and activity are dialectically connected, an explanatory model of the cognitive act is fundamental for any explanation of human activity. The model of the cognitive act which will be set out below is based on the fundamental principles of set theory. The following section will therefore involve a basic vocabulary of set theory, but it must be remembered that this vocabulary is intended only as a tool for the construction of a cognitive model, and that cognition is not in itself considered a semiological issue. It will, however, become clear that semiology is concerned with cognition insofar as it is the cognition of fundamental connections between systems of sets which form one of the fundamental issues of semiology.

The theory of cognition proposed here is that to recognize the identity of an object is to recognize it as an element of a set, or in other words to recognize it as an element of the 'extension' of a concept. Identifying an object therefore entails conceptualization.

Recognition of an object also implies referring to the properties by which it is identified, or in other words recognizing it as an element of the 'intension' of a concept. A set of objects can thus be described in terms of the objects it contains, namely as the extension of a concept, or in terms of the properties of these objects, namely as the intension of a concept.

Recognition of an object also implies recognizing its difference from all other objects constituting its 'universe of discourse', the frame of reference provided by a particular discussion. These form the complement of its set. The difference between one object and the objects forming the complement of its set is based on the differences between their properties. As difference is a reciprocal relationship, the property of an object exists in relationship to the properties of any other object. This relationship may be designated in two ways, either a, presence of a particular property in one object, or $\bar{a}$, absence of that property in another object.

A property of a given object (A) can be said to manifest itself in the set of all physiological changes which it provokes in the sensory organs of a subject. If this same property is absent from a second object (B), there would be a corresponding absence of the physiological changes provided by object A. If object B did not have positive properties of its own in addition to the above negative property, it would of course not be present to the subject.

The case may be that the positive property of object B is also present in object A. It would be only by virtue of this property common to object A and object B that one could recognize (classify) object A, as it is only by virtue of this property that it exists in a common universe of discourse with object B.

This simplified case may be used as a calculus to model perception and recognition. The recognition of an object and hence the establishment of its difference from another object may be seen as based on a calculus of two contradictory possibilities, the conclusion being which of the two possibilities applies to the object. This calculus may be written as follows:

$$ua \neq u\bar{a},$$

where a and $\bar{a}$ are the correlated positive and negative properties of the different objects, and u is that common positive property by which both objects are recognizable. The absence of a property ($\bar{a}$) can be understood as a 'zero sign': any u (for example, a moving car) with the property a (for example, flashing stoplights) is different from a u (moving car) with the property $\bar{a}$ (stoplights that do not flash).

At this point the term 'universe of discourse' may be more fully defined. An object which has been recognized, that is, has been classified as an element of a particular set, together with all those objects of which one will automatically have at least some knowledge as a result of knowledge of the recognized object and the reciprocity of the relationship of difference, form a universe of discourse. The property common to all objects in such a universe of discourse (apart from their other positive and negative properties) is defined as the common 'universal' property of the objects. This property need not necessarily be single (as u was in the example above) but may be substitutive in the sense of a logical sum ($a \cup b$). It is important to differentiate in this context between *virtual* knowledge of an 'interiorized' complete calculus of a system of sets and *actual* cognition manifesting itself in a statement attributing an object to one of the sets in a system of sets.

The example presented so far has been kept as simple as possible.

There were only two objects, a ('universal') property common to both, and one other property present in one and absent in the other.

The recognition of an object is not always based simply on the calculus of the presence or absence of only one property in addition to the presence of a universal property, as proposed in the example $ua \neq u\bar{a}$. Things are generally more complicated, as in:

$$uab \neq ua\bar{b} \neq u\bar{a},$$

or

$$uab \neq ua\bar{b} \neq u\bar{a}b \neq u\bar{a}\bar{b}.$$

Each of these possibilities confers upon an object one or more precise properties, thus distinguishing it from those objects with one or more correlated properties. These objects are in a state of logical exclusion from each other. Their logical sum forms by definition a universe of discourse.

This type of calculus can be called a system of sets. Since each set in this system is defined in reciprocity to each other set, the system forms a structure. The relation which all the sets have in common with all the other sets in the system is that of opposition. The system of sets thus forms an oppositional structure. An object is therefore recognized by establishing its membership in one of the sets of a system of sets, all of which have been subjected to the calculus of difference and identity. Knowledge of one object thus implies knowledge of all other objects pertaining to the calculus, on the grounds of the reciprocal nature of the relationship of difference.

3.4.2 From set theory to semiology: the semiotic structure

The three basic terms introduced so far have been set, complement, and universe of discourse, the latter being a logical combination of the former two. A universe of discourse was described as a system of sets existing in opposition to each other by virtue of the presence or absence of certain properties, and united at the same time by one or more 'universal' properties present in all the sets belonging to that universe. (Another term for 'set' is 'class'. The universe amounts then to what is referred to as a 'classification system'.)

The transition from this model to semiology is accomplished by asking under what conditions are what properties of objects perceived as being in a universe of discourse. What is implied here is that any such system of properties will always presuppose the particular classificatory 'point of view' of the subject by which these properties are conceptually organized into a system. This may be aesthetic, utilitarian, etc, or simply that point of view determined by the subject's physical position in space in relation to the object.

This 'point of view' in itself constitutes a universe of discourse. This is likewise composed of sets which exist in opposition to each other by virtue of the presence or absence of certain properties, but are linked by a universal property. The nature of the relationship between these two universes dictates that each set in one must correspond in a specific way to a specific set in the other. It is only as a result of this relationship with each other that the 'relevance' (or 'meaningfulness' or 'sense') of both universes can be established (in the same way that the Saussurean signifier and signified become meaningful only in relation to each other. The Saussurean model lies at the basis of this whole schema). The conferring of meaning by the one on the other is therefore reciprocal.

Prieto (1975b) called this particular reciprocal structure a 'semiotic structure'. The terminology in this book follows de Saussure (1971) and Prieto in calling the science dealing with semiotic structure 'semiology', and using the term 'semiotic' only as an adjective describing a bi-universal reciprocal structure. Why should such a structure be described as semiotic?

Any conceptual ordering of objects into a universe of discourse is always 'significative' in the sense that it presupposes another universe of discourse, the properties of each universe respectively constituting one side of a bifacial unit. This bi-universal reciprocal relationship is a relationship of signification—is semiotic—by virtue of the fact that the one is meaningful only in terms of the other, that knowledge of the one entails knowledge of the other. It may now be useful to define this semiotic structure more explicitly in Saussurean terms by calling the first universe 'the universe of the signifier' and the second (the 'point of view') 'the universe of the signified'.

In contrast to this relationship of signification there also exists the relationship of opposition between the sets within the two participating universes. What is being suggested here is a particular form of relativism which implies that an object can be known only in relation to other objects, this relation being a two-fold relation: of oppositions to objects in the same universe of discourse and of signification with objects in the correlated universe. Whereas the first relationship is based on recognition of differences between objects, the second is based on recognition of connections between objects.

In summary it seems that recognition of material reality is by its very nature significative, or 'semiotic'.

There are two possible types of semiotic structure. In the first type the relevance of one particular universe is established by another, and vice versa. The sets in both systems are in one-to-one correspondence and within each system they are in a relationship of mutual logical exclusion (if one disregards the universal property).

A simple semiotic structure of this type could take the following form:

$ua \longleftrightarrow yn,$

$u\bar{a} \longleftrightarrow y\bar{n},$

where u is the universal property of the two objects in the first system of sets (universe of the signifier), y is the universal property in the second system (universe of the signified), a and $\bar{a}$ are the positive and negative properties of the two respective sets in the first universe, and n and $\bar{n}$ are the positive and negative properties of the two respective sets in the second universe. It is this basic type which will be of interest in this book.

The second type of semiotic structure is more complex and applies to language. The universe of the signified would in this case consist of a multiple system of sets. (A third possibility would be a semiotic structure containing multiple systems of sets on both sides. This, however, would belong in the same category as the first symmetrical type.) In a multiple system the logical sum of all the sets would result in only one universe of discourse, but the relationship between them would not be one of mutual logical exclusion but partly one of logical inclusion or intersection. This second type would take the form shown in figure 3.1

$$\begin{array}{l} \text{universe A:} \\ \text{one system of sets} \end{array} \left\{ \begin{array}{l} \left.\begin{array}{l} uabc \longleftrightarrow yno \\ uab\bar{c} \longleftrightarrow yn\bar{o} \\ ua\bar{b}c \longleftrightarrow y\bar{n}o \\ ua\bar{b}\bar{c} \longleftrightarrow y\bar{n}\bar{o} \end{array}\right\} \text{system 1} \\ \left.\begin{array}{l} u\bar{a}bc \longleftrightarrow yn \\ u\bar{a}b\bar{c} \longleftrightarrow y\bar{n} \end{array}\right\} \text{system 2} \\ \left.\begin{array}{l} u\bar{a}\bar{b}c \longleftrightarrow yo \\ u\bar{a}\bar{b}\bar{c} \longleftrightarrow y\bar{o} \end{array}\right\} \text{system 3} \end{array} \right\} \begin{array}{l} \text{universe B:} \\ \text{three systems of sets} \end{array}$$

Figure 3.1. A semiotic structure in which the universe of the signified consists of a multiple system of sets.

3.4.3 The semiotic chain: (de)notation and connotation

The semiotic structure was introduced on the basis that the oppositional structuring of sets within a universe of discourse presupposes a 'point of view', namely another universe of discourse, the relevance of both sets being established by their relation to each other. The semiotic structure was thus defined as a bi-universal reciprocal relationship of signification in which the first system of sets plays the role of universe of the signifier, the second the role of universe of the signified.

The question now arises that if the principle that 'a universe of discourse presupposes a point of view' is logically valid, should this not be equally applicable to the universe of the signified? If this were not the case how would one account for its relevance, or meaningfulness, which exists prior to the particular relevance resulting from its relationship with the universe of the signifier? This could only be accounted for by reference to another 'point of view', to another universe of discourse. In this second semiotic structure this additional point of view would now play the role of universe of the signified, relegating the first to the role of universe of the signifier. The objection may now be raised as to how a particular system of sets can simultaneously function as a universe of the signified and as a universe of the signifier. The answer is straightforward. This would be the case when a system of sets is subjected twice to a calculus of its oppositional structure, that is, when it is 'classified' twice. Prieto (1975b) called the first calculus by which a universe is classified '(de)notative', as it involves a primary cognitive 'notation' which establishes it as the universe of the signifier in one semiotic structure. The second calculus by which the same universe is classified again is called 'connotative', as it involves an additional cognitive notation which establishes it as the universe of the signified in a second semiotic structure. It is important to understand that within one semiotic structure the only relationship that exists between the universe of the signifier and the corresponding universe of the signified is one of signification. There can be no connotative relationship between them, as connotation involves only one specific system, which is subjected twice to the classification process. This concept is referred to as the double calculus of the oppositional structure of a universe, or in short 'double classification'.

3.4.4 Semiology and the social sciences

So far the development of the basic theory has involved three steps. The basis was the three concepts of set, complement, and universe of discourse. The relevance of this to semiology was then shown to begin where two systems of sets are coordinated in such a way as to form a semiotic structure, the relevance of each deriving from their relation to each other and one forming the universe of the signifier, the other the universe of the signified. It was then shown, according to the logic that a universe of discourse presupposes a point of view, that the universe of the signified would likewise presuppose a point of view—another universe of the signified—and would therefore become, in the second semiotic structure, a universe of the signifier, being thus classified twice, one notatively as a signifier in one semiotic structure, the second time connotatively as a signified in another semiotic structure.

As it stands, this semiotic chain would then be recursive, each universe of the signified being subjected to double classification. In order to avoid such an infinite chain, it would be necessary to arrive finally at one universe of discourse which would in some way be an exception to the rule. As far as every subject is concerned, there is in fact a universe of discourse which fits this condition, namely a universe which guarantees the relevance of a universe of the signifier without presupposing another universe of the signified. This universe of discourse is defined for each subject by the universal property 'I–subject'. The rule that a universe of discourse presupposes a point of view is just as valid in the case of the subject, as it is for objects. In this case, however, it is the subject itself which provides the point of view by which it 're-cognizes' itself and by which its relevance is established. This does not, however, imply that there is no material reality preceding either subject or object. It simply means that objective material conditions are reflected subjectively. It was for this reason that de Saussure classified semiology as a branch of psychology, the science which studies the subjective and individual reflections of reality. However, in talking about linguistic phenomena, de Saussure was keenly aware that language ('langue') as such is intrasubjective as well as extrasubjective, individual as well as social. He therefore classified semiology as a branch of social psychology, that is, as a subjective reflection of a social reality. The re-cognition of himself which a subject carries out, is thus never independent of the society to which he belongs. If this is the case, the subject's self-concept will to a certain extent depend on the particular formation of the society (including its power structure) to which he belongs at a given moment in history. The implication of this is that his self-concept will be subjected (at least in our historical period) to the power of the dominant social classes in the society in which he has been socialized. It is at this point, where semiology is seen to be determined by class structures, that one will eventually have to move from semiology to sociology. This will, moreover, lead through sociology to political economy, at the point where class structures are seen to be determined by politico-economic factors. Under these conditions the statement that "there is no socially neutral knowledge" appears to be appropriate. Any knowledge of material reality which results from the re-cognition of identity of objects must, in the last analysis, contain a component of relevance or meaning, which cannot be derived from the properties of the object itself, but is dependent on the subject. Since this subject is a socialized being, this component of knowledge must be rooted in society and cannot, therefore, be socially neutral.

The particular set of tools (which includes housing and urban conglomerates) of a particular society, its modes of production, and all its mores and customs represent that society's way of conceptualizing ('cognizing') material reality. There is always a 'function', a practical aim, underlying the 'point of view' which establishes the relevance, or meaning, of a particular conceptualization of material reality. For example, underlying a society's or individual's conceptualization of the instruments he will use is the practical goal of production; underlying a society's or individual's conceptualization of signs lies the social aim of communication.

There is no cognition of material reality which is cognition only. The manner in which one 're-cognizes' the objects of a universe of discourse aims always at acting in a specific way on objects in another universe of discourse. Cognition always implies activity. Since activity, on the other hand, presupposes cognition of that material reality at which it is directed, one could define the object of human sciences as the different activities by which human beings act on material reality.

According to the model of cognition presented here, cognition of material reality is seen as taking place from the point of view of a subject socialized in a particular historical period of a given society. This accounts for the historical character of knowledge. Any attempt to render 'natural' the cognition of material reality, that is, to portray it as emanating 'naturally' from the object, denies the historical character of knowledge and must therefore be criticized as ideologically unsound. In denying this historical character of knowledge such an ideology would maintain conditions where knowledge and cognition are seen to be 'natural', namely fixed, and therefore not relative to historical and social conditions. Therefore the necessary and sufficient condition for a human science which takes cognition as the object of its investigation, is to recognize this historical character of knowledge.

There is a tendency among the ruling classes of different societies to attempt, more or less consciously, to conceal this historical character of knowledge. They clearly have a vested interest in maintaining the belief that the status quo derives from factors which have a natural and absolute status. There is, therefore, the concomitant tendency to resist the advancement of those human sciences which would replace existing ideologies and therefore expose them as relative and subject to change.

3.5 Application of semiology to architectural and urban activity

3.5.1 Semiology of production and consumption

The very general semiological framework which served as the basis for the model of cognition presented in the preceding section, may now be applied to the specific kind of cognition which is involved in carrying out specific types of activity. It has already been shown that semiology is geared to describe any type of human activity on the basis of its cognitive model. One element is, however, still necessary to complete the theory. Although semiology presents a model of cognition, it does not provide a general theory of activity. The cognitive model is sufficient as a prescriptive theory of activity, but is insufficient as an explanatory theory. Any theory of activity must of necessity also be explanatory. The theory of activity followed below maintains as its basic assumption that human beings (and human beings only) create their own conditions of life. This self-sustaining process is made possible by work. Work is exerted in two directions. On the one hand man creates the external conditions of his survival by working on 'objective' situations, that is, by transforming materials into objects. On the other hand he works on 'subjective' situations, by developing his human capabilities. It is work—the investment and circulation of energy—which changes 'nature' into objects and transforms the 'nature of the self'. Work presupposes on the one hand the existence of materials and objects on which to work, and on the other hand other subjects with whom to work. The latter condition derives from the fundamentally social nature of work. If it is true that work presupposes other subjects with whom to work, cooperation then becomes one of the basic issues which man has to face in order to work. The act of cooperation must be regarded as a circulation process in which men influence each other's psychological makeups and in which an interchange of skills takes place. If semiology is to concern itself with human activity, then various cooperative activities should constitute a specific branch of semiology, which could be called the semiology of cooperation. The semiology of communication, studying the circulation of signals by which subjects mutually influence each other's psychology, would fall under this section.

The supposition that work requires objects on which to work entails the question of production. In the act of production, men influence the state of nature at large and are in turn influenced by each new state of nature they produce. A circulation of energy and material takes place. Since the production of objects is not an end in itself but a condition of human survival, it entails in turn the question of consumption and the fact that man, in producing objects, creates at the same time new needs (the car, for example, has become a 'necessity'). As with cooperative activity, production and consumption should constitute a separate branch of semiology, which could be called the semiology of production and consumption. Since production and consumption necessarily entail conferring a 'sense' on the object concerned, it would appear that the semiology of production and consumption would fall under what Prieto (1975a) called the 'semiology of signification'. The semiology of production and consumption could moreover be subdivided into semiologies of different kinds of production and consumption, for example the semiology of architectural or urban production and consumption, which is the subject of this book.

This proposed subdivision of semiology presents distinct advantages. For one thing it brings out the fundamental dialectics of the semiological circulation process occurring between those involved in productive, consumptive, and cooperative activities. In these processes there is always an active and a passive partner. For example, in the semiology of communication the producer of the signals, often labelled the 'sender', plays a more active role than the consumer of those signals, often labelled the 'receiver'. The sender enters the semiotic structure by classifying meanings in the universe of the signified and expressing them by finding the appropriate signals within the universe of the signifier by which to transmit them. The receiver enters the same semiotic structure from the other end. He defines the identity of a signal by classifying it in the universe of the signifier and finds in the universe of the signified the appropriate meaning with which it is associated. The same can be said of the producer and consumer. The producer enters the semiotic structure by classifying functions in the universe of the signifier and providing appropriate tools for these functions. The consumer is faced with the task of identifying that tool (in his toolbox) by which he can carry out a particular operation, connected in principle to the tool by virtue of the same semiotic structure which the producer used. A similar parallelism is established by comparing conditions of 'breakdown' in cooperation (for example, misunderstanding in communication) and 'misuse' of tools.

Misuse of tools occurs when a particular tool is produced which is inappropriate for the purpose for which it was intended, or when a tool produced for one purpose is transfunctionalized to fulfill another. This parallelism should guide the reader through the following sections on the semiology of architecture and urban semiology. The analysis below regards the semiology of architecture first from that part of the semiotic structure which concerns the architect-producer, then from that part which concerns the citizen-consumer, and establishes finally conditions of architectural misuse. In the section on urban semiology the city is regarded as a 'box of tools' produced by architects. The same procedure is then followed as was followed for architectural semiology. The final outcome should be a thorough understanding of the semiotic structure involved in architectural and urban activity.

3.5.2 Semiology of architectural production

As far as the architect-producer is concerned, the conceptual processes involved in architectural activity entail proceeding from abstract sets of properties to a concrete design. (From this point 'conceptualization' will be used interchangeably with 'cognition', since a concept is the extension of a set.) The architect first conceptualizes one particular set of properties which are in opposition to all the other sets in the system of sets of properties which will determine what he is going to design. If he is designing a house (rather than a factory, church, school, etc) that set of properties which he will first bring to mind will be the set of properties of houses in opposition to those of factories, churches, schools, etc. He will then reconceptualize this same set, this time as a set of methods suitable for carrying out housing functions. The first classification therefore involved what he was going to design, the second how he was going to design it. The result is a set of properties describing a specific method of solving housing functions. He now has to find a housing design with a set of properties which is appropriate to this method; in other words he has to coordinate a particular set of housing functions with a particular set of design features, that is, design properties. I call this act of coordination 'composing' and the result an architectural 'composition', as these terms seem to designate well the 'putting together' of two systems of sets.

This set of design properties was, of course, conceptualized in opposition to all the other design variants which could have been composed with the method chosen to carry out the housing functions. The schema of this two-fold conceptualization of the coordination between two systems of sets is given in figure 3.2. The set of properties of the housing design and the set of properties of houses in general become relevant—that is, meaningful—in terms of their relation to each other, and together form a semiotic structure. It will be remembered that the set of properties of houses was classified twice (once in opposition to other possible building variants, and once in opposition to methods of carrying out housing functions). The second classification was connotative with respect to the first, which was denotative.

What does the system of opposed sets of properties during the connotative phase consist of? To give an exhaustive list of all the sets which are 'virtually' possible within the system is probably impossible. Nor is it necessary for the purpose at hand. It suffices to identify one major subdivision which runs throughout the system. In one possible subdivision the properties of the different methods of carrying out housing functions may be regarded as 'naturally' given. The architect will conceptualize them as those properties given in his patron's brief, which lists a set of traditional functional properties only. Because it regards what is listed in the brief simply as given and not as being in opposition to any alternative list, this method of conceptualization must be seen, in light of what has been said before, to be ideologically unsound. The composition of this set of traditional properties with a corresponding set of design features would result in a 'trivial' housing design.

An alternative to this would be the conceptualization by the architect of other methods of carrying out the housing functions in addition to those given by his patron, in such a way as to indicate the deliberately innovative character of his method. This would thus indicate the content of an intentional act of architectural conceptualization, which is lacking in the set of simply traditional functions. This second type of connotative oppositional structure could be transfunctionalized into a 'message' transmitted by the corresponding architectural design.

This design would contain signals through which the architect would attempt to express his intentions, and would be a 'nontrivial' housing design.

There is reason to hypothesize that it is this kind of housing design which would be defined as 'artistic'. If this is the case, this type of architectural activity would belong in the realm of the semiology of communication. The trivial design, on the other hand, would fall under the semiology of signification.

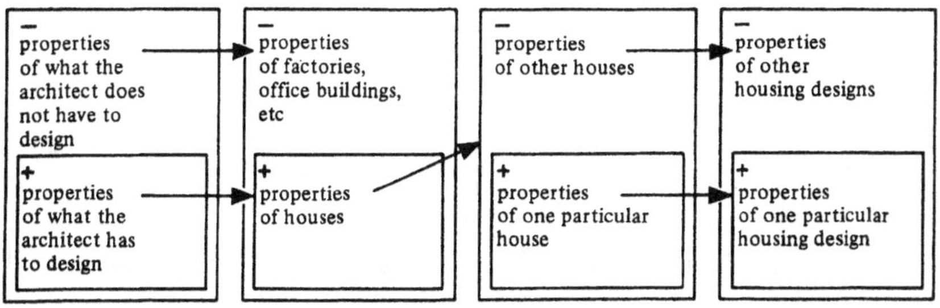

Figure 3.2. The schema of two-fold conceptualization in architectural production. First the architect conceptualizes *what* he is going to design, then he conceptualizes *how* he is going to design it.

3.5.3 Semiology of architectural consumption

As far as the 'consumer' of architecture is concerned, the conceptual process involved in architectural activity takes the form of 're-cognition' and proceeds from the realized architectural design product to abstract sets of properties. The consumer re-cognizes a housing design as a set of properties in opposition to all others within a system of sets of properties of housing design variants. He recognizes that the realized housing design, as well as its unrealized variants, is (or could be) composed with a particular set of properties of houses in opposition to all other sets of properties of houses within a system of sets of this kind. He could not, however, re-cognize a specific set of housing properties without having previously conceptualized the general set of such properties in opposition to the set of properties of other possible buildings: factories, churches, schools, etc. The schema of this process of double classification is given in figure 3.3. As was the case in the production of architecture, the set of properties of the housing design and the set of housing properties become relevant in terms of their relation to each other, as both form a semiotic structure.

Here again, as was the case in architectural production, the re-cognition of the particular set of housing properties composed with the housing design is connotative with respect to the prior denotative conceptualization of the general set of housing properties in opposition to the sets of properties of other possible buildings. Confronted with a trivial housing design, the consumer would re-cognize a set of properties in opposition to all others within a system of sets of properties of trivial housing design variants. He would recognize that the set of properties of the realized trivial housing design (and also that of its unrealized variants) is (or could be) composed with a particular set of traditional functional housing properties. This set would, of course, only be re-cognized in opposition to all other sets of properties of houses.

On the other hand, confronted with a nontrivial design, the consumer would recognize that the set of housing properties composed with the

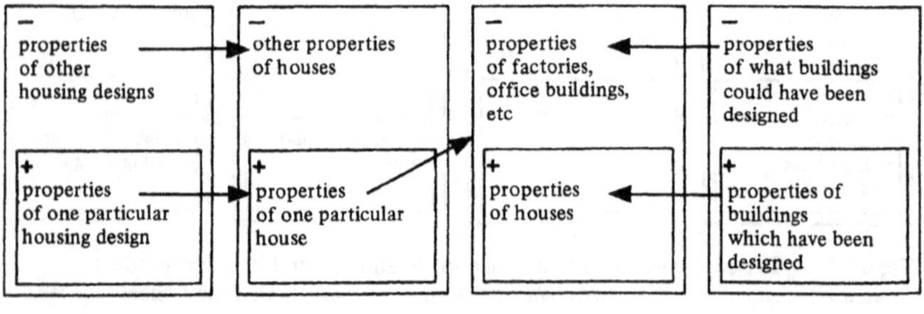

Figure 3.3. The schema of two-fold recognition in the 'consumption' of architecture. First the 'consumer' recognizes *how* a given building type has been realized, but he could not do this without having previously conceptualized, *what* kind of a building type has been realized (as opposed to any other).

realized design consists of deliberately indicative properties in addition to simply functional ones. These would place it in opposition to other sets of housing properties, notably those of the traditional functional type. These additional properties would be recognizable as indicating an intentional act of architectural conceptualization and thus as having been functionalized by the corresponding housing design to contain signals by which the architect expresses his way of conceptualizing the properties of houses as not simply given. There is reason to hypothesize that the consumer would in this case have what can be defined as an aesthetic experience, in which case this too would fall under the semiology of communication.

3.5.4 Possible discrepancies between architectural production and consumption

So far this account of architectural production and consumption has disregarded the possibility of architectural production which cannot actually be used (either in part or as a whole), and the possibility that an architectural product may be used in a different way than was anticipated by the producer. This would be the case if the sets of properties conceptualized and re-cognized during the various stages of the architectural process were not identical.

If there was no conceptual repertory of the functions of architectural products common to both producer and consumer, the consumer would be incapable of re-cognizing the conceptualizations that originally occurred in the design process. This would most likely be the case if producer and consumer came from radically different architectural cultures. This problem is mitigated to a large extent at present by the prevalent 'international style' in architecture, but may still be relevant in archaeology, where radical cultural discrepancies can result in incorrect identification of the functions of architectural artefacts. In any event the problem of different conceptual repertories points to the necessity of defining the precise position of architectural semiology in a general semiotic of culture (Lotman, 1969). Architecture would in this sense become a particular type of 'text' of particular historical origin (Agrest and Gandelsonas, 1973).

The term 'semiology of architecture' should not be understood only in terms of a professional tradition known from the time of the Renaissance, thus excluding architectural phenomena from other cultures. In a semiotic of culture, an architectural 'text' (for example, a temple), distinguishable by its material substance, may enter together with 'texts' of different substances (religious vestments, litanies, etc) into a religious cultural system as a whole (Ivanov et al, 1969). A semiology of architecture would thus be absorbed into a semiotic of cultural modelling systems. In a more limited way Raymond (1974) has pointed out the role of housing as cultural models.

Cultural differences are, however, not the only reasons for errors in identifying the functions of architectural products. There is a strong possibility that the environmental context of a specific architectural object contributes significantly to its conceptualization and re-cognition (cf Dexter and Lindsey, undated). If in the design process an architect ignores or misinterprets the context in which the design will be realized, error may be made in the subsequent identification and re-cognition of the design objects as a result of the lack of, or errors in, contextual clues. Another reason may be that the sets of properties of architectural products and the corresponding sets of properties of architectural designs are conceptualized in one way and re-cognized in another.

An architect could, for example, incorporate deliberately artistic elements into a purely functional design. These elements may not be recognized by the consumer, often for the reason that the architect-sender and consumer-receiver of the artistic message do not have a common stylistic repertory, although they live in the same culture. There is evidence that certain aesthetic experiences can be confined to a cultural elite, because the so-called lower classes are not granted the education enabling them to recognize particular artistic signals (Bourdieu, 1968). Thus class barriers, as well as cultural differences, may contribute significantly to discrepancies between architectural conceptualization and re-cognition.

The architect may be seen to be responsible for those discrepancies, in not having taken into account the possible cultural, contextual, or class differences. There may, however, be cases where the reverse is true. The consumer may reconceptualize an architectural product in a way which is intentionally different from the way he re-cognizes it to have been originally conceptualized by the architect. This is often the case when a nonsecular building is secularized, for example when a monastery is transformed into a school, or church into a museum. The casuistics of such reconceptualizations have been described by Dorfles (1958), Eco (1968), and Koenig (1970). These processes are mostly of historical dimensions, comparable to diachronic changes in language.

3.5.5 From semiology of architecture to urban semiology

Prieto (1977) has recently advanced the idea that the production of optimal urban designs may be understood as the conceptualization of a combination of architectural objects into an integrated whole. In an integrated whole the parts are recognized as objects in their own right—as universes of the signifier receiving their relevance from correlated universes of the signified. But at the same time the whole is recognized as a (composed) object in its own right —as a (composed) universe of the signifier receiving its relevance from a correlated universe of the signified. A whole would not be integrated if it did not receive its relevance from a universe of the signified of its own but consisted only of a juxtaposition of its component parts, each of the latter receiving its relevance from its own universe of the signified. Furthermore, the component parts could not be considered integrated if each did not receive relevance from a universe of the signified of its own, instead of being simply components of a relevant whole coordinated with a universe of the signified. This concept would apply to the recognition of linguistic wholes, as well as to others. Any utterance is an integrated whole; where the utterance itself has its own specific meaning and component parts, so do the individual words or monemes. Whereas a moneme is a whole, bearing its own meaning, its component parts, or phonemes, have no individual meaning but are simply components. A moneme can therefore not be called an integrated whole.

Likewise, an optimal urban design (or urban object) is an integrated whole if as a universe of the signifier it receives its relevance from a coordinated universe of the signified and if its component architectural objects (factories, office buildings, tenement houses, churches, schools) as universes of the signifier receive their own relevance from universes of the signified. With his concept of integration Prieto accentuates the optimal solution of the urban problem. A more general concept, however, would be to regard the city as a whole whether integrated or not, that is, as simply a collection or distribution of architectural objects (likewise integrated or not) existing in geographic space according to an urban conceptualization. In other words, an urban design is conceptualized in terms of a distribution of architectural objects. Whereas the ideal would be an urban design which is an integrated whole consisting of integrated architectural components, one might have integrated components but fail to integrate the whole, or alternately have an integrated urban design but fail to integrate its architectural components.

In any event, in urban design the meaning of the whole should ideally be established before one establishes the meaning of its parts, as the parts become meaningful only in terms of the whole. In this sense a semiology of architectural conglomerates would be logically superordinate to a semiology of architectural components.

With this in mind it becomes important to look diachronically at architecture (as a system of sets of architectural objects, or their properties) in order to emphasize the processes of change which occur historically within this system and the oppositional structure of sets it represents.

During a particular historical period, the architectural milieu may include temples and triumphal arches; these will in all likelihood be absent during a period which contains factories and railroad stations, which may in turn be absent in another period.

These changes in the composition and content of architectural systems are clearly not a result of factors intrinsic to architecture itself (mere stylistic changes). A diachronic comparison of different systems of sets occurring during different historical periods will give some indication of the nature of the extra-architectural causes of these changes.

The practice of compiling normative lists of architectural products was begun by Vitruvius (1960), who divided architecture into three functional categories: the making of machinery, time-pieces and buildings. Buildings were divided into public and private, and public buildings were further subdivided into three classes:

defensive: walls, towers, gates, etc;
religious: fanes, temples, etc;
utilitarian: harbours, markets, colonnades, baths, theatres, promenades, etc.

In Alberti's (1755/1965) normative list, the system of architectural products is seen to be diversified on the basis of social hierarchy. He held that there are "several divisions of human conditions whence arises the diversity of buildings" (page 64). Certain architectural products, therefore, are there for the benefit of all citizens, others for citizens of 'greater quality' and others for citizens 'of the meaner sort'. His normative catalogue looked like table 3.1.

He classed private buildings simply as town houses or country houses, both types being constructed according to the social status of the inhabitant. He recommended, however, that the "habitation of the middling sort ought to resemble those of the rich" (page 108), a recommendation which seems to be heeded occasionally by his confraternity even in our times.

Table 3.1. Alberti's (1755/1965) normative list.

Buildings for the benefit of all				
Defense	traffic and underground works			
walls	highways			
battlements	bridges			
towers	drains			
corniches	sewers			
gates	canals			
	havens			
	public squares			
'Offices' of the citizens of 'greater quality'				
Sacred	profane			
	leaders	high officials	lower magistrates	commerce
temples	fortified castles	senate house	granaries	mart
basiliques	palaces	tribunal	treasury chambers	docks
chapels		military encampment	arsenals	public stables
oratories			prisons	storehouses
cloisters				
Buildings for the 'general public'				
Hygienic	cultural			monuments
palaestrae	public schools			monuments
public baths	libraries			sepulchres
thermae	theatres			pyramids
hospitals	amphitheatres			altars
				moles

Alberti's normative inventory can be seen to be a reflection of the hierarchical structure of the social ideology of his time.

At the beginning of the nineteenth century Durand (1821; 1823; 1825) drew up a normative inventory (complete with plans and elevations) which excluded defensive or nonsecular buildings and in which no account was taken of hierarchization:

barracks,
bathing establishment,
city hall,
courthouse,
customshouse,
fair building,
high school,
hospital,
hotel,
institute,
lighthouse,
library,
market hall,
museum,
observatory,
palace,
post office,
prison,
registrar's office,
slaughterhouse,
stock exchange,
theatre.

Included in public buildings was the category of monuments, which contained 'tomb' and 'triumphal arch'. The category of private buildings contained 'dwelling houses', 'tenement houses', 'villas', and 'farmhouses'. A plan was also given for 'public squares'. The standard character of these types becomes especially evident if one examines the plans and elevations, which are reproduced visually in figure 3.4. This absence of hierarchical structuring in Durand's normative inventory parallels the general absence of hierarchical structures within social ideologies after the French Revolution. New types such as factory, office building, railway station, airport terminal, low-cost housing have of course emerged since the Industrial Revolution, whereas others such as palace or triumphal arch have disappeared.

These lists suggest that the reasons for the emergence or disappearance of certain architectural products from an architectural system (or normative inventory) lie not in phenomena intrinsic to architecture, but can be explained in terms of changes in 'urban ideologies', namely changes in the meaning of the urban environment as a whole. In other words, if there is a change in the universe of the signified correlated with the universe of the signifier which is the city, there will be a consequent change in the repertory of architectural objects as universes of the signifier and their coordinated universes of the signified. (This would not be altered if one were to include negative space, such as streets and squares, and other urban phenomena, such as traffic, in the category of urban production and consumption. However, in order to simplify matters as much as possible, these have been excluded and the system of architectural products has been regarded simply as 'real estate'.)

(text continued on page 88)

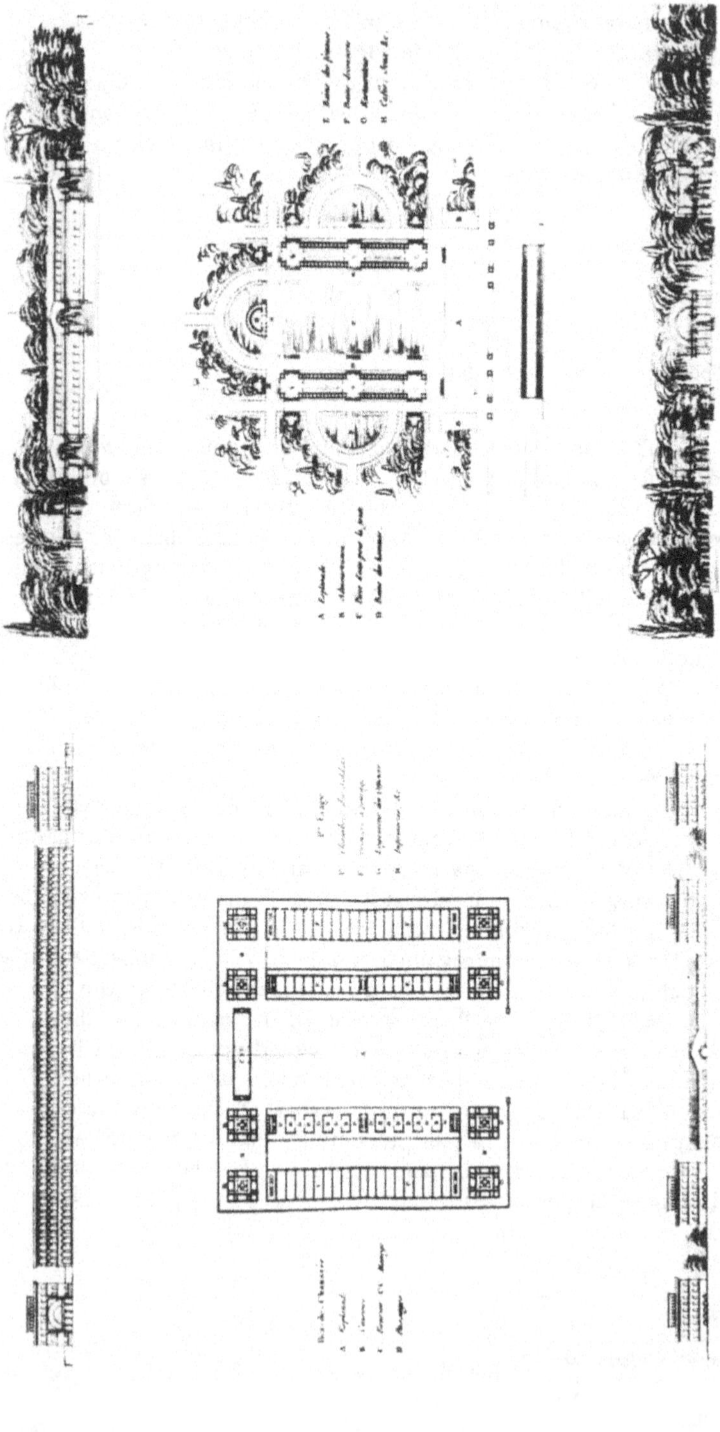

Bathing establishment

Barracks

Figure 3.4. Durand's (1821; 1823; 1825) 'normative' inventory of architectural products. These may be considered parts of the city, which in turn may be considered as a whole.

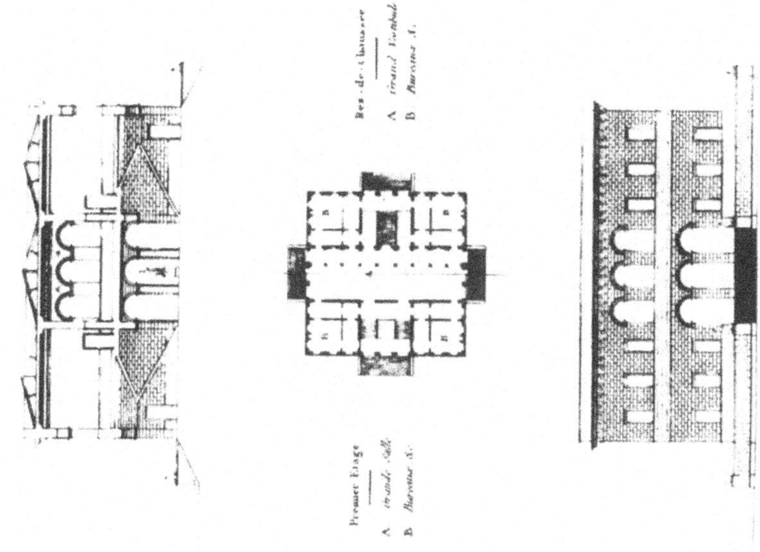

Town Hall

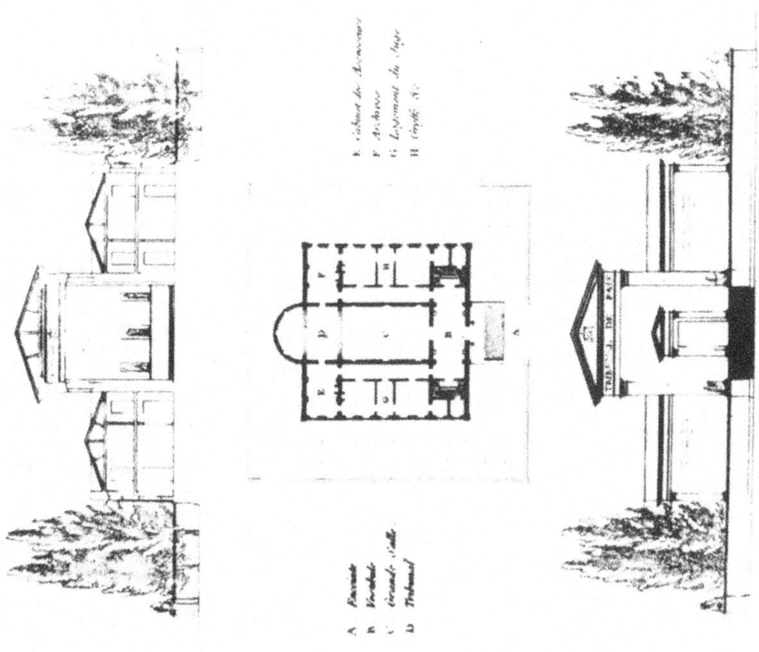

Office of the justice of the peace

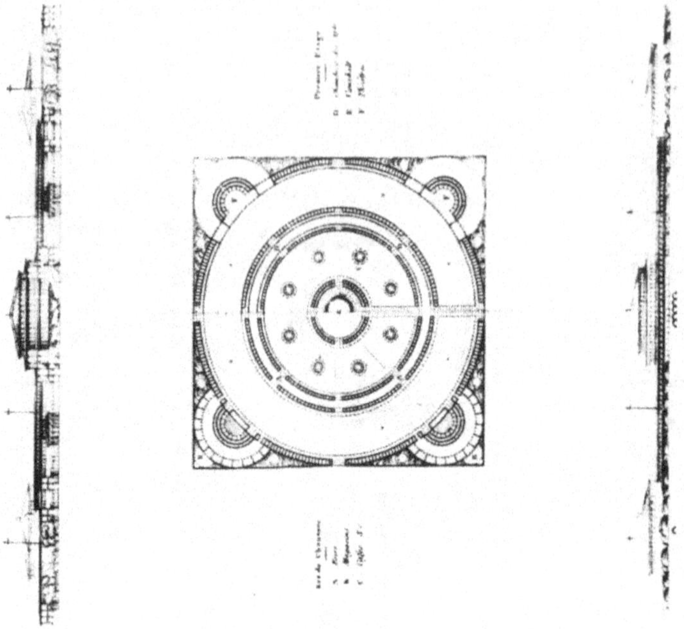

Trade-fair building

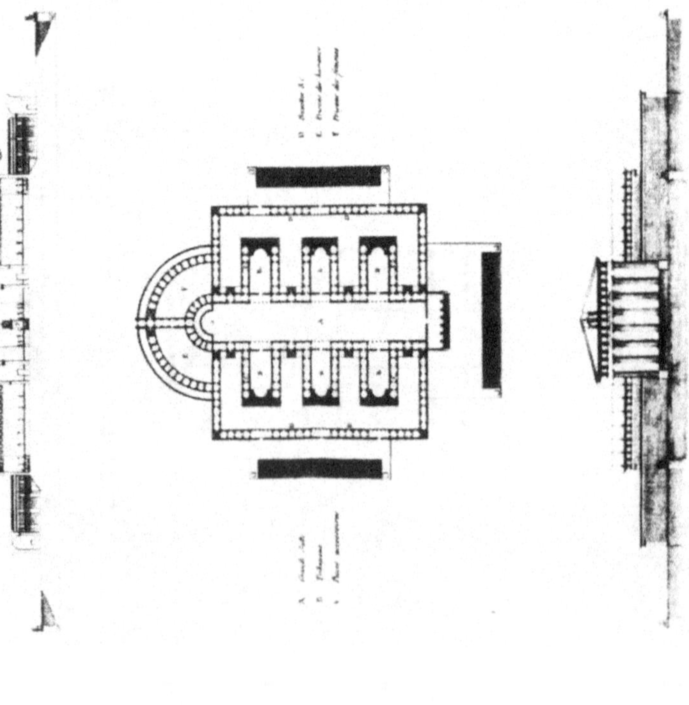

Courthouse

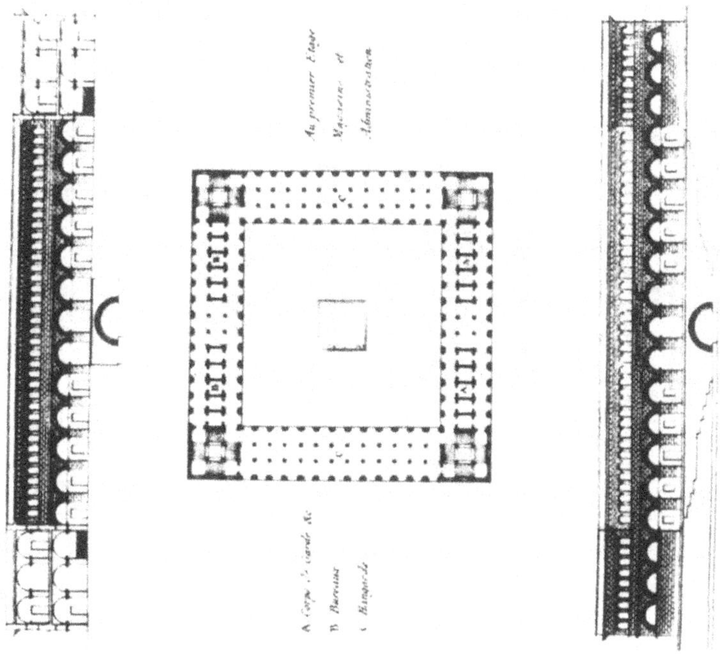

Customshouse

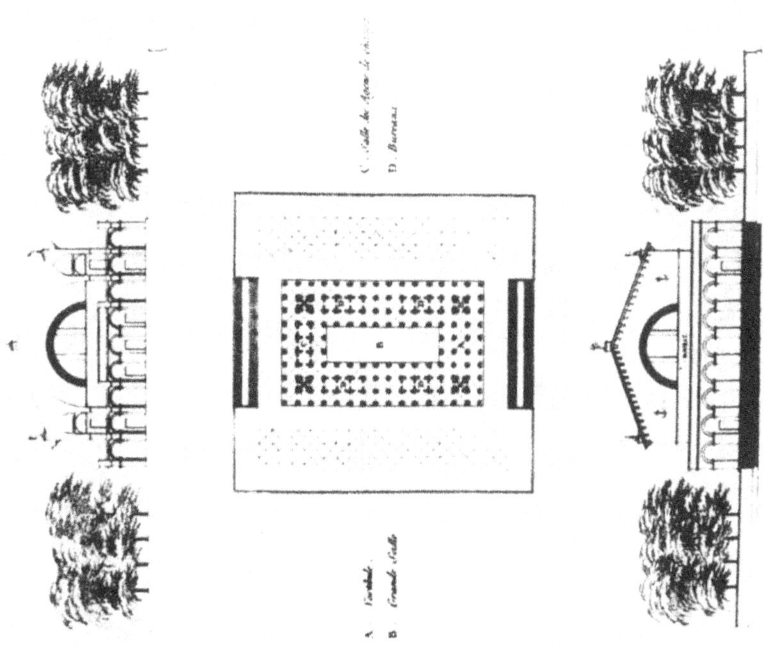

Stock exchange

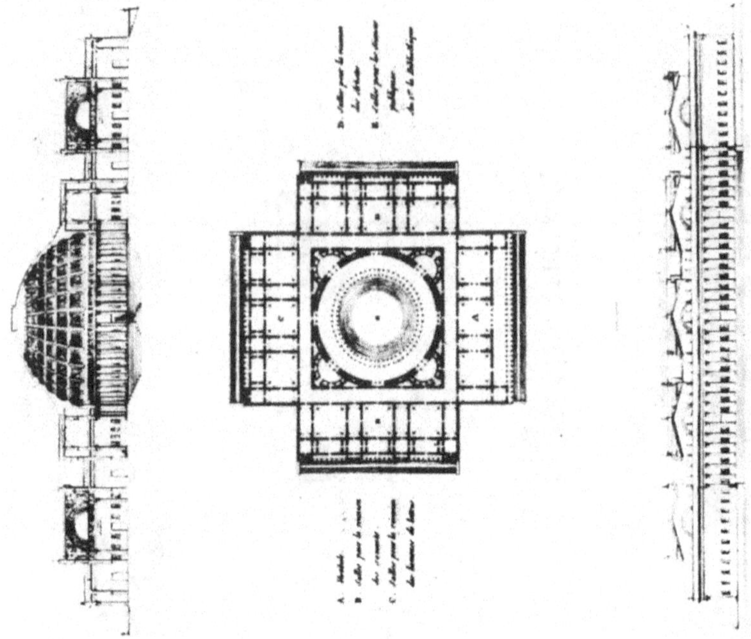

Institute

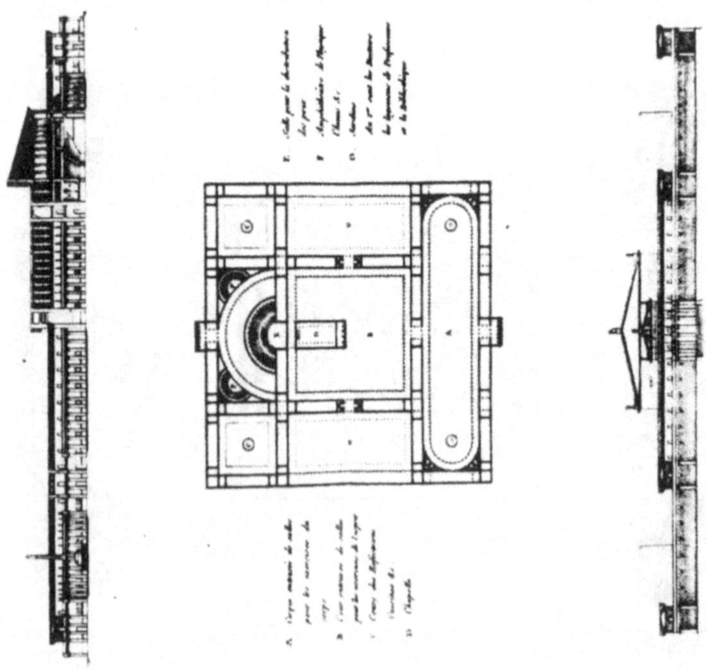

High school

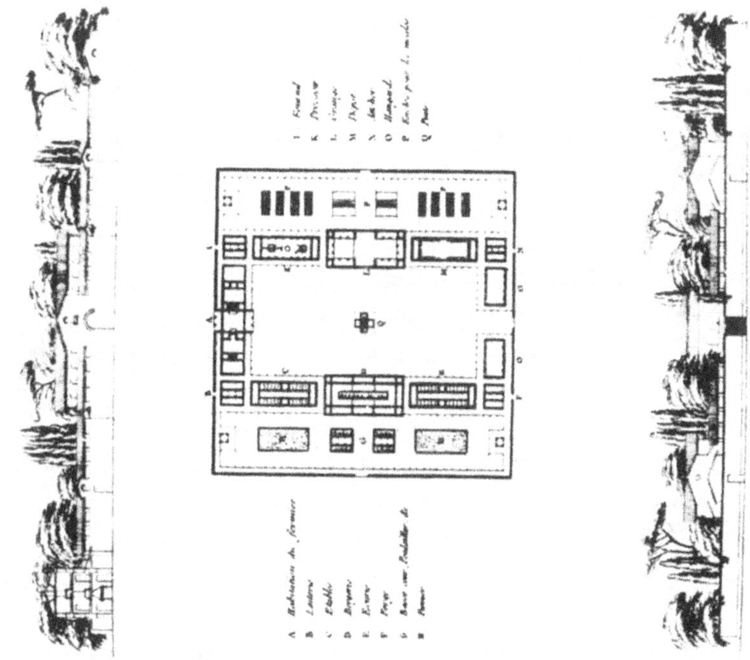

Farm

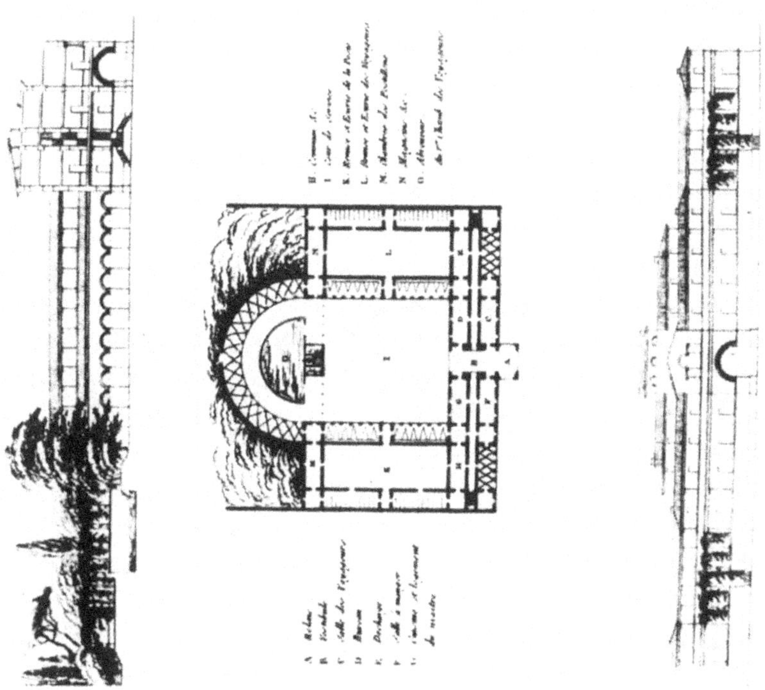

Hotel and Post Office

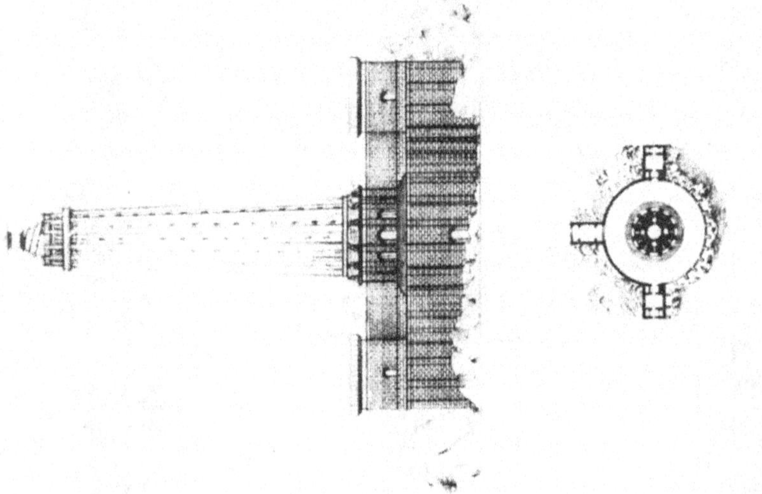

Lighthouse

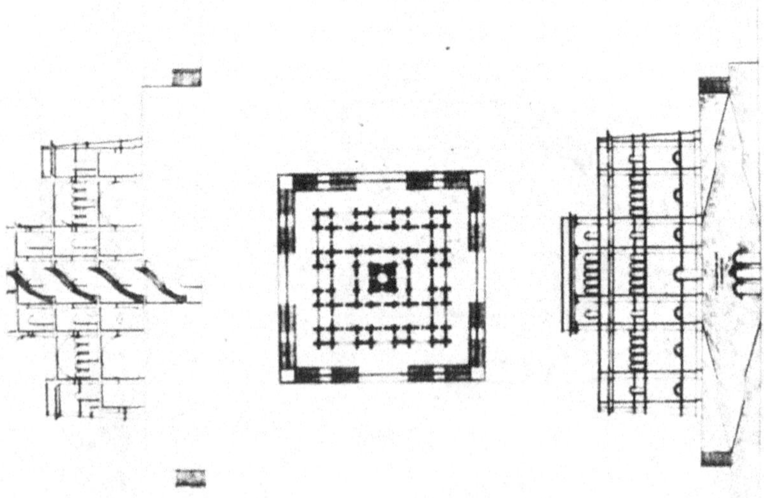

Observatory

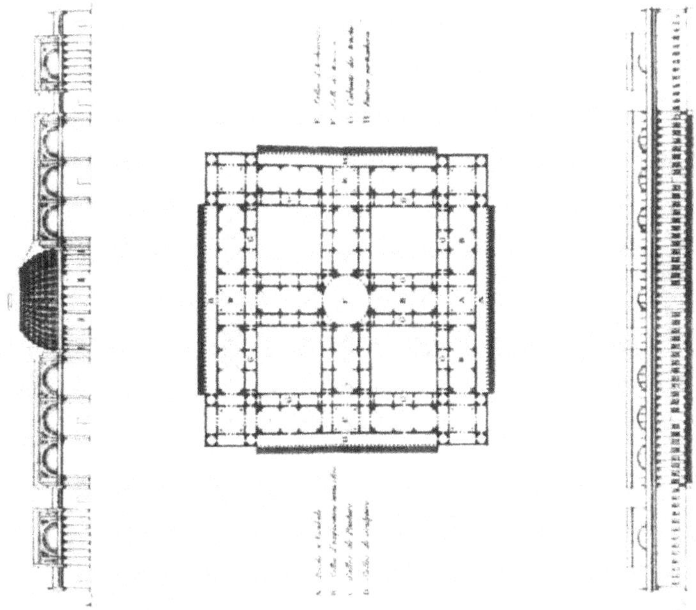

Museum

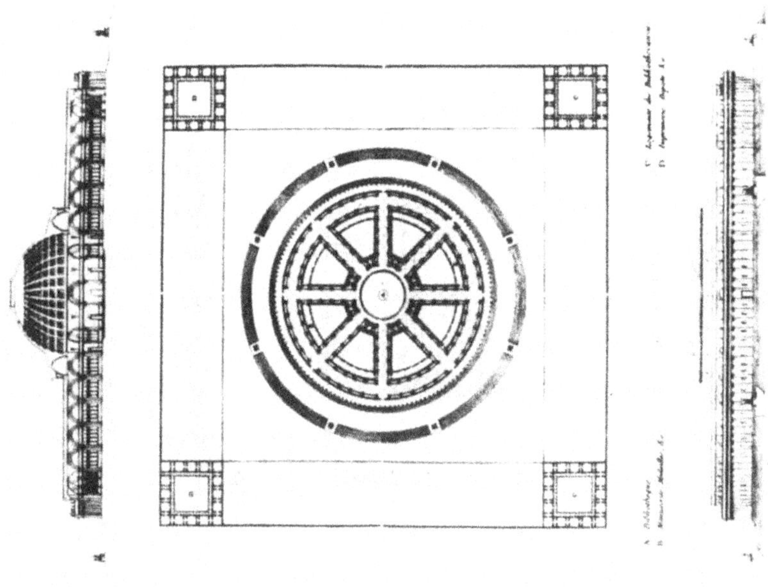

Library

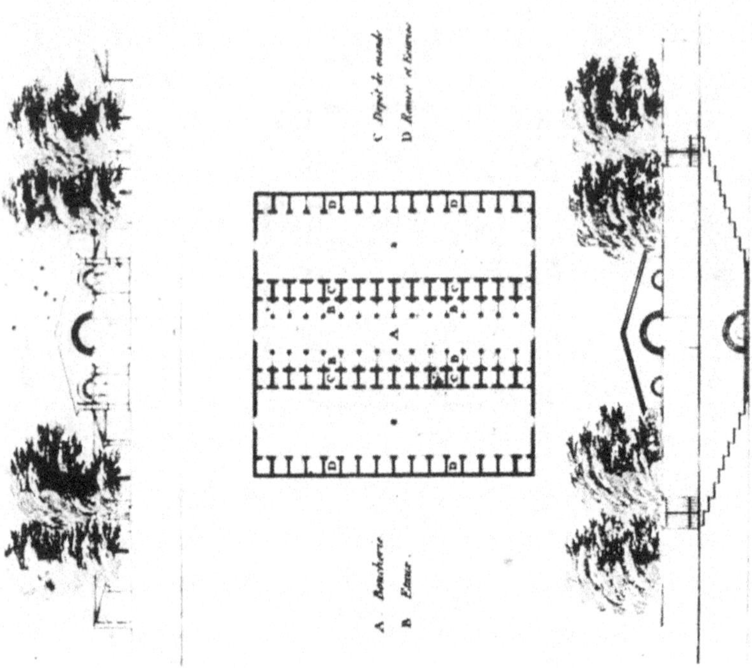

Slaughterhouse

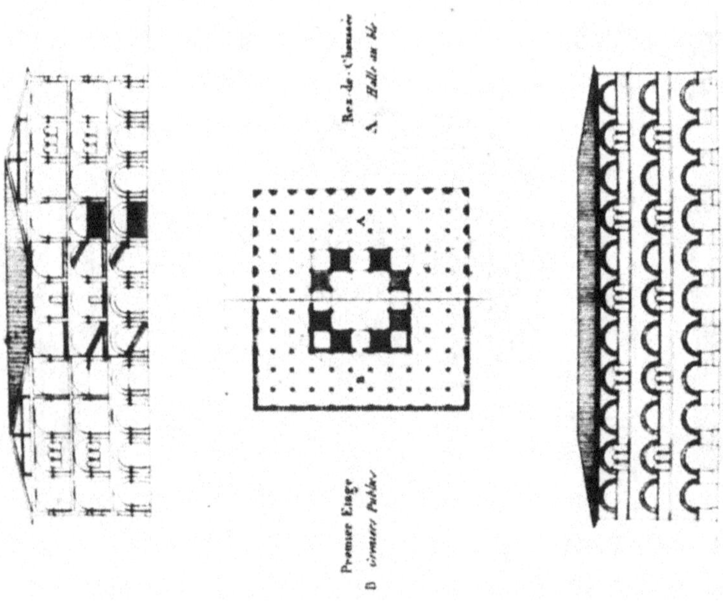

Market hall

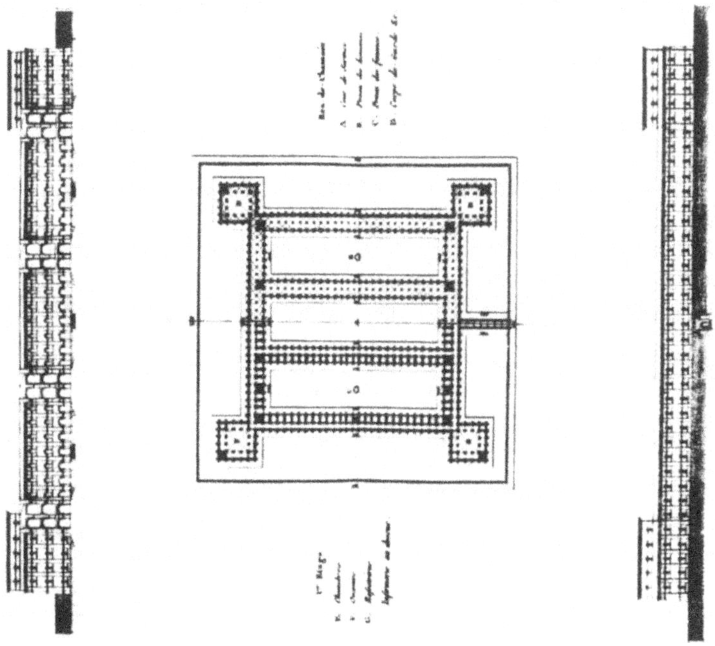

Prison

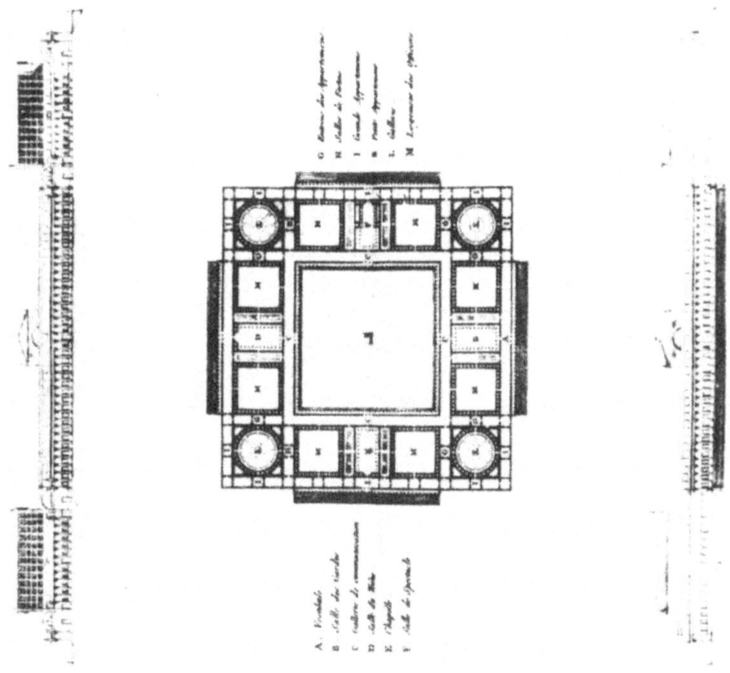

Palace

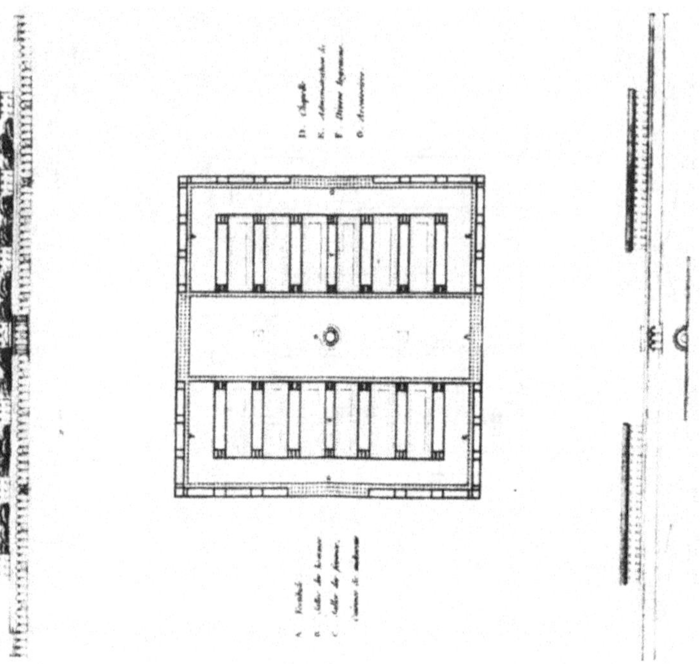

Hospital

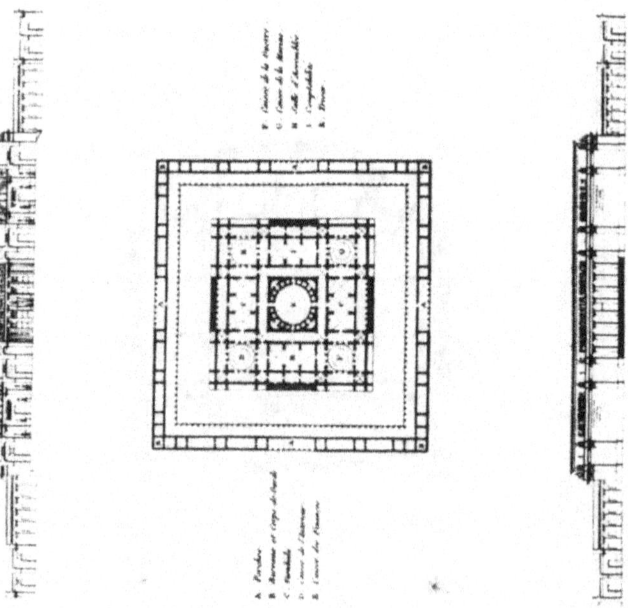

Tax Office

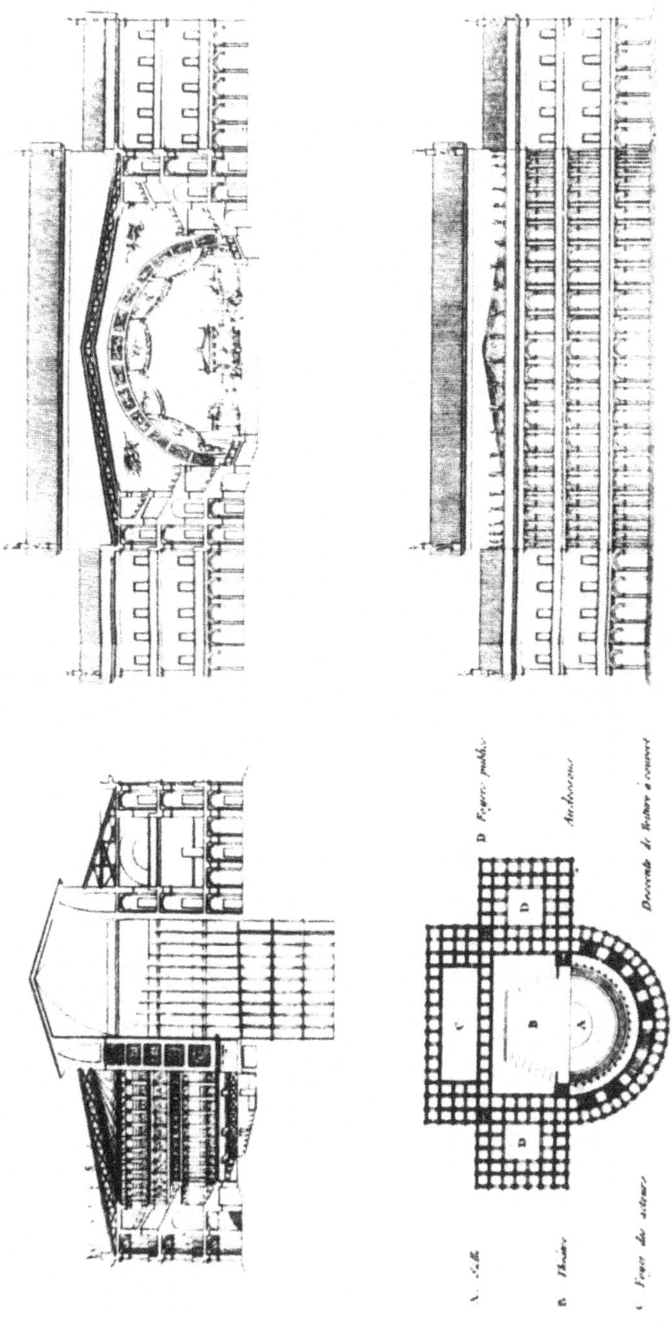

Theatre

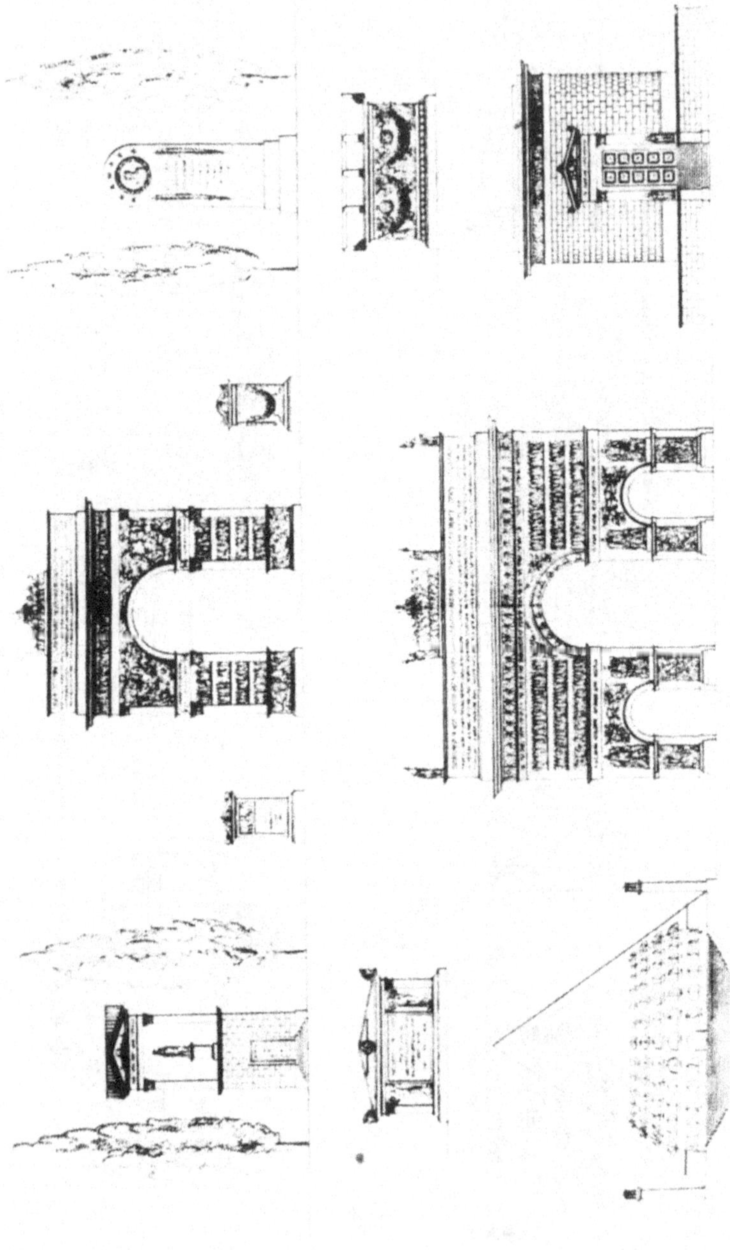

Tombs and triumphal arch

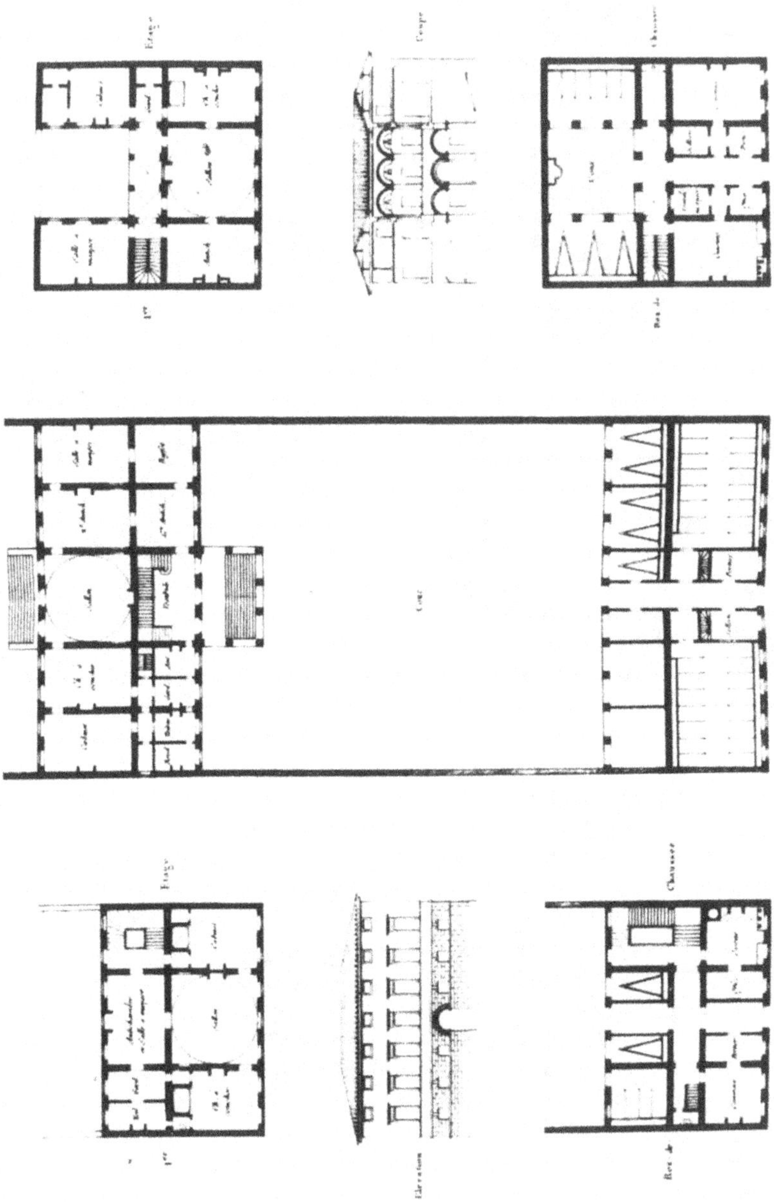

Townhouse

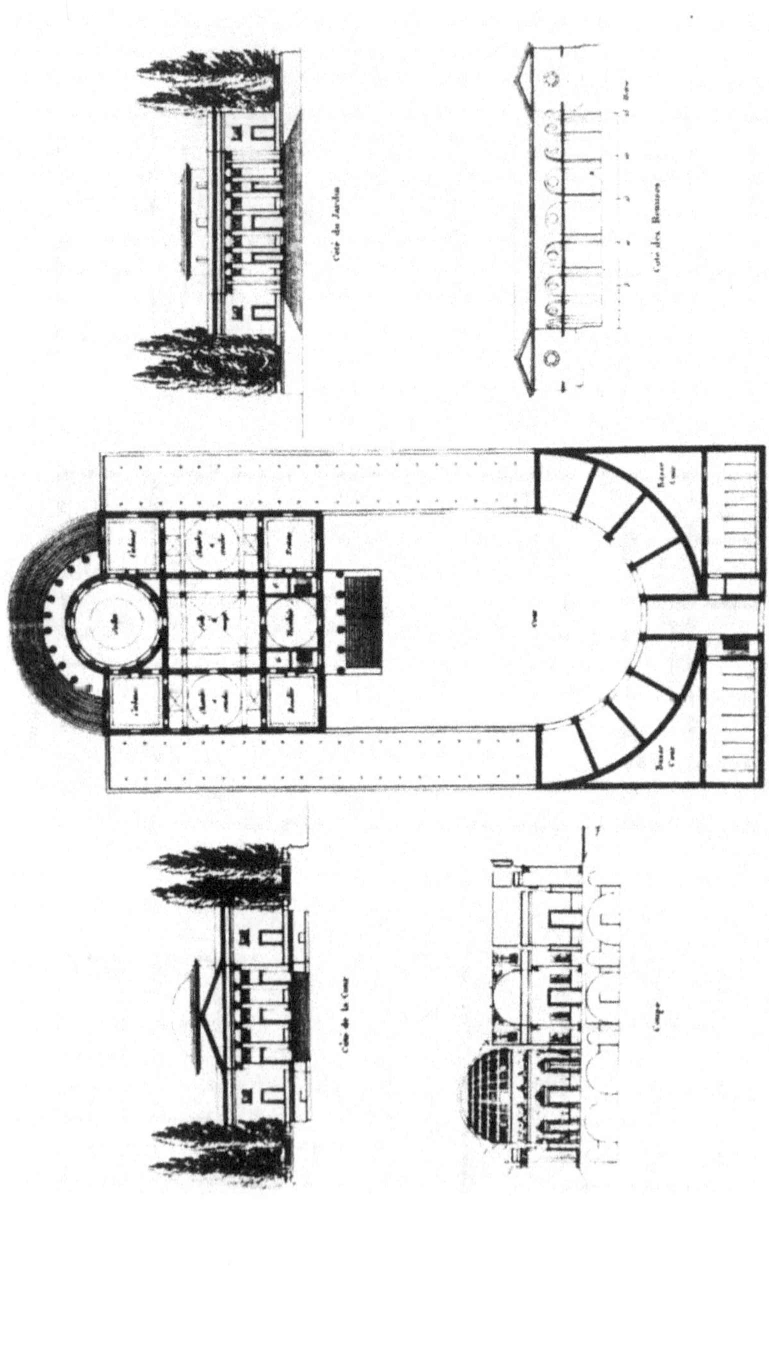

Town and country house

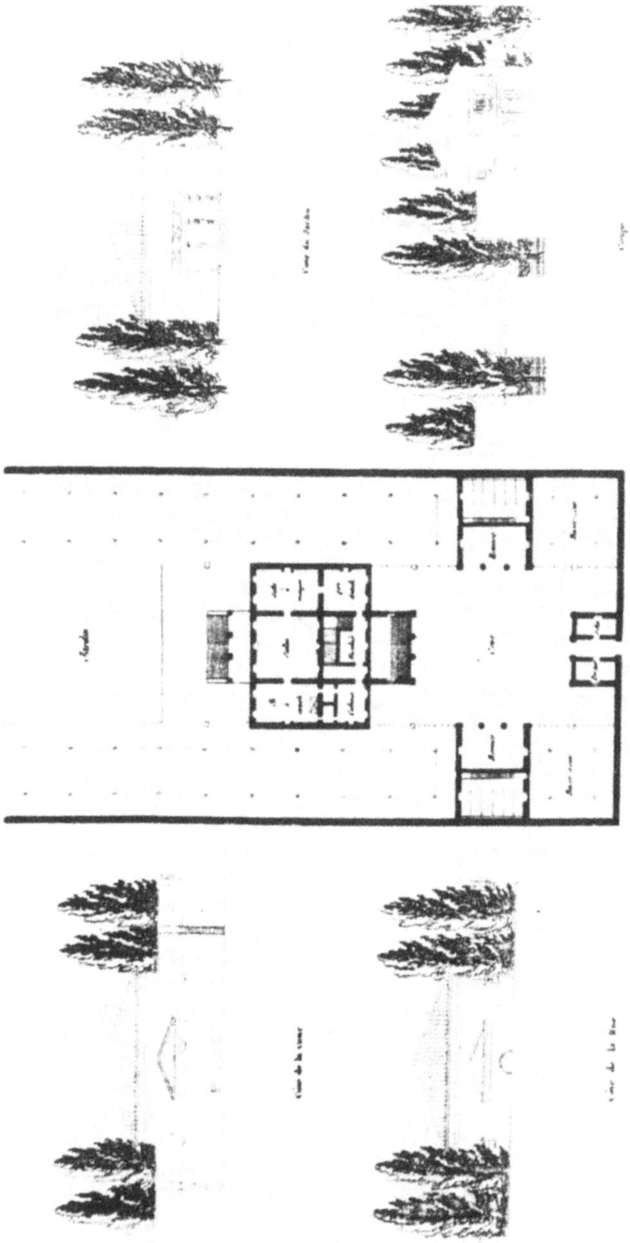

Town and country house

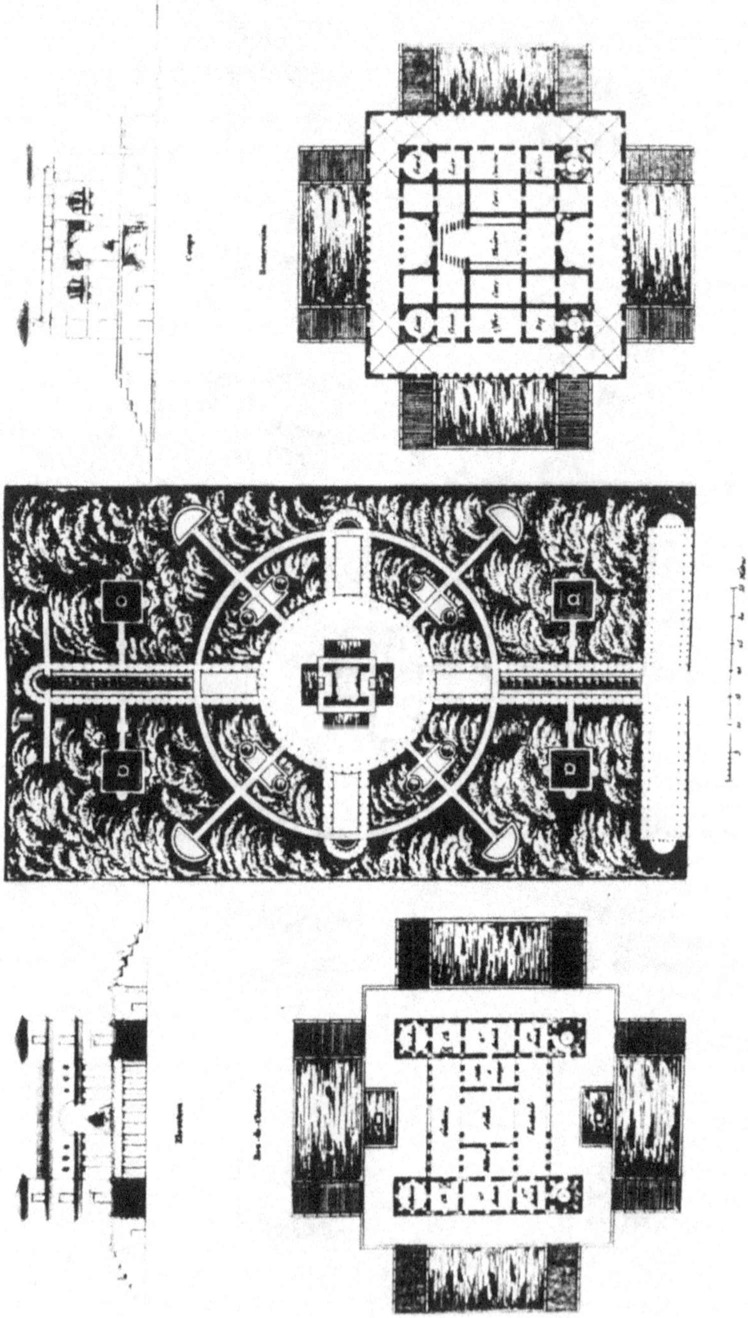

Country house

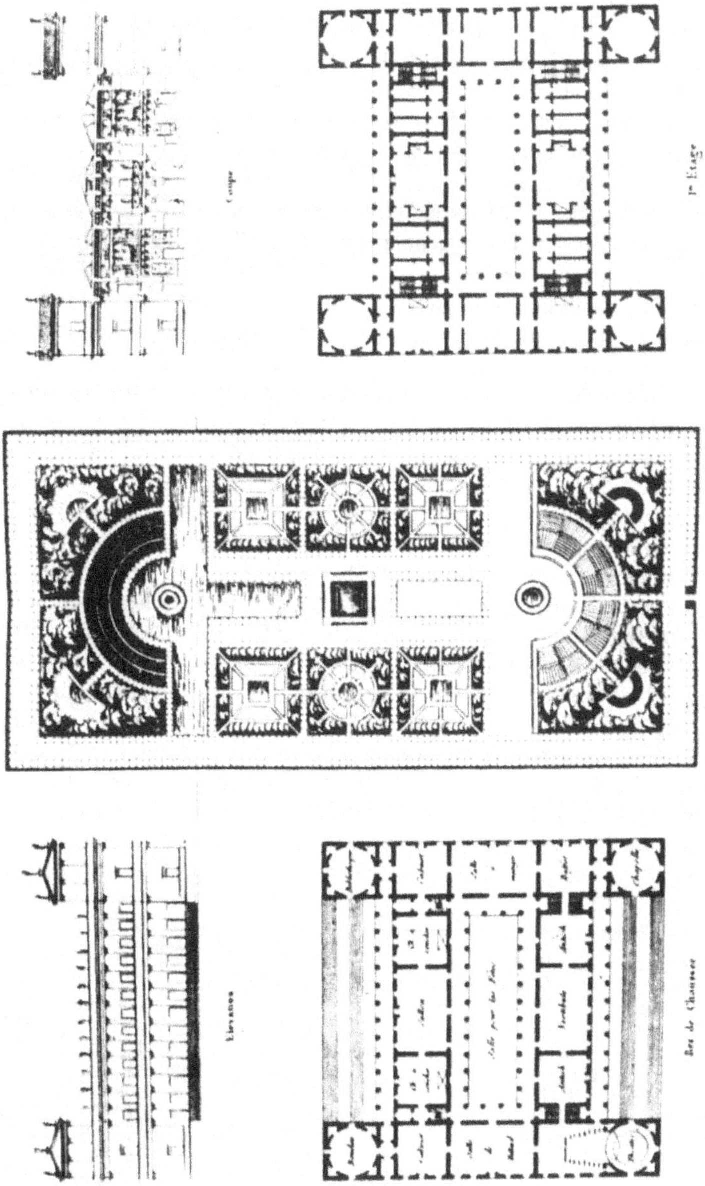

Country house

3.5.6 Semiology of urban production

Urban production involves precisely the same conceptual processes as did architectural production. The urban designer goes through the same procedure of double classification (figure 3.5), conceptualizing first that set of properties which will determine the content of what he will design, reconceptualizing that set in terms of the design method he will employ, and finally composing it with a set of urban design features. In this case, however, the set of properties of the first conceptualization refers to possible urban agglomerations differentiated by factors such as geographic site, size, and function, for example a port of 500000 inhabitants as distinct from an inland industrial city of 300000 inhabitants. As with architectural production a semiotic structure is formed, and the second phase of the conceptual process is connotative with respect to the first.

This connotative stage can be seen to fall, in the same way as in architectural production, into two categories, resulting on the one hand in an urban design which indicates intentionality and originality on the part of the urban designer, and on the other hand a lack of intentional conceptualization. It would appear that in this connotative phase the designer may be guided by some ideology (cf Berndt, 1968; Choay, 1973b; Ostrowetsky, 1972), for example, the 'hygienist' ideology of the last century (Benevolo, 1968), which may have ultimately given rise to the concept of the urban agglomeration as a 'garden city'. The functionalist ideology of the Athens Charter sees the city in terms of a subdivision of the functions work, transportation, and recreation. A widely accepted ideology is that the city grows as an organism and must be conceptualized accordingly (Berndt, 1968; Geddes, 1949). During the last decade an ideology of 'high density' agglomeration has also had some influence. A number of these city 'ideals' have been compiled by Schoof (1965).

Whatever the ideology, a particular design method can be functionalized by a corresponding urban design to become a signal by which the particular character of the design method is expressed. The design is composed with a

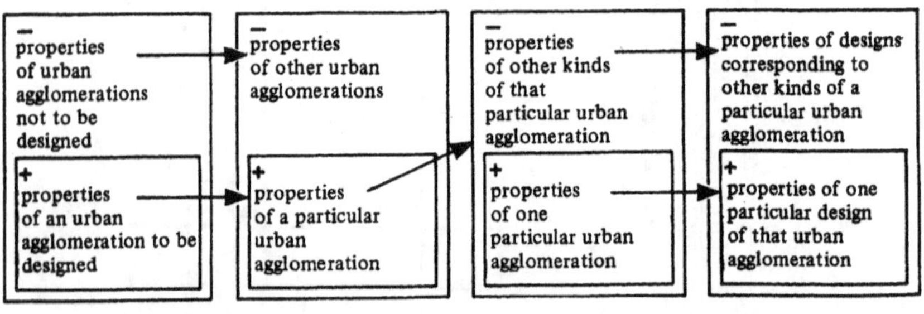

Figure 3.5. The schema of two-fold conceptualization in urban production. First the urban designer conceptualizes *what* type of city he is going to design (depending on site, size, function, etc). Subsequently he conceptualizes *how* he is going to design it (as a 'garden city'', as a 'high-density agglomeration', etc).

set of both functional and ideological properties, resulting in what could be called an ideological design, as distinct from one which lacks such signals. If this is the case, the urban design process may become, via ideological functionalization, part of the semiology of communication.

3.5.7 Semiology of urban consumption

Urban consumption similarly involves the same conceptual processes as did architectural consumption, the citizen proceeding in both cases from a concrete object to an abstract set of properties. The citizen-consumer recognizes an urban design product by reference to its set of properties (the notative universe of the signifier), which are in opposition to those of other design variants, and recognizes that these properties receive their relevance in terms of their relation to the properties of the coordinated urban design method (the connotative universe of the signified). In doing this the citizen-consumer re-cognizes the semiotic structure which the urban designer originally operated with.

As it is impossible to re-cognize a design method without re-cognizing at the same time the particular design problem of which the design method is the solution, the consumer-citizen therefore re-cognizes an urban design method at the same time, or even prior to, referring to the particular urban design problem related to it. In other words the citizen puts the urban design problem through the process of double classification, once as a connotative universe of the signified belonging to one semiotic structure, and once as a notative universe of the signified belonging to another. The schema of this process is given in figure 3.6.

Because urban production may entail either an urban design which indicates that it involved an intentional act of conceptualization, and therefore has an ideological basis, or one that lacks these features, the citizen-consumer may or may not experience the city as an ideological message.

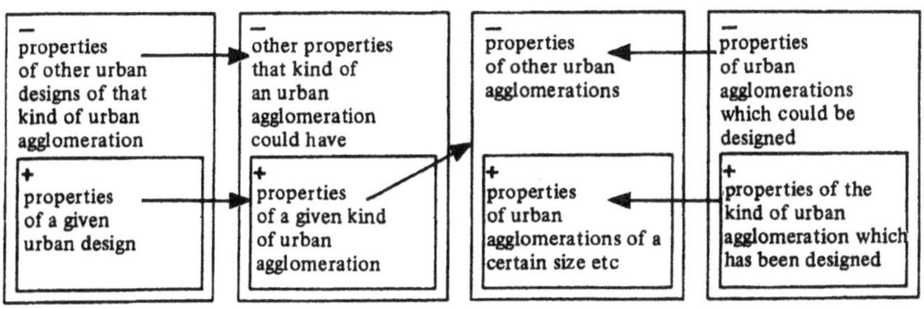

Figure 3.6. The schema of two-fold recognition in urban consumption. First the citizen-consumer recognizes *how* a particular urban design has been conceptualized (whether it has been designed in any particular way as opposed to any other). He could not, however, recognize this without having previously conceptualized *what* particular type of urban problem had to be solved by the urban designer.

3.5.8 Possible discrepancies between urban production and consumption

Discrepancies may occur between urban production and consumption for the same reasons as were discussed in the section on architectural production and consumption. The possibility that the consumer may not recognize properties which are in fact present in the object may give the impression that what is involved is cognitive subjectivism as a form of subjective idealism, as though the object is created by the cognitive act. In order to avoid this, the conditions for re-cognition must be made clear. The citizen-consumer

may re-cognize the set of properties of an urban agglomeration and the set of properties of the corresponding urban design as they were conceptualized by the urban designer,

may not re-cognize the above at all,

may conceptualize the above in a way which was not intended by the urban designer,

cannot conceptualize a set of properties of an urban agglomeration and a set of corresponding urban design properties which are simply not there.

One measure which has been proposed with increasing frequency in an attempt to alleviate the problem of conceptual discrepancies is the indirect participation of the consumer in the process of production by participating in discussion groups and workshops. Although this may seem to be the only answer to the problem, its practical implementation gives rise, in turn, to obvious problems as regards division of labour. If the burden of the cost of this participation fell on the consumer, that is, if it were done without remuneration in the consumer's spare time, the situation would arise where the working hours and corresponding remuneration of the designer would remain the same, whereas the consumer's working hours would in effect increase without an increase in remuneration. Another obvious problem would be the radical differences in professional qualifications and competence of the designer and consumer, a problem which would in effect finally render the idea impracticable. It would seem then that in a given historical situation this notion of participation would become simply another urban ideology.

3.5.9 Critical comparison of the semiotic and semiological approaches to architecture and the city

A critical evaluation of the utility and ideology of the semiotic approach was presented in section 2.3. Its utilitarian value was seen to lie in the ability of its classification system to order data in the planning phase of the design process. An inconsistency was, however, apparent within the school itself as regards the scope of the approach, with Blomeyer and Helmholtz (1976) maintaining that semiotic terminology concerns only the planning phase, whereas Bense (1971c) claimed that it involved the creative process as well. An ideological weakness of the approach was seen to be the mandatory inclusion of pragmatism into the trichotomous schema, whereas it should ideally remain ideologically open and concern itself only with the nature of the semiotic structure itself and not its ideological relation to the interpretant.

Semiology, in comparison, has distinct advantages. Its utility lies in the fact that, although its model is not geared to the actual classification of objects or potential objects, it nevertheless recognizes that cognition is an act of classification. Instead, semiology sets up procedures and sequences for different activities by means of the concept of double classification.

Another advantage is its careful analysis of the different ways in which the producer and consumer approach the architectural and urban object, and its anticipation of the potential discrepancies possible between the two approaches.

More important, however, are its ideological advantages. Semiology operates by reducing signs to objects, rather than by reducing objects to signs (which tends to undermine the material substance of the object). Semiology is capable, however, of a precise description of the mechanics whereby a building or city becomes a message (over and above its signifying potential as an instrument).

The semiological model of cognition is based on psychology rather than on epistemological philosophy, although its relation to extrasemiological issues can be spelled out precisely. Since psychology is an empirical science, these hypothetical internalized semiotic structures must be open to empirical verification, and psychological analysis would have to be able to bring these internal structures into the open and make them accessible to empirical examination and verification. Semiology concentrates on the functioning of semiotic structure. That an analysis of semiotic structure would itself never be devoid of an ideological point of view is not denied by semiology, but the specific nature of the ideology is regarded as an extrasemiological issue. Semiology is therefore an ideologically open system, which is subordinate and not superordinate to a guiding ideology and epistemology.

The actuality of the advantages of semiology must now be demonstrated by practical application in empirical research.

Part 2

Empirical studies in the psychosemiology of architecture

4

Psychosemiology and the recognition of different functional building types

4.1 Psychosemiology

The question now arises as to how the theoretical constructs proposed in the preceding section are to be verified. In order to do this the material status of both universe of the signifier and universe of the signified should be established. Since neurophysiology has not yet produced a physiological model capable of describing and explaining these processes, the only option is to proceed by developing an intervening hypothetical model. These universes were said to result from calculi, acts of classification, conceptualization and re-cognition. Signification or meaning would therefore have to be regarded as basically psychological in nature. Semiotic structures are cognitive structures. The question, however, is how these structures are made up. Since this investigation is not neurophysiological in scope, it is simply assumed that the universe of the signifier involves some sort of reflection in the brain of visually selected properties of material objects. The universe of the signified, however, is conceptual in nature, that is, is a schematic memory trace, abstracted from previous perceptual information derived from the different senses. This perceptual information would include not only information derived from the visual channel (which would include graphic language) but also information derived from the acoustic channel (which would include spoken language).

Whether one depicts this internalized cognitive semiotic structure in terms of 'associations', 'mediating processes' (Osgood, 1958), cognitive processes (Neisser, 1967), or facilitated synaptic networks (Hebb, 1961) etc, is for the moment not really the issue. It is sufficient for this investigation that the visual perception of objects activates certain thought processes (the nature of which can be deduced from observable behaviour), or vice versa, that certain thought processes activate selective responses in the visual perception of objects. The universe of the signifier, namely visual images in the perceiver's mind, may activate the universe of the signified, namely specific memory traces, and vice versa. (This is simply a reformulation in psychological terms of the earlier model which stated that each universe supplied the relevance of the other.)

In analogy to the term 'psycholinguistics' I have called the psychological study of internalized semiotic structures psychosemiology (although the psychological status of semiology was clearly present in its original definition by de Saussure).

The study may proceed by three methods:

1 focusing on the content of and processes involved in memory;
2 focusing on the content of and processes involved in perception;
3 focusing on concomitant variations in memory and perception, for example, by means of a commutation test.

Prieto (1975a) called attention to an analogous tripartition in linguistics by referring to phonology as the study of the level of expression in what Wells (1958) called 'quasi-total abstraction' from the level of content. Prieto called the study of the level of content in quasi-total abstraction from the level of expression 'noology' (it is more often referred to as 'semantics'). The third discipline would involve the study of signs, that is, the study of the bifacial units which comprise semiotic structures or codes. This would be called semiology. The two studies in this book are concerned with this last discipline.

In the first study subjects were required to classify a number of photographs of buildings in accordance with given categories (in the form of labels) of architectural types (differentiated on the basis of function). They were thus required to proceed on the basis of their own conceptions of the functions of buildings, to the classification of the photographs. In the second study the subjects were required to express their emotional responses to given photographs. In both cases an attempt was made to reconstruct the semiotic structures embedded in the subjects' minds.

4.2 The re-cognition of different functional building types

4.2.1 Introduction

Definition of the problem

The object of the following exploratory study is to establish the minimal critical properties necessary for a subject to be able to classify an architectural design as belonging to a particular functional type. A further question is whether the ability to classify a building type varies according to the amount of detail available in the design.

The conceptualization of the properties of building types of different functions is logically prior to the conceptualization of the properties of building types of the same function. In other words, before one can re-cognize stylistic differences between churches, for example, one must be able to re-cognize the difference between churches and other building types.

In the terminology of psychosemiology the question is which sets of visual properties of different architectural designs would be activated by which corresponding sets of properties connected to verbal labels such as 'factory', 'office building', 'church', 'school', etc.

Background of the problem

Different authors have attempted to establish what they call 'architectural meaning' (cf Dexter and Lindsey, 1976) or 'architectural codes'. According to Eco (1968) architectural codes can be divided into the two categories of 'syntactic' and 'semantic' codes (cf pages 23, 24). The elements of syntactic codes follow an exclusively technical structural logic: beams, roofs, pillars, vaults, etc, defining the structural conditions necessary for the denotation of space. He divided semantic codes into two categories—element codes and type codes. Element codes contain primary function elements (roof, terrace, staircase, window, etc); secondary function (symbolic) elements (gable, tympanum, metope, etc); and primary function spatial programmes, which may also connote ideologies of habitation (common room, dining room, living room, etc). Type codes are subdivided into social types (hospital, villa, school, castle, station, etc) and spatial types (religious buildings with circular, open, or labyrinthine plans, etc). He also stressed the importance of anthropological codes that take form in different spatial arrangements and vary from culture to culture: fixed-feature space (for example, layout of a tribal village or city), semifixed feature space (for example, 'sociofugal' or 'sociopetal' arrangements of furniture), and informal space (which finds expression in the different intimate, personal, social, and public spatial codes which people adhere to 'automatically' in different cultures). Most of these anthropological codes have been introduced by Hall (1966).

Although it remains open to debate whether the term 'code' should refer only to semiotic structures in the semiology of communication or not, the questions raised in this particular investigation concern what Eco calls 'social type' codes.

All the architectural semiotic structures or architectural codes proposed as yet in the literature on the subject lack empirical verification. The question is whether people actually perceive architecture in these terms and which semiotic structures they utilize in their dealings with architecture.

Since connotative aspects of architecture attach themselves to aspects to which they are subordinate, these connotative aspects may be investigated once the superordinate aspects have been defined, but not vice versa. It is therefore one of the most urgent issues in the semiology of architecture to clarify the relationship of signification existing between the specific functional (architectural) types and corresponding differences in design. The question is simply which properties are critical for the identification of the specific function of the building.

In order to answer this question, the following investigation was carried out to compare the recognition performances of architectural and non-architectural students under different viewing conditions (a brief exposure of a number of slides, and a leisurely examination of a number of photographs). (The investigation was carried out in 1970 by the author and a group of architectural students at the University of Stuttgart, FRG. The students were W Auer, Heike Füßmann, F Gschwind, C Hirtschulz, H Müller, and U Oppenländer.)

Slides of the façades of six individual houses, six tenement houses, six office buildings, six factory buildings, six school buildings and six churches (a total of thirty-six) were shown to groups of twenty architectural and nonarchitectural students. In addition photographs of the same buildings were viewed individually by six architectural students and four nonstudents in random sequence. Both groups were required to write down what they thought the function of each building was and to name the features which had helped them identify this function.

These features were then sorted by the experimenters into the six major categories which the subjects seemed to have used in the identification process (table 4.1). There was a significant correlation between the number of features listed by the subjects and the mode of presentation. Fewer features were produced in response to the brief exposure of the slides than for the unlimited viewing time of the photographs. While this is in itself unremarkable, a noteworthy difference became apparent in the number and kind of features regarded as relevant by the nonarchitectural students and those chosen by the architectural students. In the leisurely examination of the photographs the nonarchitectural students chose as relevant, significantly more of the features attached to buildings (curtains, lettering, etc) than did the architectural students. In response to the brief exposure of slides, on the other hand, the architectural students listed significantly more environmental features (trees, cars, etc) than did the nonarchitectural students. On the whole, however, the difference in length of exposure seemed to be more important than the difference in architectural training. This is not surprising as the task was the correlation of function with architectural features and not the differentiation of styles.

In the case of the photographs, the frequencies of the feature categories chosen by both groups constituted the following rank order:

façade > overall building design > general impression > building elements > attached features > environmental features.

Table 4.1. The six major categories used in the identification process.

Environmental features
fixed: garden, trees, etc
mobile: cars, commercial vehicles, etc

Features attached to the building
interior: curtains, Venetian blinds, etc
exterior: lettering, TV antennas, etc

Building elements
balcony, roof, staircases, etc

Design of the façade
structural and material composition
elements of the façade itself: windows, doors, etc
composite elements: a row of windows, a shop, a garage, the façade as a whole

Overall design of the building
building parts: entrance hall, ground floor, etc
kind of building: skyscraper, tower, etc
dimensions of the building: height, width, mass, etc

General impressions
style, emotional reactions evoked, general concept, etc

In the case of the slides the following rank order appeared:

façade > general impression > overall building design > building elements > attached features > environmental features.

It appeared that the subjects (regardless of the degree of architectural training) used more global features to identify a building's function when allowed only a brief exposure of the stimuli, and used more detailed information when given unlimited viewing time. It was probably in regard to these more detailed categories that the architectural and nonarchitectural students differed.

This preliminary investigation showed that viewing conditions have a significant influence on the kinds of features people draw on to identify a building as a particular functional type. It seemed important, therefore, to investigate this influence in greater depth and to control for features not contained in the architectural design. If only a minimum amount of detail of a building were accessible to observation, which design features would be used to identify it? What would happen if more detail were added? Could one, by successively adding more detail, establish the 'critical detail' responsible for the 'stereotype' of a building? Indications of an answer to these questions were sought in the main study.

Hypotheses

This study is basically inductive, as a taxonomy of the critical design features pertinent to the identification of building types was lacking at the outset. There were, however, two guiding hypotheses.

1 From the theoretical framework developed thus far it would be predicted that the subjects' task of classifying certain pictures under certain labels would involve the following internalized semiotic structure. The given set of verbal labels ('factory', 'church', 'school', etc) would correspond to a system of sets of functional categories ('for production', 'for worship', 'for education', etc) which are established as concepts in the subjects' minds. These sets of functional categories correspond to sets of design properties also established in the subjects' minds. This would then activate the search for corresponding sets of visual perceptual properties in the different architectural designs. Some of these properties will be relevant, some will not, the relevance being established by the sets of properties established in the subjects' minds and connected to the labels.

2 The fewer visual design properties the subjects have to choose from, the less successfully will they be able to carry out the above procedure. Conversely, the more complete the design properties are, the more successful will be their performance.

4.2.2 Methods

Stimuli

The same thirty-six pictures were used in the beginning of this study as were used in the preliminary study (at a later stage, one of the pictures had to be dropped because of incomplete data). There is a preliminary question to be answered as regards the status of the pictures in relation to realized architectural products. One would normally refer to the picture's properties as iconic 'metaproperties' with respect to the properties of the actual architectural design (cf Prieto, 1975a). It was assumed here, however, that the effect of the presence or absence of such iconic metaproperties would be similar, if not identical, to the effect of the corresponding properties of the architectural design. This assumption was made on the basis of Gibson's (1954) theory of 'surrogates', which asserts that the patterns of visual stimulation resulting from 'surrogates' produce similar or, under special circumstances, identical patterns of stimulation on the retina as the original environmental stimulus. In the theory developed for this study the term 'architectural product' was used interchangeably with the term 'architectural design', the latter referring to features which can be recognized either in an actual building or in an appropriate 'iconic' representation of a building. The substitution of pictures of buildings for the buildings themselves is further supported by Bortz's (1972) study and Shafer and Richard's (1974) study, both of which will be discussed later.

Three tracings, each with a different degree of detail, were made of the photographs by superimposing drafting paper of 10 cm x 15 cm on the 9 cm x 12·5 cm photographs. The first minimal level of detail contained only the general outline of the building, with the occasional inclusion of an object such as a person or car as a scale reference. The second level indicated the number of storeys present in the building and in some cases indications of windows, entrances, and staircases. The third added details of the shape, number, and subdivision of the windows and their distribution on the façade. The original photographs constitute the fourth maximum level. (All 3 x 36 = 108 drawings were made by A Lienhard, a former student of architecture at the University of Stuttgart.) A total of 144 stimulus drawings (108 drawings + 36 photographs) would have been used by each subject, except that one of the pictures was accidentally left out of some of the packs, and so had finally to be dropped from all. The four groups of thirty-five pictures used in this study are given in figures 4.1–4.4.

Experimental subjects

The subjects who participated in the entire experiment were thirty-six French-speaking architectural students of both sexes from the School of Architecture of the University of Geneva, Switzerland. The data derived from one of the subjects had to be dropped later as he had not followed the experimental instructions. Some of the subjects took part in the experiments on the first levels of detail but could not be contacted for the later levels. Because their data were independent of that of the later stages, it was used for the preliminary study establishing the frequency ranks of the specific levels of detail they had used, but was excluded from the analysis for the second, more complete, stage.

The choice of architectural students to simulate the re-cognition processes proper to the consumer of architectural products is assumed to be justified on several grounds. Architectural students cannot be considered fully fledged professionals, but are simply initiates in a long process of professional socialization which only at its completion would be assumed to result in perceptual sophistication. More important, however, is that from the theoretical point of view on which the study is based, there is no crucial difference (apart from very general cultural factors) between the process of conceptualizing architectural products carried out by the architect-producer and that carried out by the citizen-consumer. The choice of subjects will become critical only when the question of connotation arises, namely the question of re-cognizing stylistic differences between buildings of the same functional category. Any subject therefore, regardless of training, could be assumed to be capable of classifying buildings into functional types, just as any English-speaking subject, regardless of linguistic training, would understand what meanings correspond to different English phonations, whereas one would not necessarily expect him to appreciate stylistic differences in different ways of expressing the same meaning.

General procedure

Each of the subjects was approached and asked if he or she wanted to participate in an experiment attempting to establish which architectural features people use to classify a building as a functional type. If he agreed, the first level of detail was presented to him in the form of a well-shuffled deck of thirty-five (thirty-six) pictures. The subject was asked to sort the pictures at his leisure into the following categories: factory,. office building, tenement house, church, school, individual house. He was left free to attribute to each category as many pictures as he thought belonged to it. He was then asked to rank the pictures in each category from the most to the least typical (for example, the most typical office building, second most typical, ..., least typical). He was finally asked to write down which features he had used to identify the most typical building of each category. This procedure was repeated four times with each subject until he or she had been exposed to all four levels of detail. One week was allowed to elapse between each exposure to ensure that the subjects would not remember details of the previous exposure. (Most of the data were collected by P Lazaridis, who also helped considerably in the first stage of the analysis.)

Analysis of the data

In the first stage of data analysis, in each category only the picture which was most frequently ranked first by the subjects was chosen to be used as data by the experimenter. The experimenter then listed all the subjects' verbal comments describing the features they had used to identify the buildings. The purpose of this stage of the analysis was simply to give a

(text continued on page 108)

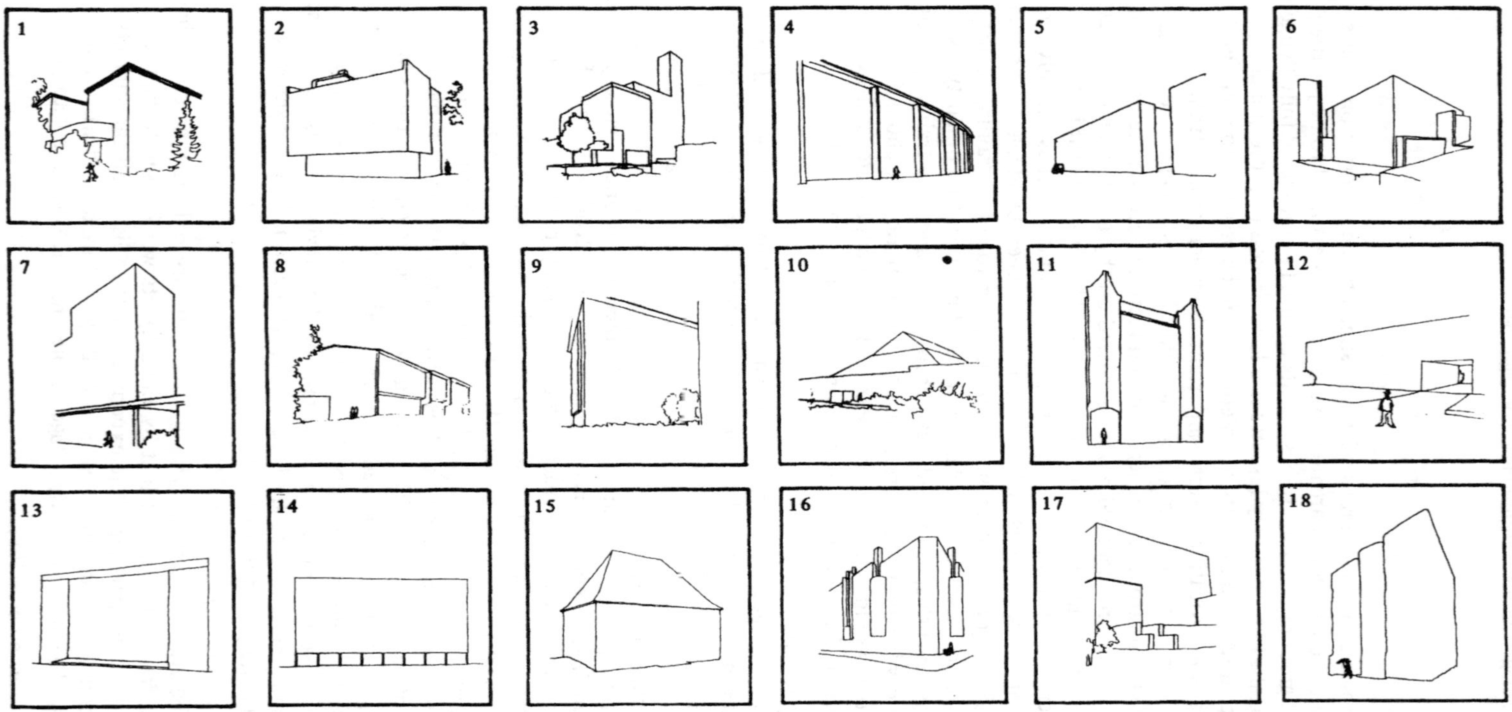

Figure 4.1. The first series of thirty-five building pictures used in the investigation of the re-cognition of different functional building types. These pictures were obtained by tracing photographs, retaining only the general outlines of the buildings, with occasional objects or people as scale indicators—a minimal level of detail.

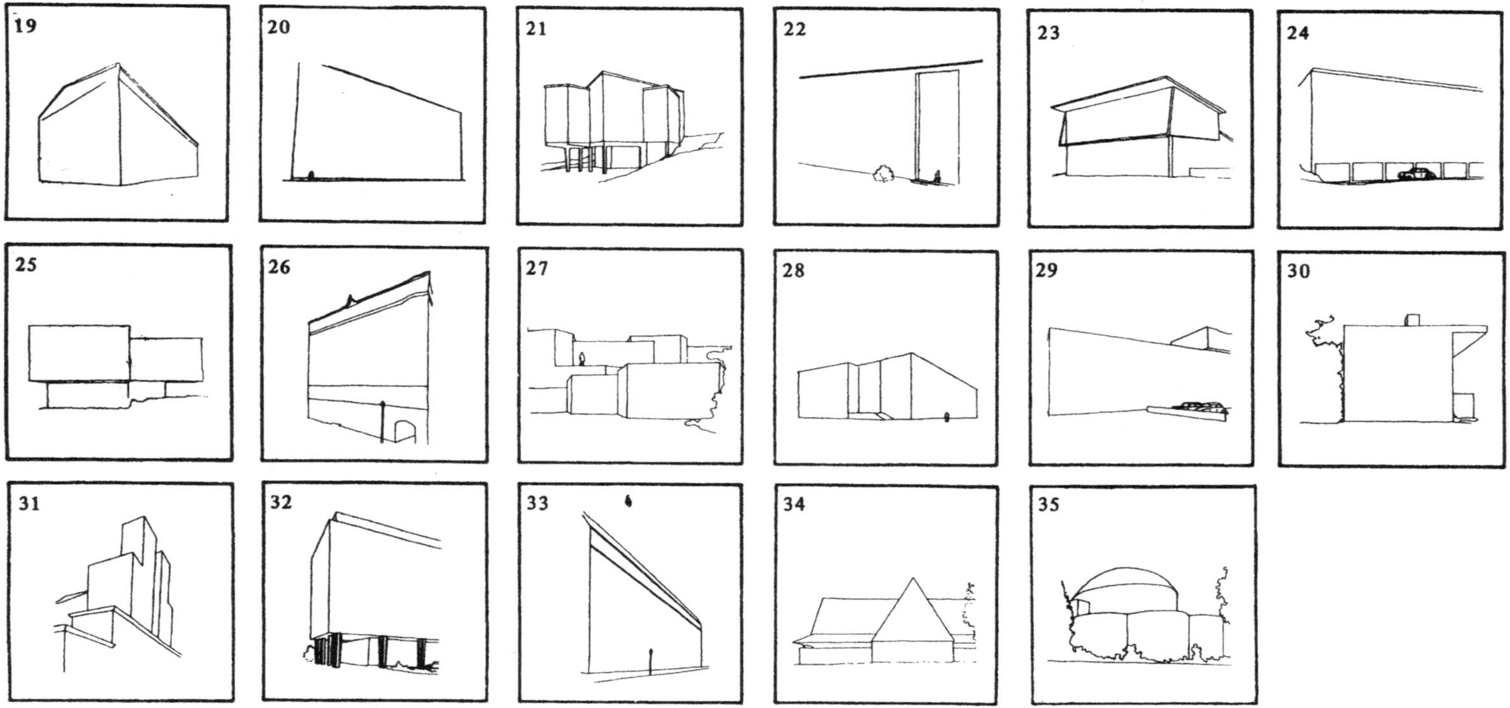

Figure 4.1. (continued).

Figure 4.2. The second series of thirty-five building pictures used in the investigation. At this second level of detail the number of storeys of the buildings was indicated. In some cases additional detail of windows, entrances, and staircases was given.

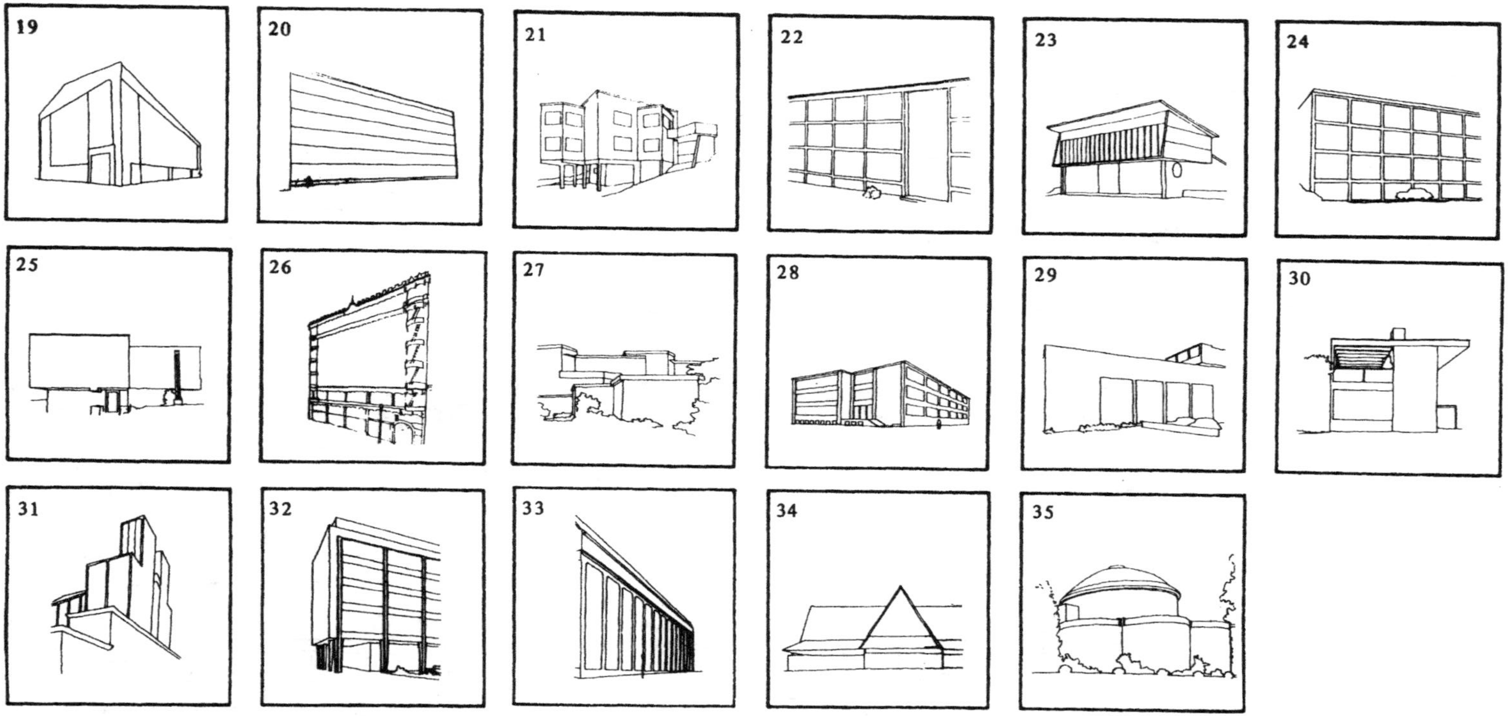

Figure 4.2. (continued).

Figure 4.3. The third series of thirty-five building pictures, featuring the third level of detail—detail of the shape, number, and subdivision of the windows and their distribution on the façade.

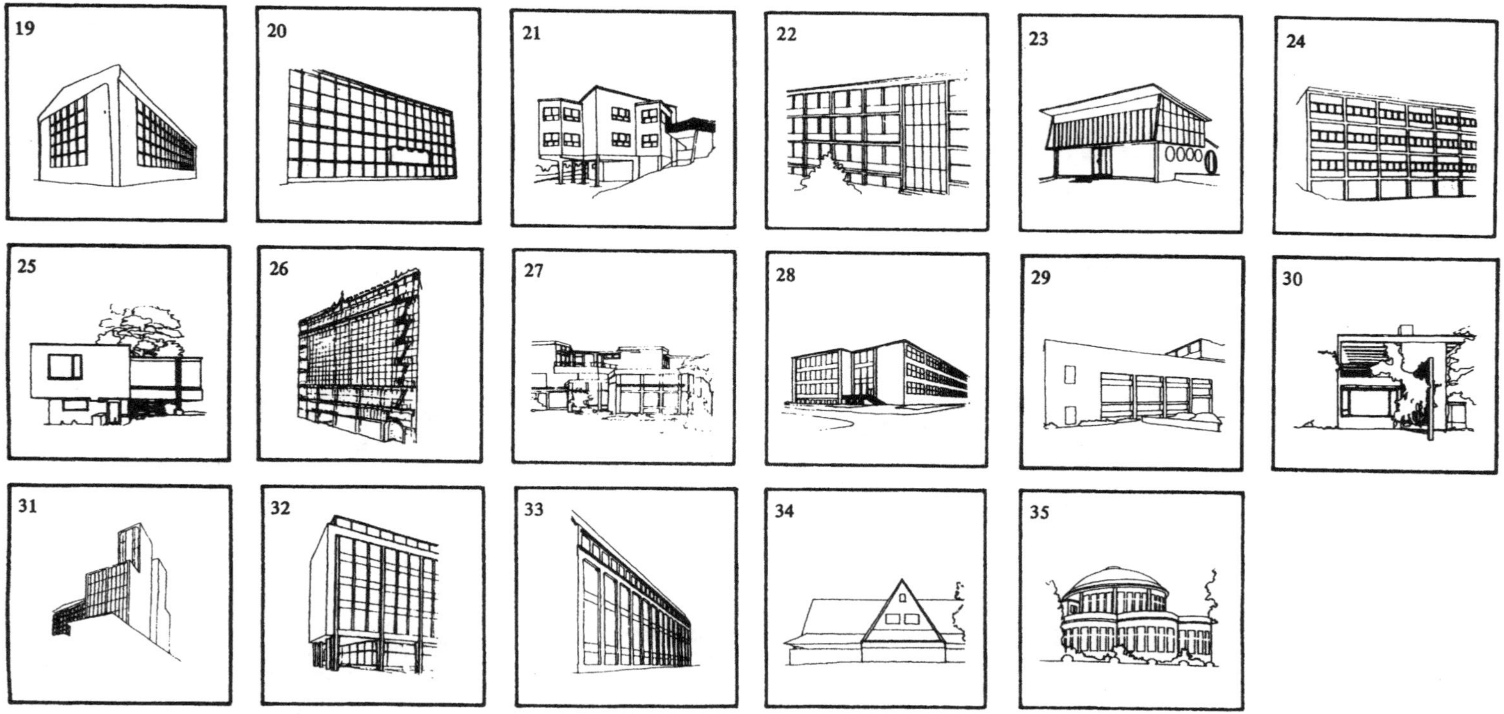

Figure 4.3. (continued).

Figure 4.4. The fourth series of thirty-five building pictures, photographs presenting the maximum level of detail. The three preceding series were traced from these photographs.

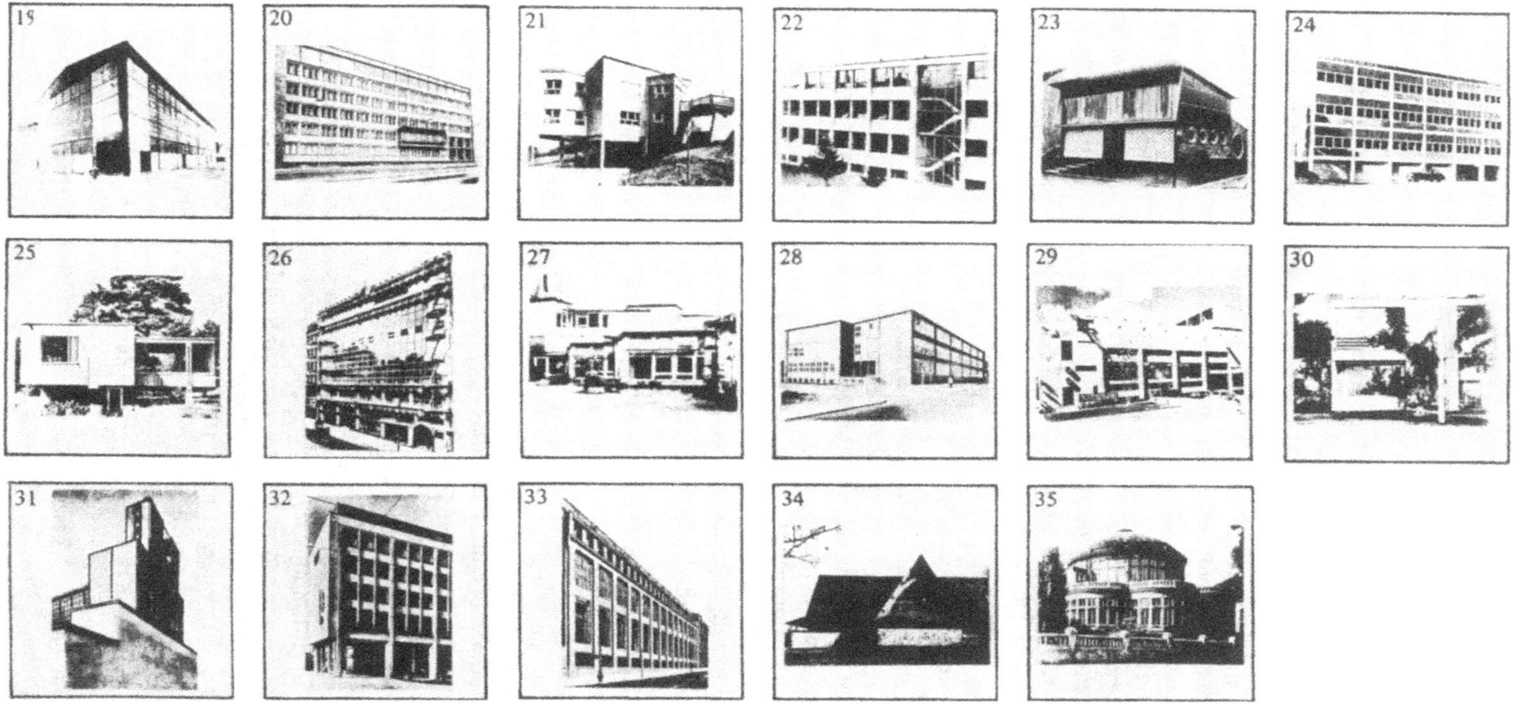

Figure 4.4. (continued).

general sense of how the subjects had interpreted their task and what global results were to be expected.

In the second stage of the analysis the information contained in the data was analyzed in far greater depth. The analysis was based on the assumption that the subjects would in effect act as 'synthesizers' of the sets of properties of building functions and the corresponding sets of design properties. This assumption lends itself to testing by the empirical methods developed in the psycholinguistic study of word meaning.

According to Miller (1967) no objective methods for the direct appraisal of the 'semantic content' of verbal stimuli have yet been devised. There have, however, been indirect approaches by way of investigating 'semantic distances'. As Miller points out:

> "Distances can be related to similarity, and judgements of similarity of meaning are actually easy to get and to analyze."

The empirical procedure would involve three steps:

1 One would first obtain estimates of semantic similarities between lexical items from a group of subjects. This data forms a symmetric item-by-item matrix containing similarity frequencies.

2 One would then investigate the structure underlying the matrix by means of factor analysis or cluster analysis.

3 One would finally identify the resulting factors or clusters by appropriate names, which represent the 'meaning' of the respective lexical items judged to be similar.

Psychologists have used four methods to accomplish step 1:

the scaling of similarity,
the association of other words to a given stimulus word,
the substitution of a given word within different contents,
the classification of words by similarity.

Several methods of cluster analysis have been developed. The method which concerns us here was developed by Johnson (1967) for use in psychological experiments where subjects are asked to sort objects on the basis of similarity. This method may be described best by reference to its practical application in Miller's (1969) paper. The object of Miller's study was to investigate semantic similarities between different lexical units. Forty-eight English nouns were typed on separate cards and a group of fifty subjects were asked to sort the cards into different piles on the basis of similarity of meaning. They were left free to form as many piles as they wished and to place as many items as they wished into each pile. The clusters formed by each subject were recorded and then converted into item-by-item matrices. The individual matrices were then pooled, and the resulting matrix was used as the data for further analysis. The number of subjects placing a pair of items in a cluster was taken as the measure of similarity. These similarity measures were then processed by Johnson's method.

The effect of this method is to produce what is called a *hierarchical clustering scheme* on the basis of the conversion of similarity measures into ordinal distances, that is, distance $a >$ distance $b >$ distance c, etc. Items are ranked along the horizontal axis of the hierarchy and the distance indices along the vertical axis. Miller's study therefore had forty-eight items along the base of the hierarchy and the corresponding distance scale was established by means of converting the similarity measures into distance measures according to the number of subjects placing the same items in a cluster. [The mathematics involved in this type of conversion is fully explained in Johnson's (1967) paper, and will not be gone into here.] The first level of the hierarchy represents the 'weakest' clustering, where each object constitutes a cluster, so with forty-eight words there will be forty-eight clusters. This level is given the 'value' or 'rating' 0·00 on the distance axis.

The procedure is as follows. The smallest distance measure, or metric (representing in this case the largest number of subjects placing two items together) is read, and the two items to which this smallest metric refers are merged into one cluster. These two items are then regarded as one unit, and a new, smaller matrix of distances is then formed in terms of the new metric, which is now rated as 0·00. (Since the items are now regarded as one unit, there can be no distance measure between them.) This new, smaller matrix is then put through the same procedure. The smallest distance is found, the items to which it refers are merged, and a new, smaller matrix formed in terms of the merger. The first sequence of clustering is recorded graphically in the hierarchical scheme (cf figure 4.7) by drawing a vertical line from each item in the cluster and connecting the lines at a node corresponding to the appropriate level on the distance axis. The next sequence is recorded by drawing a vertical line from the node of the subordinate cluster and connecting it at a new node to the vertical lines drawn from the other items in this second cluster. This procedure is repeated for each sequence, and results finally in a hierarchical tree where the peak of the hierarchy represents the 'strongest' clustering, namely, all the items merged in one cluster, this cluster including successively 'weaker' clusters down to the 'weakest' level where each cluster has only one item. As Miller (1969, page 177) puts it:

> "A hierarchical clustering scheme consists of a sequence of clusterings having the property that any cluster is a merging of two or more clusters in the immediately preceding clustering."

This procedure is based on the assumption that the distance between the merged items (called here for the sake of simplicity a and b) is 0·00 and that consequently the distances ak and bk (where k is any other item in the cluster) will be equal. In practice, however, sorting data will generally be noisy, so that the distance between two merged items will often not be exactly 0·00, and the distances between each merged item and another item in the cluster will not be exactly equal. The question when this happens is what distance to assign to the cluster. Johnson's method is to solve the problem twice, first taking the minimum distance—its distance to the nearest member of the cluster—and then the maximum distance—its distance to the furthest member of the cluster.

Johnson's method will be applied to the data of this study, but not at this stage. His method presupposes a square, symmetrical, item-by-item matrix, whereas the data obtained from the study at this stage form a rectangular matrix of 6 (functional categories) x 35 (pictures of buildings) for each of the four levels. This is because the subjects were not free to order the pictures of architectural designs into as many piles as they wished, as Miller's subjects were able to do with lexical items, but were required to distribute them into six prescribed categories. The limitation of this 'forced choice' method, in not being truly representative of the subjects' reactions to the stimuli, is compensated for by the fact that there could be no confusion as to the object of the task. The labels made it clear that the criterion of similarity was similarity of function and not of style, aesthetic preference, etc.

There is, however, a method of investigating the structures underlying data matrices of similarity measures which is not restricted to square, symmetrical matrices. This method, called 'smallest-space analysis', was developed by Guttman and Lingoes (Guttman, 1968; Lingoes, 1973). By this method coefficients of similarity between pairs of objects are converted into distances between points within a coordinate system. Here clusters of similar objects may be portrayed as configurations of points in a scatter diagram or plot. This technique has been applied in studies of architectural psychology to a number of problems involving perception of similarity, from similarity of living-room styles (Laumann and House, 1970), of room functions (Canter and Tagg, 1972; Tagg, 1974), of distributions of furniture in rooms (Canter and Lee, 1974) to that of the functional proximity of departments in a hospital layout (Tagg, 1971). The configurations produced by this technique need not be represented on two-dimensional surfaces only, but in certain circumstances can be represented in spaces of any number of dimensions.

Increasing the number of dimensions results in a more accurate correlation between the information contained in the input matrix and the output configuration, as the constraints of a two-dimensional schema do not reflect the complexity of all the similarity relations involved. The configurations can be successively improved by a process of iteration. The discrepancy between the input matrix and the output configuration is indexed after each iteration by a coefficient called the 'coefficient of alienation' (Lingoes, 1973). As the discrepancy between the input and output decreases, so does the coefficient of alienation. At a level below 0·15 the coefficient arrives at progressively more acceptable values, thus reducing the necessity for further iteration (Laumann and House, 1970).

The computer program SSAP-I (Lingoes, 1973) (from a program series suitable for smallest-space analysis of different kinds) operates on rectangular matrices and is perfectly suited to process the kind of data produced at this level of the study. It puts the categories (in this case building functions) and the pictures of the buildings into a joint space. The categories become reference points situated in various parts of the plot, around which points representing the pictures are scattered. The proximity of a point representing a picture to a point representing a category is determined by the frequency with which the picture was placed into that category by the subjects. The proximity to each other of points representing categories is determined by the relative frequency with which the same pictures were placed in them. This type of joint-space analysis was used by Canter and Tagg (1972; 1974), Canter and Lee (1974) and Tagg (1974) to investigate the distribution of furniture or activities per room in a given house. The rooms would be represented as category points (living room, kitchen, bedroom, etc) and the furniture (chairs, cupboards, etc) or activities (smoking, eating, reading, etc) as object points clustered round the category points. (The present study could not have been carried out without knowledge of these studies and consultation with S Tagg.)

This technique unfortunately also has drawbacks. One problem is how to decide where the boundaries of a particular scatter configuration begin and end. If a point is located some distance from an otherwise dense cluster, it becomes difficult to say whether it belongs to that cluster or not. This becomes even more difficult when a point is located somewhere in the middle between two clusters. This difficulty can be overcome by putting the output of the SSAP-I program as an input into the Johnson method. As well as printing out scatter plots, SSAP-I calculates 'derived coefficients' of similarity between all the objects and categories for every iteration. A derived coefficient is the numerical expression of the approximate degree of similarity arrived at after each iteration. The mathematics of this procedure is given in detail by Lingoes (1973). The coefficients of similarity put out by SSAP-I are calculated for each pair of objects and each pair of categories and are thus printed out in the form of a square, symmetrical matrix. Now that the initial rectangular matrix of 35 pictures x 6 categories has been transformed into a square, symmetrical matrix of similarity coefficients, it can be fed into the Johnson routine. And since the output of Johnson's cluster analysis separates individual clusters into precise hierarchical levels, it can be used to delineate the boundaries of each hierarchical level in the SSAP-I plot. [For the purpose of this study the Johnson (1967) method was transformed into a computer program (with the use of SSAP-I produced coefficients of similarity as an input) by M Sund, formerly of the computing centre of the University of Ulm, FRG.]

Another drawback of the SSAP-I method is that, since the discrepancy between the informational constraints of the input matrix and the output configuration decreases with an increase in the number of dimensions used, solutions may often take the form of a scatter of points in three-dimensional space, that, like an architectural problem, can be represented only by three separate 'views' (plan, front elevation, side elevation) which have to be examined separately.

One way of rendering such an output more easily intelligible is to transform it into a perspective drawing of the three-dimensional space. For this study, a computer program controlling an automatic plotter for the drawing of perspectives of molecular structures was adapted to the input of three-dimensional solutions produced by the SSAP-I program. (This program was produced by N Krier, formerly of the computing centre of the University of Ulm, FRG.)

In the resulting perspectives the functional categories are represented in three-dimensional space as small spheres around which smaller spheres representing the pictures are scattered, in much the same way as planets are dispersed around their stars.

The same technique of representing three-dimensional results was applied to the results of Shepard's (1962a; 1962b) analysis of the speech sound data (Shepard, 1966) also used by Johnson (1967). Lingoes (1970) himself represented in a perspective space the configuration of stratigraphic sites and archaeological pottery types. These representations are given in figures 4.5 and 4.6.

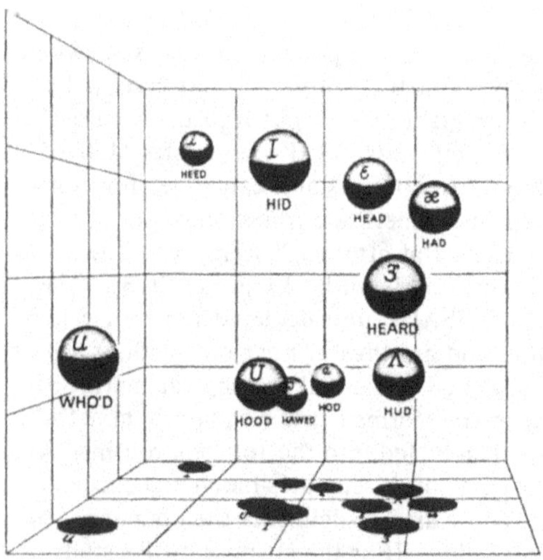

Figure 4.5. Perspective representation in subjective space of the proximities ('nearness' or 'farness') among ten vowel sounds, indicated at the left by phonetic symbols. Test subjects heard these vowels in words like *heed, had, hid,* etc, and were asked to identify them. Pairs of vowels frequently confused are considered 'near' each other; pairs which were seldom confused are regarded as 'far apart'. This computer solution displays these spatial relationships relating each vowel to every other vowel, in a form easily visualized (Shepard, 1966).

The special feature of the perspective program used in the present study is, however, that all small spheres of great proximity are connected to their corresponding categories by means of small bars (cf figure 4.59). Any sphere which does not satisfy a certain requirement of proximity either is connected to another cluster, or remains isolated.

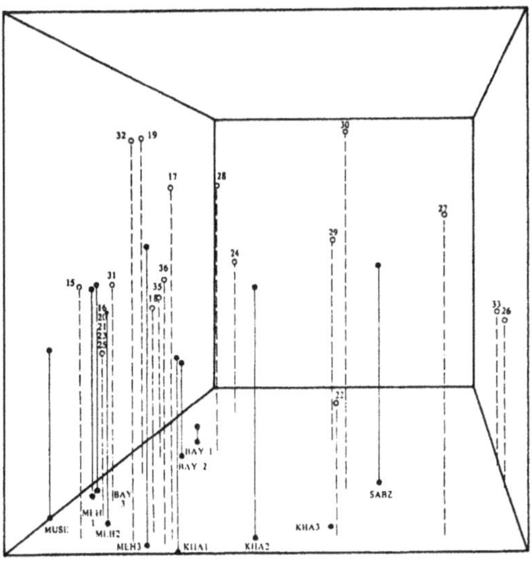

Figure 4.6. Perspective representation of three-dimensional results of a smallest-space analysis. This diagram shows distances and relationships of stratigraphic sites and pottery types found in archaeological excavations (Lingoes, 1970).

Inference of the oppositional structure of design features

The data analysis described thus far resulted in a clustering of building pictures of similar functions. This brings out only half of the semiotic structure, as the particular design features which determined the subjects' choice of which pictures to put into which categories have not yet been established.

The next step is to infer the visual system of design properties which corresponds to the cognitive properties of the building functions within the semiotic structure. This can be done with the aid of the spatial plots of the pictures for the four levels of detail. So far the analysis has established which pictures belong to which categories with which degree of proximity (similarity). These pictures can now be analyzed, in turn, with the focus on the particular design features common to all the pictures in one category and those opposed to all those in the other categories. As more design features were added for each successive level, each level defines which additional features could have been used by the subjects to classify a particular picture.

On the *level of minimum detail* the possibilities open to the subjects were mainly impressions of size (the scale reference being additional details such as people or objects), façade outline, building volume (mostly drafted in perspective), in some instances vertical subdivisions of the façades, the forms of the roofs, repetition of buildings as in row houses, a general impression of complexity or simplicity, and additional cues such as pillars or cues inherent in the perspective. Distinctive volume features, for example would be:

large volume,	low volume,	additional vertically extended volume (a church tower),
small volume,	horizontally extended volume,	vertical volume,
high volume,	horizontal volume,	form of the roof.

On the *second level* the possibilities open to the subjects were mainly horizontal subdivisions of the façades, sometimes with summary indications of windows. Some distinctive features here would be, for example:

ground floor only (hall),	few storeys,	many storeys.

On the *third level* additional possibilities were mainly the number, size, and form of the windows, sometimes in conjunction with indications of doors, balconies, bays, etc. Examples of these would be:

few windows,	different window forms,	few balconies (bays, etc),
large windows,	uniformity of windows,	many balconies (bays, etc).

On the *maximum level* additional possibilities would have been any external properties so far absent in the preceding levels, particularly cues as to the internal structure and function of the storeys. At this level the subjects would have almost complete 'transinformation', as Kiemle (1967) called those cues in the exterior of a building which give information about the internal function of the building. According to Kiemle (1967, page 85) these cues are of three kinds: where one can see directly what is happening on the inside of a building, where a particular function is associated with a particular form (by frequency of common occurrence), or where a particular function is represented by a symbol (religious functions by a dome or a cross; cf Wallis, 1973). Kiemle concluded that the experimental subject can help define what information the exterior of a building provides about the function of the building, and in what measure a subject's uncertainty about this function decreases during the perception of a building. Theoretically, transinformation can be regarded as that information which becomes available when two sets X and Y are interdependent in such a way that events in X do not occur independently of events in Y, and vice versa. A subject's uncertainty about what happens in X would be maximum if he had no knowledge of Y and would decrease with increasing information about Y. This uncertainty is therefore a dependent uncertainty. As regards this study, knowledge of the external appearance of a building would decrease uncertainty about the particular function of that building.

At this maximum level the features available to the subjects for final classification would be:

regular subdivision of storeys (transinformation about regular subdivision of interior),

curtains (transinformation about living styles),

horizontal form of the storeys (transinformation about lack of subdivision of storeys into separate rooms).

Each cluster of pictures around a category point would thus be described on each level in terms of the presence or absence of certain features common to all the pictures within that category (or at least substitutive for each other and thus forming a logical sum) and opposed to those features common to all the pictures in the other categories.

The less detail available to the subjects, the more random will their classification of the pictures be. Because of this random nature of the initial classification, analysis of the data in the early stages will result not in six distinct categories, but rather in a number of logical sums of categories. For example, if an equal number of subjects put one picture into two categories, those two categories will form a logical sum as regards that picture. The analysis of the distinctive design features which correspond to a particular building function can thus only be considered complete at a certain level. This level must furnish enough visual features for the subjects to be able to construct a system of design features which would be in exact correspondence with the system of sets of building functions. A cluster analysis of this final level would show six clearly distinct clusters of scattered points.

Testing of classification

As has been shown, the distribution of building pictures into categories by the subjects can be analyzed by certain clustering techniques. Subsequently, pictures within the clusters may be described in terms of a hypothetical system of features proposed by the investigator. It then remains to be established whether this proposed system of features corresponds as a whole to those features which the subjects used initially to categorize the pictures. The system must thus be verified as being completely relevant in terms of the subjects' actual performance as well as being internally coherent.

To answer these questions one must critically examine the proposed system of design features. The hypothesis that this system of features was used in the subjects' classification process would be verified if it were possible to reconstruct more or less exactly by a purely mathematical analysis the same clusters of the design features which the subjects originally produced. This system of features could then be considered coherent and complete.

Some archaeologists, working with computer documentation techniques, have shown that relatively exhaustive descriptions can be made of artefacts by using their distinctive features. Archaeologists are generally interested in exhaustive description and documentation of human artefacts so as to be able to store and compare the information they have collected. Gardin (1958) has given examples of such exhaustive descriptions of metal tools, containers, and ornaments, and proposed a method for coding iconographic material. Lagrange (1973) proposed a method for testing whether any description, such as may be given by an art historian, is exhaustive and coherent. Her aim was to construct a standardized documentation language for architectural form. The need for such a language is obvious, as technical and aesthetic terminologies are generally mixed in descriptions of architecture, particularly in texts on the history of art. Lagrange tried to develop a method whereby proposed classification systems can be tested and possibly improved. Such a method should discover the relationship between description (for example, an inventory of characteristics of a group of buildings) and interpretation (operations of comparison, selection, weighting of such characteristics).

An exhaustive description requires a set of descriptive elements and a listing of classes, categories, or sets. The descriptive elements are a result of three levels of analysis. First the main categories or sets are defined. According to Lagrange, in the case of buildings, such categories could be:

environment,	elevation (façade),	form of the openings,
volume,	roofing,	form of the roof.
plan,	material of the walls,	

The subdivisions of the main categories are then listed. These are characterizations of the categories. If 'façade' is the category, then its characterizations may be:

material	proportion of height to width,	form of the doors.

Finally minimal mutually exclusive, descriptive features are found. In the case of material (as a characterization of a façade) such minimal features could be:

wood panelling,	bricks,	stone.

Once the descriptive elements have been defined, the categories are listed, either by name or a number, by noting down the presence or absence of characterizations in them, or by enumerating the minimal descriptive features. These categories are then put into relation with objects existing in time and/or space, for example: class C characterized by features f_1, f_2, f_3 contains objects O_1, O_2, O_3, etc.

The test of the coherence of a classification system involves reconstructing the classification by purely mathematical procedures, by using the knowledge of the levels of analysis and of the presence or absence of distinctive features.

The test would reveal (a) whether all the elements of description were used, (b) whether these elements were detailed enough, and (c) whether the sequence of classes in the listing has any morphological significance as a result, for example, of historical or geographical factors.

If a concrete classificatory text is to be tested, care should be taken that it is homogeneous, simple, and brief. Analysis would proceed first by an inventory of all the classes or sets mentioned in the text and an inventory of the elements of description. The results can then be represented in an objects-by-features matrix. The elements of the matrix can be arranged without altering the relational content of the matrix. This rearrangement aims rather at altering the distribution of the signs indicating the presence (+) or the absence (−) of a feature in such a way that the distance between the classes and between the descriptive features becomes (visually) clear. According to Lagrange the best way of doing this is with the scalogram. The scalogram renders visual the distances between the three main positions represented by the two extremes and the centre of the grouping of signs. The objects at the extremes have no features in common whereas those at the centre have a maximum number in common.

Testing the coherence of a classification thus involves the question of the covariation of multiple elements. The question is, which features are so regularly connected to certain objects or classes that they can be used to distinguish these objects or classes from one another; and which objects or classes are so regularly connected that they could be regarded as forming a class in turn.

The different mathematical techniques by which this problem can be solved tend to have several characteristics in common.

- The point of departure is always an objects-by-features matrix.
- The original matrix is generally then transformed into one or two similarity matrices, for which an index of similarity is necessary.
- Among the high-, medium-, and low-value coefficients of similarity randomly arranged in the matrix, an order has to be established which is inherent to the matrix content.
- The different mathematical techniques stress one or other aspect of this order.
- These aspects fall into two major groups: classification and seriation.
- Classification techniques group objects and classes which have many common features.

Seriation establishes the proximity or distance between classes and objects. On the whole the mathematical techniques used in this context are either techniques of seriation by rearrangements of matrices, cluster analyses, or forms of multivariate classification (factor analysis, distance analysis).

In this study the coherence of the proposed system of features can be tested by the same method that was applied in the first stage of the analysis. The pictures would be clustered on the basis of similarity by the Guttman–Lingoes technique (Lingoes, 1973) by using the SSAP-I program.

The Johnson method would be applied to establish the precise boundaries of the clusters. Whether these clusters were identical with, or at least similar to the clusters originally produced by the subjects could then be tested by calculating a coefficient of contingency for the two clustering solutions.

There are, however, some important differences between the Lagrange classification test and that which is applicable here. Whereas the former is applied to verbal material produced by individuals, the latter applies to sets of nonverbal material (in this case iconic representations of buildings) and their inferred nonverbal features. Moreover, the aim of the former method is to test a classification system and to arrive finally at a standardized documentation language for architecture. The latter method is used to infer the visual properties that may have been used to classify a building function and to arrive finally at a reconstruction or simulation of (subjective) recognition processes.

If the goal of this study was to propose a systematic classification of architectural form, one would have to proceed by listing classes of architectural main forms (square, cylinder, etc) in groups (horizontal, vertical, etc) and subgroups (long, short, etc). The main section and position (horizontal, vertical, etc) of secondary architectural forms of first order (balconies, bays, etc), second order, and third order would have to be described in a similar way. This method could possibly lead to an exhaustive description of architectural products. Such a method has been proved successful with manufactured artefacts of great complexity (Zimmermann, 1966).

In contrast to this goal of a rational description of artefacts such as buildings, this study proposes a type of computerized model of the complex and perhaps not always 'rational' process by which people re-cognize building functions. Whether this model would actually have relevance beyond the present investigation is simply a matter of empirical research which would take the proposed model as an hypothesis.

A model of architectural re-cognition

Since any act of cognition concerning objects requires a double classification (notative and then connotative) of the universe of the signified, this must hold true, as was shown above in theory (pages 62–65), for architectural objects. The notative classification system should contain properties characteristic of buildings of different functions and should arrange these into an oppositional structure. The connotative system should contain properties characteristic of buildings of the same function and should arrange them into an oppositional structure of what might be termed 'styles'. In this model the characteristics of the system pertaining to the notative phase will be labelled by capital letters, those of the connotative phase by small letters. (This does not imply any difference in value.)

It will be recalled that Johnson's hierarchical clustering technique used in conjunction with the SSAP-I program transforms an item-by-item matrix of similarity coefficients into a hierarchical tree. The lower branches of this tree structure form clusters of similar objects or frequently co-existing properties. A simplified example of such a hierarchical tree structure is given in figure 4.7.

This simplified structure shows four clusters of two similar objects (from left to right): 1 and 3, 7 and 6, 5 and 4, 2 and 8. On a higher (and more general) level of the hierarchy there are two clusters of four objects: 1, 3, 7, and 6; 5, 4, 2, and 8. If this simplified example is taken as a model for the re-cognition process concerning a series of eight architectural objects, it shows that objects 1 and 3, for example, are more similar than objects 1 and 7, or objects 1 and 6, or objects 3 and 7, or objects 3 and 6, respectively. It also shows that objects 1, 3, 7, and 6 are similar to each other, and different from objects 5, 4, 2, and 8, which are in turn similar to each other.

The concept of similarity (or difference) implies that objects with like properties are classified as belonging to one set and objects which do not exhibit these properties as belonging to its complement. One would regard the property common to objects 1, 3, 7, and 6 as differentiating them from objects 5, 4, 2, and 8 by its absence from these latter objects. Similarly, objects 5, 4, 2, and 8 should exhibit a common notative feature not present in objects 1, 3, 7, and 6. This system of notative features could refer, for example, to four office buildings exhibiting the notative feature A for 'large' or to four individual houses exhibiting the notative characteristic B for 'small'. The system could be presented as in table 4.2.

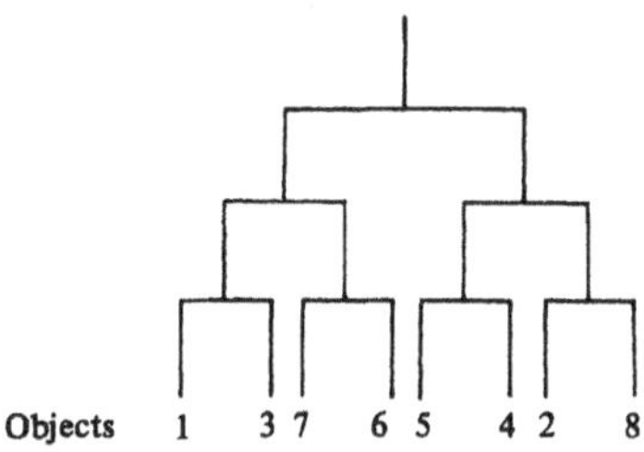

Figure 4.7. Hierarchical tree structure representing successive levels of proximity or similarity between objects or properties. Objects 1 and 3 are more similar than any of the pairs, objects 1 and 7, 1 and 6, 3 and 7, or 3 and 6. Objects 1, 3, 7, and 6 are similar, and different from objects 5, 4, 2, and 8, which in turn are similar to each other.

Table 4.2. System of two 'notative' distinctive features applicable to eight architectural objects.

	Object							
	1	3	7	6	5	4	2	8
Notative characteristic								
A	+	+	+	+	–	–	–	–
B	–	–	–	–	+	+	+	+

On the lower level of the hierarchy there are, within clusters exhibiting the same notative features, objects which differ by other connotative features. For example, objects 1 and 3 have at least one common property which is absent from objects 7 and 6, while objects 7 and 6 may have a property which is absent from objects 1 and 3. The same is valid in the other half of the tree for objects 5 and 4 versus objects 2 and 8. These features which differentiate objects with common notative features must, according to the theoretical framework used here, be connotative. This system of connotative features could refer, for example, to two office buildings exhibiting the connotative feature a for 'vertical', and to two others with the connotative feature b for 'short horizontal rectangle'; the connotative features of the individual houses could be c for one storey and d for two storeys, respectively. The entire system can be represented as in table 4.3. 'Large' and 'small' are notative and 'vertical' and 'horizontal' connotative as the categories 'large' and 'small' are superordinate to the categories 'vertical' and 'horizontal'. An office building will always be larger than a house, but either can be horizontal or vertical in form.

In this study an attempt is made to discover which notative and connotative features people may use to differentiate buildings of different functions. At this point of the investigation the attempt will be subjectively biased by the investigator. In the present study the investigator's bias may be somewhat reduced by the fact that the subjects were asked to name the features which had helped them to identify the most typical pictures. A further question then arises as regards the labelling of the features. The labelling of features common to various objects is not unusual in the social sciences. It is frequently applied in factor analysis when a factor is named on the basis of properties which are exhibited by objects with high loadings on that particular factor. Nevertheless it would remain to be established whether the labels used are consistent with their use in everyday language.

Table 4.3. System of two 'notative' and four 'connotative' distinctive features applicable to eight architectural objects.

	Object							
	1	3	7	6	5	4	2	8
Notative characteristic								
A	+	+	+	+	–	–	–	–
B	–	–	–	–	+	+	+	+
Connotative characteristic								
a	+	+	–	–	–	–	–	–
b	–	–	+	+	–	–	–	–
c	–	–	–	–	+	+	–	–
d	–	–	–	–	–	–	+	+

Limitations of the model of architectural re-cognition and the concept of 'fuzzy sets'

At this point some further critical comments on the model are necessary. The model differentiates various building functions on the basis of notative characteristics and simultaneously on the basis of connotative characteristics. This does not mean that the notative and connotative characteristics discovered by means of the model are absolute or ahistorical. For the purpose of a synchronic study of a certain corpus of buildings one must, however, claim a relative validity for them just as one would do for distinctive features discovered in a synchronic study of a sample of recorded English speech.

Another problem of the model is that the notative and connotative characteristics tend to shift easily from one category to the other. There are well-known examples of churches built in the style of a factory, universities built like Gothic cathedrals, and individual houses built like Renaissance castles.

In this context the concept of 'fuzzy sets' (Bellman et al, 1964; Zadeh, 1965; Volli, 1974a) becomes very useful. Zadeh (1965, pages 338–340) remarks, in introducing the concept:

> "Clearly, the 'class of all real numbers which are much greater than 1', or 'the class of beautiful women', or 'the class of tall men', do not constitute classes or sets in the usual mathematical sense of these terms. Yet, the fact remains that such imprecisely defined 'classes' play an important role in human thinking, particularly in the domains of pattern recognition, communication of information, and abstraction... the notion of fuzzy set provides a convenient point of departure for the construction of a conceptual framework which parallels in many respects the framework used in the case of ordinary sets, but is more general than the latter and, potentially, may prove to have a much wider scope of applicability, particularly in the fields of pattern classification and information processing. Essentially such a framework provides a natural way of dealing with problems in which the source of imprecision is the absence of sharply defined criteria of class membership rather than the presence of random variables."

Zadeh goes on, defining the concept of a fuzzy set as follows:

> "Let X be a space of points (objects) with a generic element of X denoted by x. Thus $X = (x)$. A fuzzy set (class) A in X is characterized by a membership (characteristic) function $f_A(x)$ which associates with each point in X a real number in the interval 0, 1, with the value of $f_A(x)$ at x representing the 'grade of membership' of X in A. Thus, the nearer the value of $f_A(x)$ to unity, the higher the grade of membership of x in A. When A is a set in the ordinary sense of the term, its membership function can take on only two values 0 and 1, with $f_A(x) = 1$ or 0 according as x

> does or does not belong to A. Thus, in this case $f_A(X)$ reduces to the familiar characteristic function of a set A."

On the relationship of the concept of fuzzy set to probability theory Zadeh states:

> "... although the membership function of a fuzzy set has some resemblance to a probability function when X is a countable set (or a probability density function when X is a continuum), there are essential differences between these concepts ... In fact, the notion of a fuzzy set is completely nonstatistical in nature."

This latter assertion is based on "the rules of combination of membership functions and their basic properties".

This lengthy quotation from Zadeh's (1965) article speaks for itself as far as the applicability of the concept of fuzzy set to the problem of architectural re-cognition is concerned. Architectural objects with a certain 'grade of membership' may belong in one set on the basis of certain features while at the same time having a certain 'grade of membership' in another set on the basis of certain other features. The 'normal' set theoretical notion of intersection between two sets is complicated here by the 'grade of membership' concept. But it is intuitively clear that this complication does not invalidate but rather forces a generalization of the 'normal' operation of logical intersection. The mathematical details of this and other generalizations from set theory to the theory of fuzzy sets are dealt with in Zadeh's article. For the purpose of this study and other investigations of this kind the concepts of 'fuzzy set' and 'grade of membership' may prove to be of important explanatory and heuristic value as well as for semiotic studies in general (Volli, 1974a).

4.2.3 Results of the study on the re-cognition of building functions

The first stage of data analysis: the pictures chosen most frequently as representative of a building type and the subjects' written comments

The first stage of the analysis dealt only with those pictures chosen by the majority of the subjects as the most representative of the six building functions. These pictures are given in figure 4.8.

For the office building category the first choice, a building of vertically structured volumes, remained the same for the first three levels of detail. However, when the original photograph of the building was presented for the fourth level, new details such as church bells cancelled it out as a possibility.

For the factory category a different picture was chosen for each of the four levels, the only common feature of these pictures being horizontal building volumes. At the second level, where indications as to the number of storeys of the buildings were introduced, 50% of the subjects chose one picture and 50% chose another. One of these exhibits complex vertical additions to the main volume, ending in what was probably perceived by the subjects as smokestacks.

The first choice for the first level of the tenement house category was a building on 'piloti' (pillars), possibly reminiscent for the subjects of Le Corbusier's "Unité d'Habitation" in Marseille. A picture of four-storey row housing with balconies was chosen for each of the other three levels.

For the first two levels of the church category the subjects chose a building with a central cupola surrounded by what must have looked to them like lateral naves. A building with the more conventional hall and steeple was chosen when more detail was added for the other two levels.

A building of three identical, low, horizontal volumes was chosen for the first three levels of detail of the school category. At the fourth level 50% of the subjects chose a picture showing a loose configuration of one-storey or two-storey buildings, and 50% chose a four-storey modular curtain-wall façade on 'piloti'.

A building with a gabled roof was chosen as the most typical individual house for the first level of detail. 50% of the subjects chose this again for the second level while the other 50% chose a building with a flat roof. Most of the subjects then chose the latter for the next two levels.

It will be recalled that the subjects were required to give reasons for their choices of the most representative buildings. Not all the subjects did this, often because of an apparent inability to articulate the reasons for their choices. The thirty-five subjects together produced a total of 347 comments, which covered the four sessions necessary for the four different levels. This means that only about fourteen of the thirty-five subjects actually gave comments, although the comments for each category or level of detail were not always made by the same subjects. An English translation of the original French comments is given in the appendix (table A1). This table will only be intelligible if one realizes that not all the comments listed refer to the pictures which were the most frequent first choices. Some of these comments refer to pictures which were one subject's first choice but not the majority's. The subjects also sometimes commented on more than one choice or on their general reactions to all the buildings they had placed in one category.

The comments on the most representative office building for the first three levels stressed height, size, and the (vertical) articulation of the different building volumes. The latter probably also provoked the comments 'Manhattan', 'New York', and 'skyscraper'. At the third level some mention was made of the 'continuous windows' and 'glass surface'. At the fourth level persistent reference was made to the uniformity or modularity of the façade, whereas the comment 'height' disappeared at this stage. Transinformation (cf pages 114, 115) also became relevant at this level in the references to (uniform) 'white curtains', 'subdivision of the interior space', and 'lack of balconies'.

In the comments on the most representative factory, stress was laid for all four levels on monotony (uniformity) and size. The subjects expressed negative responses to all the levels in comments like 'ugly', 'prison', 'sad',

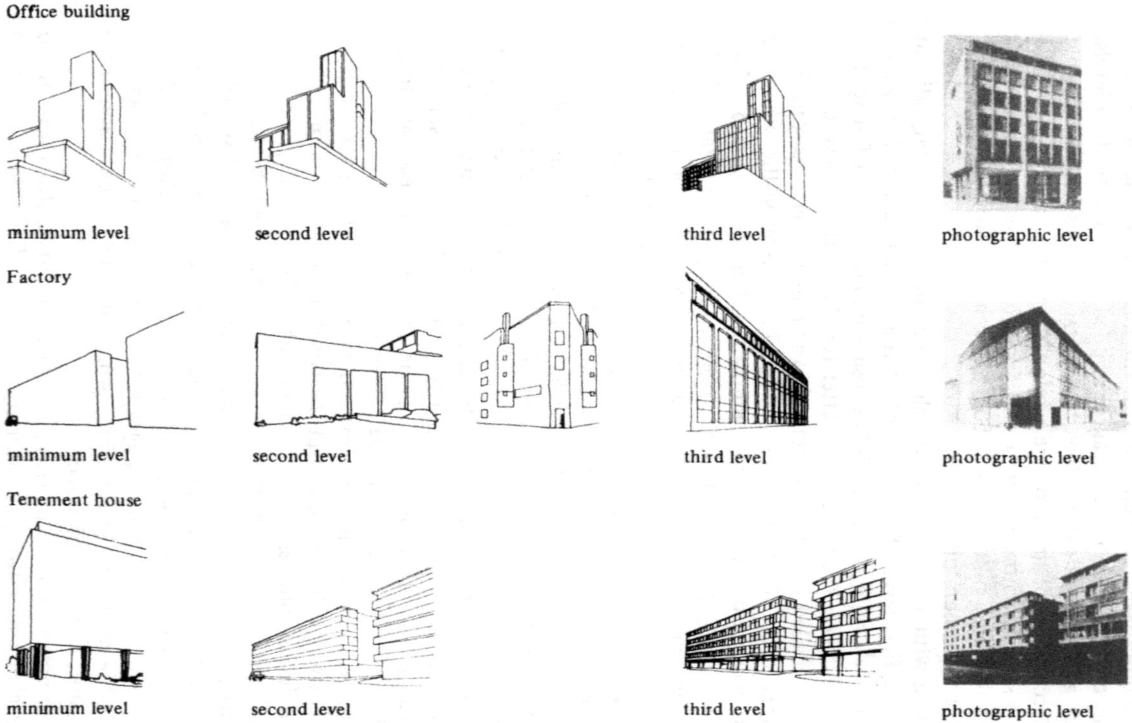

Figure 4.8. The pictures chosen by the majority of subjects as the most representative of the six building functions—office, factory, tenement house, church, school, individual house—at the four levels of detail. Where two pictures were chosen equally often, both are shown.

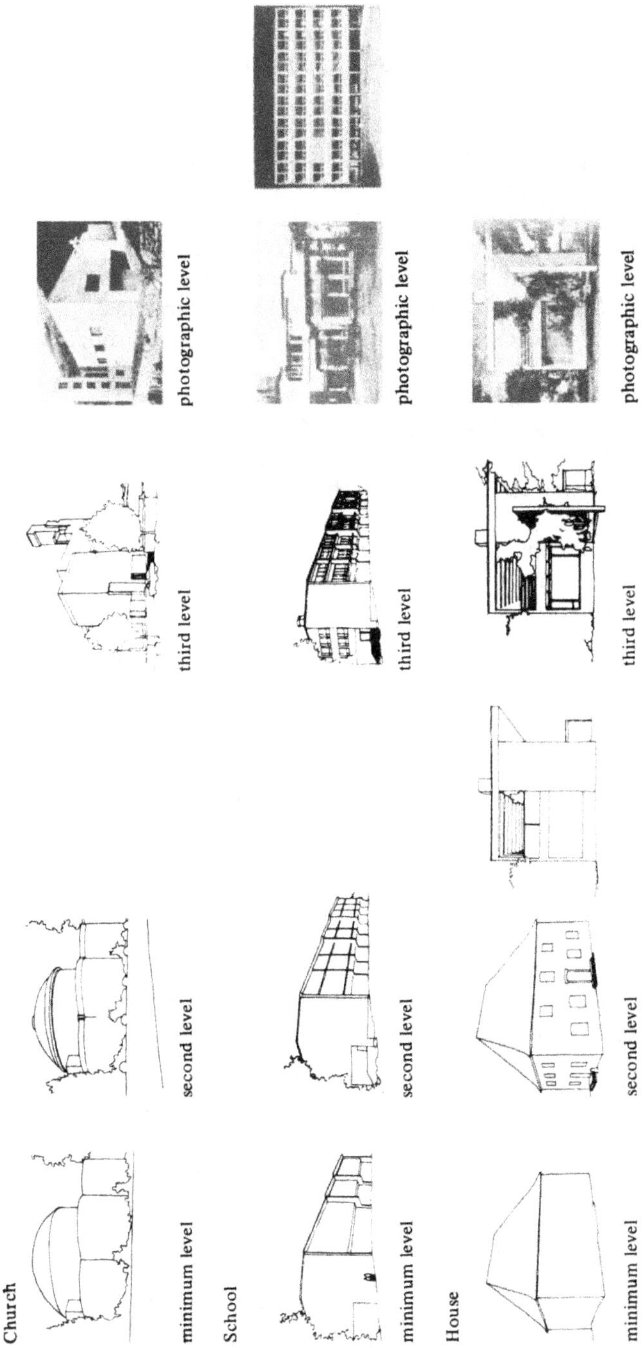

Figure 4.8. (continued).

and 'cheap'. Reactions to the window size were diverse: on the second and third levels mention was made of 'small windows' and on the third and fourth levels of 'large openings' and 'transparency'.

For the first level of the most representative tenement house the suspension of the building on pillars, as in Le Corbusier's "Unité d'Habitation", was most frequently emphasized. On the other three levels the balconies or loggias were mentioned thirty-three times in fifty-six comments. At the fourth level the subjects again seemed to use trans-informational cues. Curtains were mentioned several times as a possible indication of the internal function of the building. Television antennas, the relation of window to surface, and the presence of ground-floor shops were also referred to.

The comments on the most representative church concentrated at the first level on the term cupola and form (circular). The steeple was mentioned at the second level and again in nine of thirteen comments at the third level. Reference was also made to the freedom and variety of volume composition.

At the fourth level predominant emphasis was placed on the cross as a symbol, although the bells and steeple were also occasionally mentioned. Reference to the cross would be an example of symbolic transinformation (cf page 114).

The comments on the most typical school were largely the same for all the levels. Reference was made on every level to the possible relation of the exterior subdivision of façades into large modules, to interior classroom size. The large openings and glass surfaces provoked references to an effect of transparency and luminosity. Mention was also made of the open ground floor, the long, low proportions of the building and the surrounding greenery.

In the comments on the most typical house, reference was made on every level to the (small) size of the building. The gabled roof was mentioned at the first level and the surrounding greenery ('nature') at the other levels. Mention was also made of the patio (claustra), terrace and picture window, and at the third and fourth levels the variety and freedom of composition and use of material were emphasized.

Discussion: results of the first stage of the data analysis. The results at this stage provide some preliminary insights into the re-cognition behaviour of the subjects. One result is that a specific building stereotype generally emerged from the pooled data for each functional category. Another result is that the features chosen for comment by the subjects tended to fulfill the experimenter's expectations (pages 114, 115). In addition to these relevant features, however, the subjects also gave indications of their general impressions and emotional responses to the stimuli. The general nature of these impressions and reactions is perhaps an indication that despite the fact that they were architectural students, the subjects were still largely unaware of precisely which distinctive features they were selecting in order to classify a building. Environmental features included at the first level as scale references possibly played a decisive role in the re-cognition process, for example a truck as a cue for 'factory', a group of school children as a cue for 'school'. If this were the case it would support the hypothesis of the importance of context in the classification process.

A number of general features characteristic of buildings of different functional types gradually emerged from the pictures and comments. An office building is characterized by its height, size, and occasionally by its vertical volume structure. The arrangement of windows is generally uniform as are the curtains, the façade usually has no balconies and there is sometimes an open ground floor.

A factory is characterized by a large, horizontal volume structure, standardized windows resulting in a monotonous façade, use of bricks as building material, environmental features such as trucks and a general impression of ugliness.

One of the most important characteristic features which distinguish a tenement house from office buildings of the same shape is the presence of balconies. Other characteristic features are shops on the ground floor, different curtains in the windows of each apartment, TV antennas, and possibly gardens or trees in the immediate vicinity.

A church is characterized by a cupola or steeple, bells, a cross, side chapels, the form of the roof and occasionally greenery in the vicinity. Dome, steeple and ground plan of the Christian temple are ancient symbols of heaven, gate to the heavenly city and cross (Wallis, 1973) and thus deeply rooted in the minds of people of Western civilization.

The characteristic features of a school building are large windows forming the module of the façade, these windows corresponding to the size of the classrooms, an open ground floor, large staircases visible from the outside, and a large entrance. It is generally a low building organized at different levels (floors) and is often situated in open grounds.

The typical house has a saddle roof (gable) or flat roof with a pergola or balcony, curtains, picture windows, and a characteristic use of wood, glass, metal, and stone as surface materials. The most important feature, however, is its situation in a garden.

These results indicate that the subjects selected different distinctive features to classify buildings of different types. Where little information was available, there was a tendency to act on general impressions, probably on the basis of the overall shapes of the buildings and contextual clues. Where more detail was available, use was made of the form and distribution of windows, features such as the presence or absence of balconies, and attached features such as curtains, antennas, crosses, etc. One of the most important features is the configuration of the façade and the possible transinformation it provides about the internal distribution of space. A tentative list of these distinctive features is given in table 4.4.

Table 4.4. Distinctive features of the most frequently chosen representatives of six building categories.

	Large	Uniform	Many storeys	Balcony	Steeple	Large module	Greenery
Office	+	+	+	–	–	–	–
Factory	+	+	–	–	–	(+)	–
Tenement house	0	0	0	+	–	–	–
Church	0	–	–	–	+	–	(+)
School	0	0	–	–	–	+	(+)
Home	–	–	–	(+)	–	–	+

The second stage of data analysis: the distinctive design features which correspond to different building functions

Preliminary remarks. As this study consisted as a whole of four substudies corresponding to the four sessions necessary for the examination of each level of detail on the building pictures, the results, which generally take the form of computer printouts, will likewise be presented in four sections.

For each of the four levels the results of the initial sorting of the pictures into categories will be presented as an output of the Johnson hierarchical cluster analysis. The results of the smallest-space analysis prior to this, which translates the rectangular input matrix into a symmetrical square matrix of derived coefficients, will be presented not in terms of derived coefficients but as scatter plots. Then, since these clusters are by themselves not particularly instructive, their boundaries are derived from the Johnson schema and printed in. In this way one has two different representations of the results—a hierarchical tree and a 'map' with boundaries. In order to render the results more accessible, those represented in the printouts as clusters of numbers are also represented as clusters of pictures.

The process of architectural re-cognition which should have gone on in the mind of the subjects according to our model of architectural re-cognition (cf pages 118–120) is presented following the results on each level of detail. This attempt to reconstruct the subjects' perception of building functions is made by postulating certain design features they must have used in their classification. These features are then fed into the same computer programs as were used to analyze the subjects' data and are treated in the same way. The results of the reconstruction tests are likewise represented by a hierarchical clustering scheme and a scatter plot with boundaries. The reconstructed model and the subjects' performance are now available for comparison for each level of detail.

Results of substudy 1: the subjects' classification of the building pictures that show only the outlines of the building volumes or façades—the first level of detail. The results of the smallest-space analysis are given in figures 4.9 and 4.10 together with the boundaries derived from the

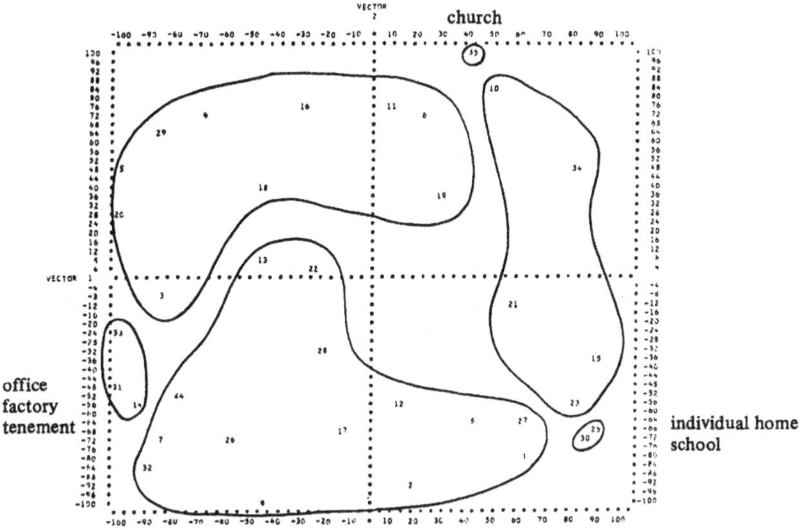

Figure 4.9. Smallest-space analysis 1, substudy 1 (minimum level of detail). This figure shows the similarities between the thirty-five building pictures (represented by the numbers 1-35) resulting from the subjects' sorting them into the six categories of building function. The boundaries around the clusters are added by using the results of the 'maximum solution' of the Johnson hierarchical cluster analysis shown in figure 4.12.

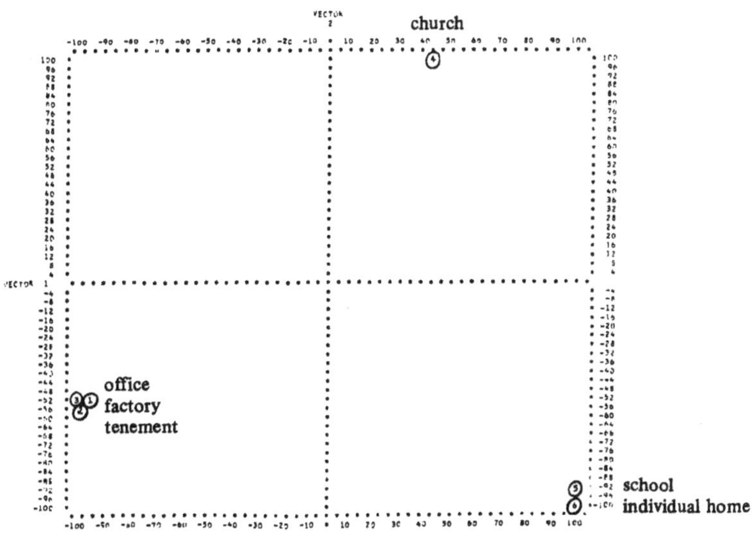

Figure 4.10. Smallest-space analysis 2, substudy 1. This figure shows the similarity between the six categories of building functions (numbers 1-6) used in this study. The subjects were not able to distinguish, at this minimum level of detail, between offices, factories and tenements, nor between schools and individual houses.

previous Johnson hierarchical cluster analysis. The difficulty the subjects had in classifying the pictures can be inferred from the almost homogeneous distribution of the numbers (representing pictures) over the four vectors of the SSAP-I plot. Were it not for the boundaries, hardly any cluster would be distinguishable. The results of the minimum and maximum solutions (cf page 110) of the Johnson method are given in figures 4.11 and 4.12. The maximum solution resulted in a subdivision of the pictures into six clusters on two hierarchical levels, while the minimum solution showed no clustering at all.

The pictures represented by the number clusters are given in figures 4.13–4.18. Examination of these pictures confirms the general impression. The largest cluster contains fourteen pictures, although they clearly cannot all belong to one category. Some of these are outlines of low buildings

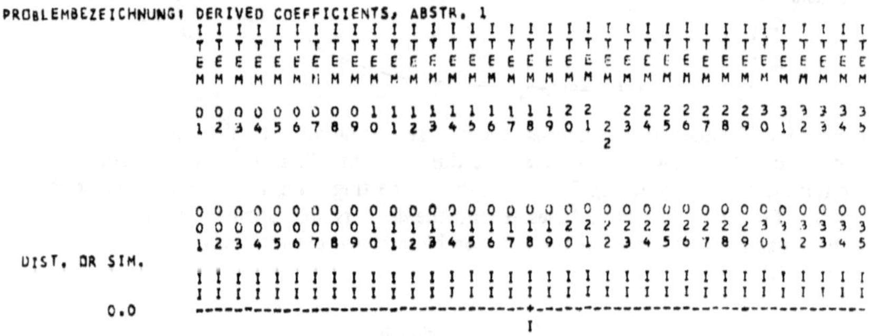

Figure 4.11. Johnson hierarchical cluster analysis, minimum solution, substudy 1 (minimum level of detail). The numbers at the ends of the branches represent the thirty-five pictures. In the minimum solution of the analysis, the subjects were not able to form any hierarchically clustered structure of building pictures.

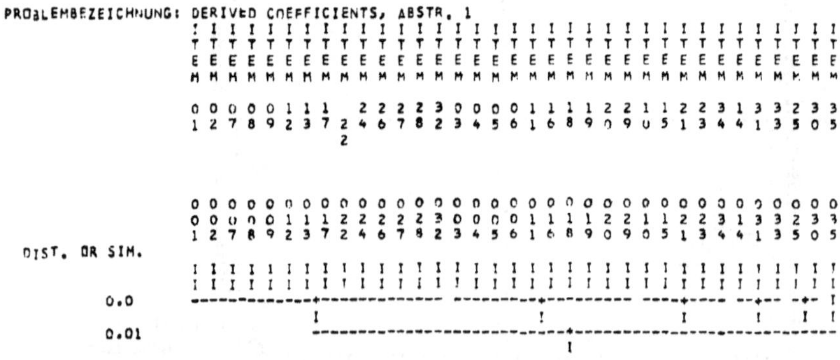

Figure 4.12. Johnson hierarchical cluster analysis, maximum solution, substudy 1. A vague hierarchical structure of two levels is shown, reflecting the subjects' attempt to sort the thirty-five building pictures featuring a minimum degree of detail into six categories of building types. The groupings indicated by this solution were used to draw the boundaries in figure 4.9.

extended by addition or repetition of volumes and could be either houses or schools. Others have large entrances and could be either schools or offices. A third group of tall buildings of simple shape and occasionally with a marked ground floor could be either offices or tenement houses. This cluster thus represents an intersection of four categories: school, office building, tenement house, and house.

The second-largest cluster contains ten pictures which similarly cannot all belong in one category. Some of these are simple in structure and horizontally rather than vertically elongated. These would possibly be factories. Others with vertical elements which could be interpreted either as steeples or chimneys could be either churches or factories. This cluster thus possibly represents an intersection of the categories factory and church.

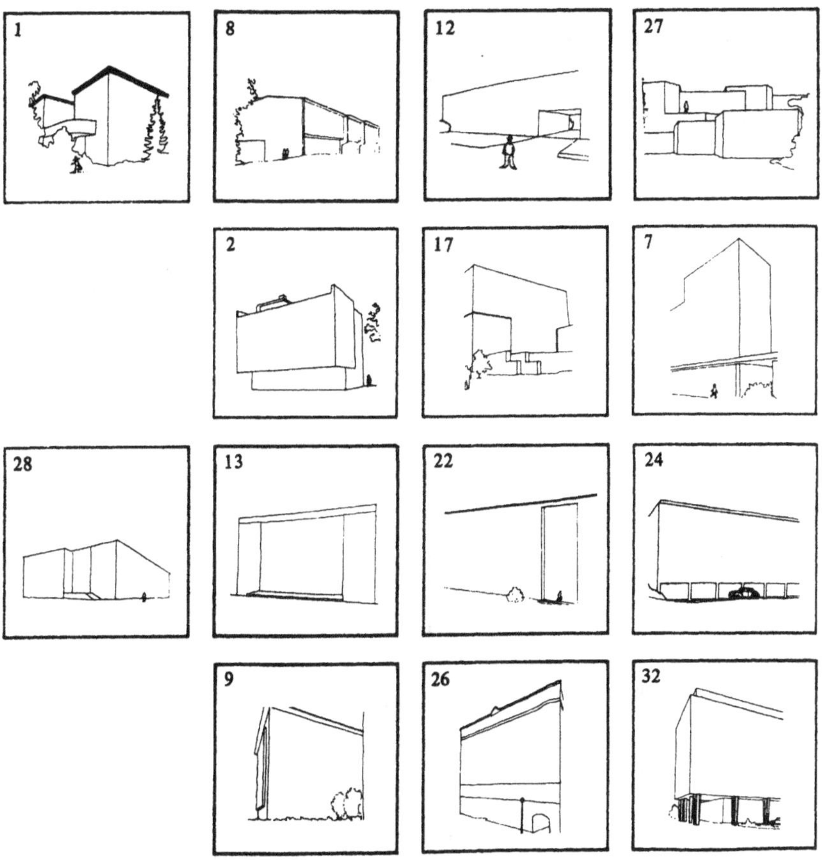

Figure 4.13. Substudy 1: the fourteen pictures in the largest cluster. This cluster appears to represent an intersection of four building categories: school, office building, tenement house, and house.

The smaller clusters are easier to interpret. The largest of these contains what are obviously low buildings with a specific roof structure or complex overall structure. These would seem to be houses, although two could possibly also be churches. The buildings in the second largest of the smaller clusters would seem to be office buildings: the skyscraper and the horizontal box-shaped type. Two other buildings which would seem to be modern-style houses with flat roofs form another cluster, and in a category all by itself is the building with the dome, probably recognized as a church by the majority of the subjects.

It could therefore be concluded that the larger clusters contain those pictures about which there was least consensus, whereas there was relatively more agreement about the pictures in the smaller clusters. The latter would apply to less than one-third of the pictures.

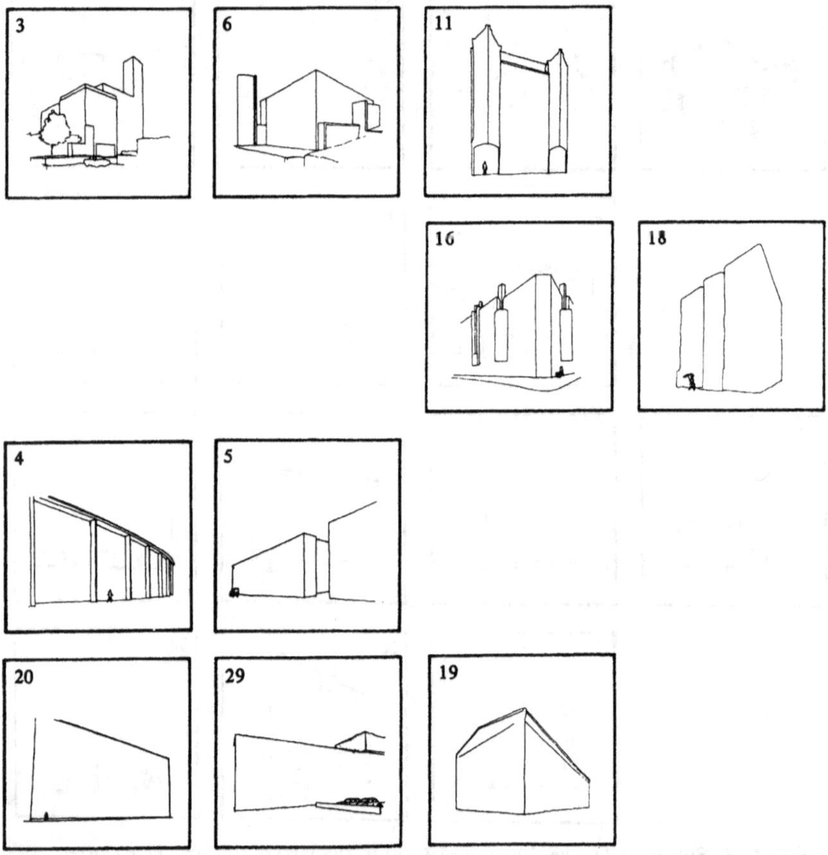

Figure 4.14. Substudy 1: the ten building pictures in the second-largest cluster. This cluster possibly represents an intersection of the categories factory and church.

While it is clear that the subjects were unsuccessful in classifying the pictures at this level into six distinct categories, two types of clusters did at least emerge—two large clusters containing a common mixture of design features (horizontal, vertical, high, low, complex, simple) opposed to four

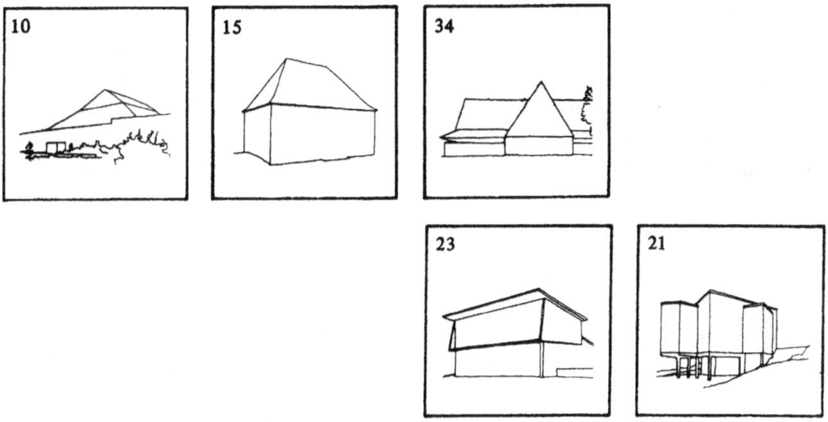

Figure 4.15. Substudy 1: five pictures of low buildings with a particular roof or complex overall structure—houses and/or churches.

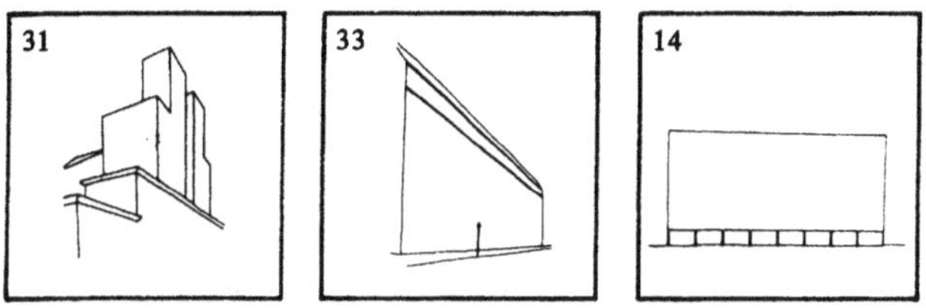

Figure 4.16. Substudy 1: three pictures of tall buildings, one a skyscraper, the others box-shaped buildings which are probably offices.

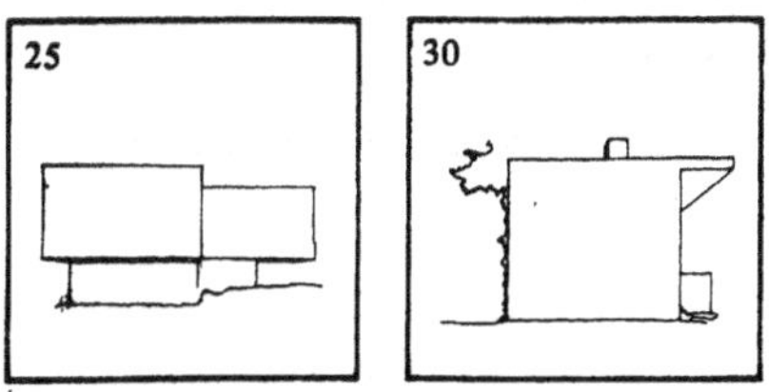

Figure 4.17. Substudy 1: two pictures of small houses. These are probably modern-style houses with flat roofs.

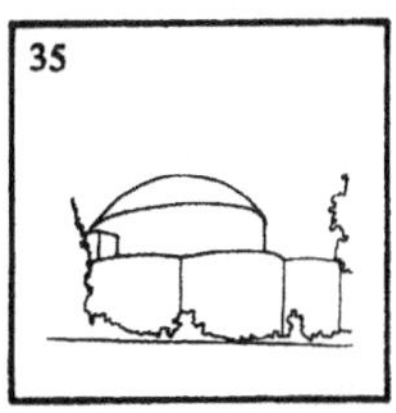

Figure 4.18. Substudy 1: one picture of a building with a dome. This picture was recognized as a church by the majority of the subjects.

small ones with relatively precise features:

a group of small buildings with a specific roof structure: houses,

a group of large buildings, a skyscraper and two horizontal box-shapes: offices,

a group of small, flat-roofed buildings: houses,

a round building with dome: church.

The characteristic of this initial clustering is thus random versus nonrandom. The random group is divided in turn into two further random variables. The nonrandom group differentiates between small and large buildings and the small buildings are in turn divided into those with a specific structure and those without. The building with the dome forms a distinct category of its own.

Reconstruction of the classifications obtained from substudy 1 by means of the re-cognition model. By following the logic of the re-cognition model the two largest clusters can now be described in terms of the common notative property of random grouping (notative property A), while the smaller clusters can be described in terms of the absence of this property ($\overline{A}$). The two large clusters can then each be seen in terms of a second notative property (B) which is present in the larger and absent in the smaller and by which they are differentiated. (Examination of the pictures could lead one to suppose that the notative property B is 'smaller buildings', although it is not actually necessary at this stage to assign any specific content to the properties.) The second-largest cluster would then also be described in terms of an additional property C which is that positive notative property differentiating it from the largest cluster. Although it is not necessary, this property could be inferred to be 'large buildings'. Interpreting B and C respectively as 'smaller buildings' and 'larger buildings' is further supported by the fact that in substudy 2, where the number of storeys of the different buildings was included, one of the two largest clusters, featuring an average per building of 1·82 storeys, was more or less identical to the largest cluster of substudy 1. Similarly, the large cluster in substudy 2 with an average of 2·73 storeys (therefore higher buildings than the first) was more or less identical to the second largest cluster of substudy 1 (cf table 4.5).

Table 4.5. Comparison of the common building pictures in the two largest clusters formed by the subjects in substudy 1 and substudy 2.

Substudy 1	Substudy 2		
	cluster 1	cluster 2	cluster 3
Cluster 1	1, 9, 26, 32	2,8,12,13,17,22,27,28	7, 24
Cluster 2	3, 5, 6, 11, 16, 19	4, 29	18, 20

It is now possible to assign to the buildings in the two random clusters certain connotative features such as 'short horizontal volume', 'vertically extended volume', 'complex nesting of volumes'. These features are selected simply on the basis of the experimenter's personal reactions to the pictures. This procedure should simulate the random choices the subjects made as evidenced in their comments. Care has to be taken, however, to ensure that the features chosen as representative of each distinctive cluster are mutually exclusive, so as to eliminate the possibility of an overlapping of clusters.

The nonrandom section of the hierarchical tree may now be treated similarly. Here the larger cluster can be definitely identified as having the common notative property 'small' (D). In addition, all except one of the buildings in this cluster have the connotative property 'specific roof form' (either gabled or 'pulpit'). The buildings were classified by the majority of the subjects as belonging in the house category.

The next cluster can be described in terms of a connotative property 'horizontal versus vertical extension of large buildings'. Two of the three remaining pictures were tentatively classified on the basis of the connotative property 'very small' and the last on the basis of the property 'presence of a dome'.

The proposed notative and connotative properties of all the clusters are summarized in table 4.6. These data were then put through the SSAP-I procedure, the results of which are given in figures 4.19 and 4.20. The coefficients of similarity derived from the SSAP-I program are then fed into the Johnson routine, the results of which are given in figure 4.21.

A comparison of the hierarchical trees and vector plots resulting from the subjects' data (figures 4.9–4.12) with that produced by the reconstruction model show identical clusters. There are, however, differences in the concentration of the contents of the clusters. The cluster points derived from the data of the model have a correspondingly high density in comparison to those derived from the subjects' data. This can be explained by the fact that the model was operated systematically according to prior knowledge of the results, whereas there were obviously large variances between the subjects' performances, as a result of the lack of precise design cues. The level of randomness in the reconstruction was reduced by using only three random characteristics (A, B, and C).

The same effect is apparent in the hierarchical tree. If the two maximum solutions are compared, it can be seen that the tree resulting from the model is articulated on eight levels, the tree from the subjects' data on only two. The same reason holds here as for the SSAP-I results. It will be seen below that the discrepancy between the subjects' performance and that of the model decreases as more design detail is added.

Table 4.6. Substudy 1: fifteen properties assigned to the six subjective clusters for their reconstruction by means of the re-cognition model.

		A	B	C	D	a	b	c	d	e	f	g	h	i	j	k
Probably smaller buildings	1	+	+	−	−	−	−	−	−	−	−	−	+	−	−	−
	2	+	+	−	−	−	−	−	−	−	−	+	−	−	−	−
	7	+	+	−	−	−	−	−	−	−	−	−	−	−	−	−
	8	+	+	−	−	−	−	−	−	−	+	−	−	−	−	−
	9	+	+	−	−	−	−	+	−	−	−	−	−	−	−	−
	12	+	+	−	−	−	−	−	−	−	−	−	−	−	−	−
	13	+	+	−	−	−	−	−	−	−	−	−	−	+	−	−
	17	+	+	−	−	−	−	−	−	−	−	+	−	−	−	−
	22	+	+	−	−	−	−	−	−	−	−	−	−	+	−	−
	24	+	+	−	−	−	−	−	−	−	−	−	−	−	−	−
	26	+	+	−	−	−	−	+	−	−	−	−	−	−	−	−
	27	+	+	−	−	−	−	−	−	−	+	−	−	−	−	−
	28	+	+	−	−	−	−	−	−	−	−	−	−	+	−	−
	32	+	+	−	−	−	−	+	−	−	−	−	−	−	−	−
Probably larger buildings	3	+	−	+	−	−	−	−	−	+	−	−	−	−	−	−
	4	+	−	+	−	−	−	−	−	−	−	−	−	−	−	−
	5	+	−	+	−	−	−	−	−	−	−	−	−	−	−	−
	6	+	−	+	−	−	−	−	−	+	−	−	−	−	−	−
	11	+	−	+	−	−	−	−	−	+	−	−	−	−	−	−
	16	+	−	+	−	−	−	−	−	+	−	−	−	−	−	−
	18	+	−	+	−	−	−	−	−	−	−	−	−	−	−	−
	19	+	−	+	−	−	−	−	−	−	−	−	−	−	−	−
	20	+	−	+	−	−	−	−	−	−	−	−	−	−	−	−
	29	+	−	+	−	−	−	−	−	−	−	−	−	−	−	−
Individual homes 1	10	−	−	−	+	−	−	−	−	−	−	−	−	−	+	−
	15	−	−	−	+	−	−	−	−	−	−	−	−	−	+	−
	21	−	−	−	+	−	−	−	−	−	−	−	−	−	−	−
	23	−	−	−	+	−	−	−	−	−	−	−	−	−	+	−
	34	−	−	−	+	−	−	−	−	−	−	−	−	−	+	−
Office	14	−	−	−	−	−	+	−	−	−	−	−	−	−	−	−
	31	−	−	−	−	−	+	−	+	−	−	−	−	−	−	−
	33	−	−	−	−	−	+	−	−	−	−	−	−	−	−	−
Church 2	25	−	−	−	−	+	−	−	−	−	−	−	−	−	−	−
	30	−	−	−	−	+	−	−	−	−	−	−	−	−	−	−
Church	35	−	−	−	−	−	−	−	−	−	−	−	−	−	−	+

A random grouping

B probably smaller buildings
C probably larger buildings
D small
a very small

b horizontal versus vertical
c short horizontal, high

d vertical addition of vertical volume
e thin vertical volume added
f row of small volumes
g complex nesting of volume

h small secondary volume (balcony)
i large entrance

j particular form of the roof
k cupola

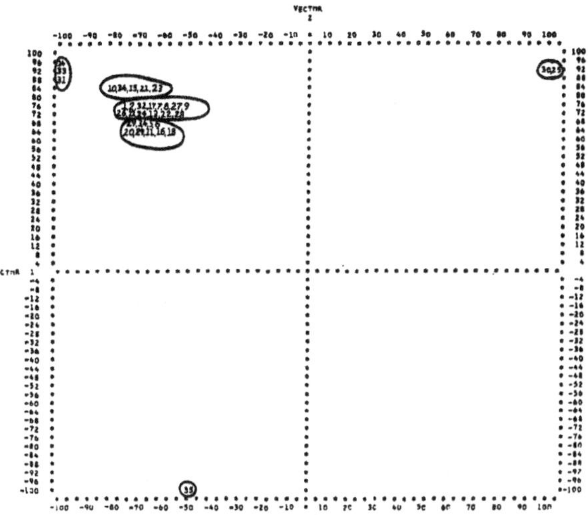

Figure 4.19. Reconstruction of the classifications in substudy 1 by means of the re-cognition model, smallest-space analysis 1: the similarities between the thirty-five building pictures that resulted from assigning properties to the subjective clusters according to the model of architectural re-cognition (pages 118-122). The boundaries around the clusters are added by using the results of the 'maximum solution' of the Johnson hierarchical cluster analysis shown in figure 4.22.

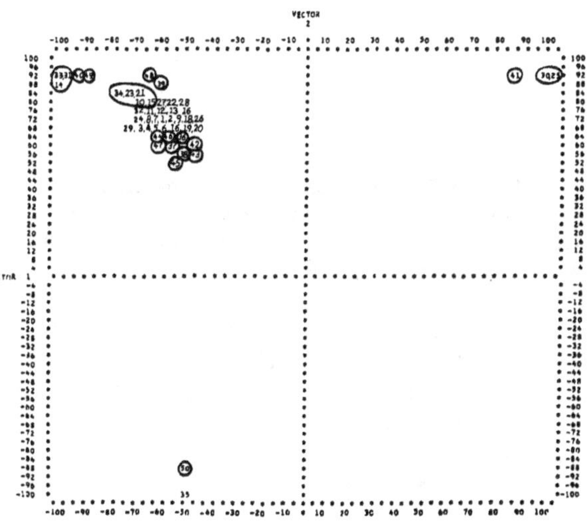

Figure 4.20. Reconstruction of the classifications in substudy 1 by means of the re-cognition model, smallest-space analysis 2. The thirty-five building pictures (1-35) and the fifteen properties assigned to them (36-50) are shown in a joint two-dimensional space. The fifteen properties and their distribution over the various buildings are shown in table 4.6.

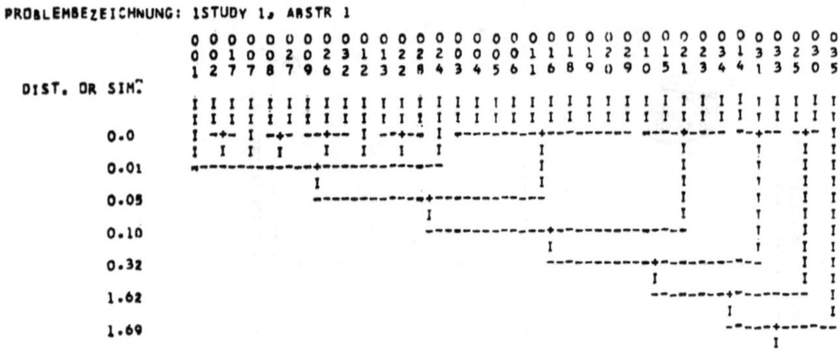

(a)

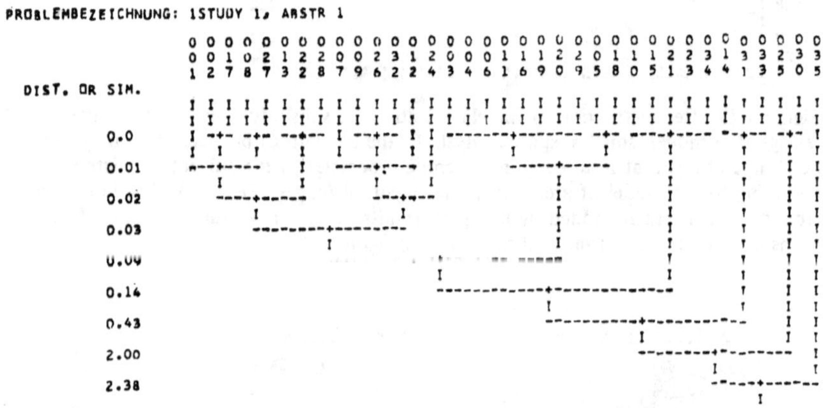

(b)

Figure 4.21. Reconstruction of the classifications in substudy 1 by means of the re-cognition model: (a) 'minimum solution' of the Johnson hierarchical cluster analysis; (b) 'maximum solution' of the Johnson hierarchical cluster analysis. This solution was chosen to draw the boundaries in figure 4.19.

Results of substudy 2: the subjects' classification for the second level of detail, where the number of storeys was added. The results of the smallest-space analysis conducted on this level are given in figures 4.22 and 4.23, together with the cluster boundaries derived from the Johnson analysis. The hierarchical trees from the Johnson analysis are given in figures 4.24 and 4.25 and the pictures represented by the five clusters in figures 4.26–4.30.

As was the case in the first substudy, no clusters emerged from the minimum method. The maximum solution therefore had to be chosen so that at least some clusters could be worked with and boundaries drawn

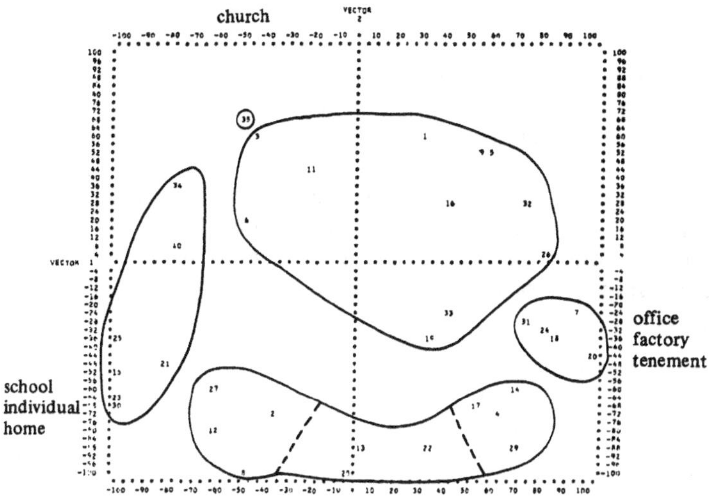

Figure 4.22. Smallest-space analysis 1, substudy 2 (second level of detail). The similarities between the thirty-five building pictures are shown, reflecting the subjects' attempts to sort them into six building categories. The boundaries around the clusters are derived from the results of the 'maximum solution' of the Johnson hierarchical cluster analysis shown in figure 4.25.

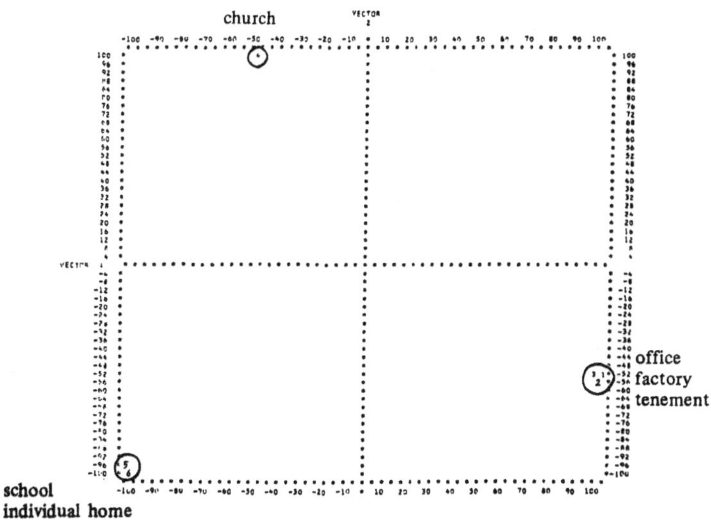

Figure 4.23. Smallest-space analysis 2, substudy 2. The similarities between the six building categories are shown. The indication of the number of storeys in the pictures did not help the subjects to distinguish between offices, factories, and tenement houses, or between schools and individual homes.

into the vector plot. Each of the two largest clusters contains eleven pictures. The pictures in the first of the large clusters could represent any of the categories except houses and perhaps schools. There are steeples on some and balconies on others and different numbers of storeys. This cluster could therefore represent an intersection of the categories office, factory, tenement house, and church.

The second large cluster contains low buildings extended by addition or repetition of volume. Some have large entrances, some have a simple structure, others a complex structure. This cluster could therefore represent an intersection of the categories office, factory, tenement house, school, and possibly house.

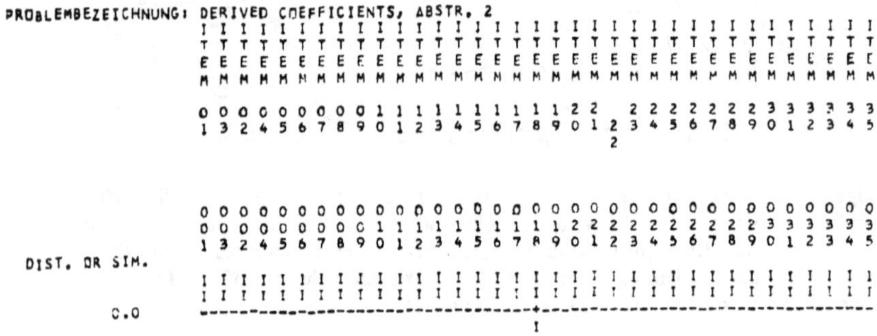

Figure 4.24. Johnson hierarchical cluster analysis, minimum solution, substudy 2 (second level of detail). In the minimum solution of the analysis, no hierarchy of the building pictures was formed. The addition of the number of storeys of a building was not a sufficient clue for the recognition of the six building types.

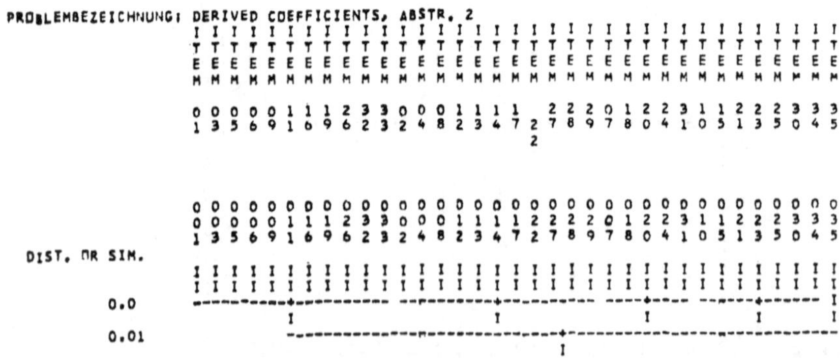

Figure 4.25. Johnson hierarchical cluster analysis, maximum solution, substudy 2. A vague hierarchical structure of two levels is shown. The groupings indicated by this solution were used to draw the boundaries in figure 4.22.

The third large cluster contains low buildings, some with a particular form of the roof. This cluster would probably represent the house category, although some of the buildings could possibly also be churches.

Office buildings and perhaps tenement houses seem to comprise the next cluster, and the building with the cupola is again in a cluster on its own.

It can thus be seen that no appreciable difference was made by indicating the number of storeys of each building. The two largest clusters appear to be as random as in the previous study, while the smaller clusters again seem open to more precise classification by examination of the pictures.

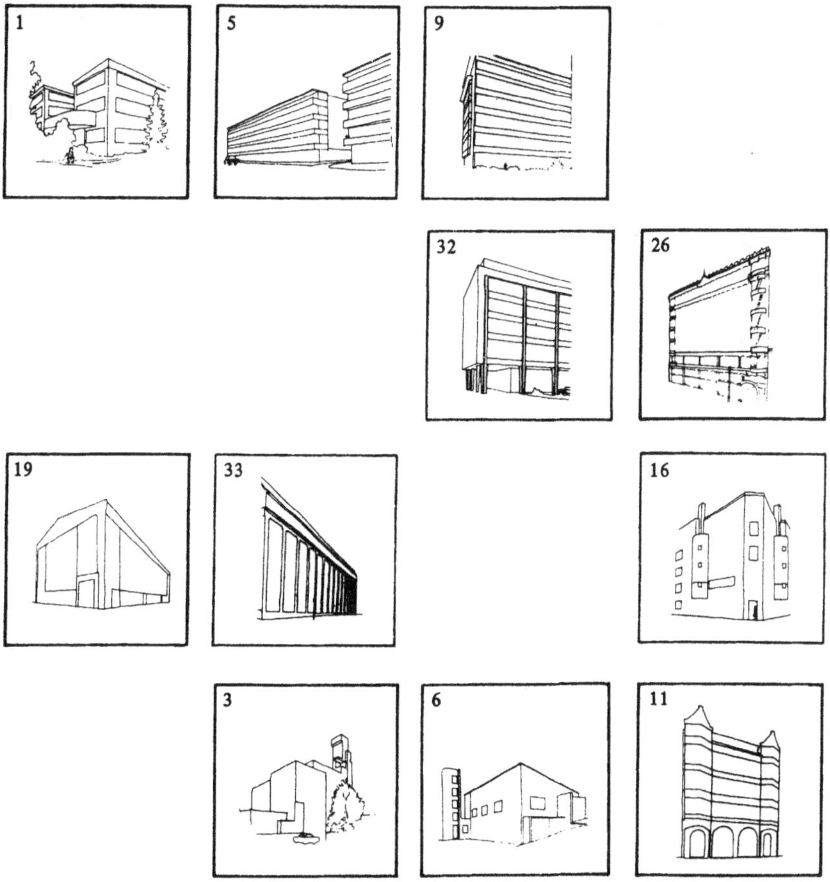

Figure 4.26. Substudy 2: the eleven building pictures in the first of the two large clusters. This cluster could represent an intersection of the categories office, factory, tenement house, and church.

These three clusters can be seen as:

a group of large, vertically extended or box-shaped buildings: office buildings;

a group of small one-storey buildings, most of which have a characteristic roof form: houses;

a round building with a dome: church.

The only difference between the first two studies is that in the second the small buildings are grouped together regardless of form and the sequence of the two large clusters is reversed in the Johnson hierarchy.

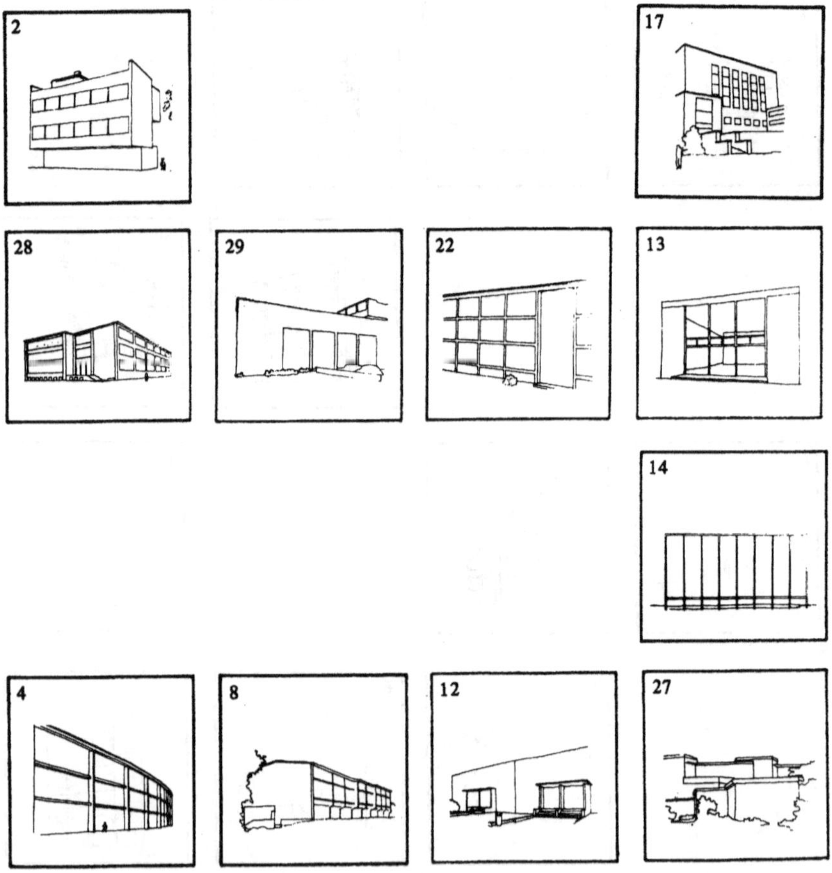

Figure 4.27. Substudy 2: the eleven building pictures in the second of the two large clusters. The low buildings in this cluster could be offices, factories, tenement houses, schools, or possibly individual houses.

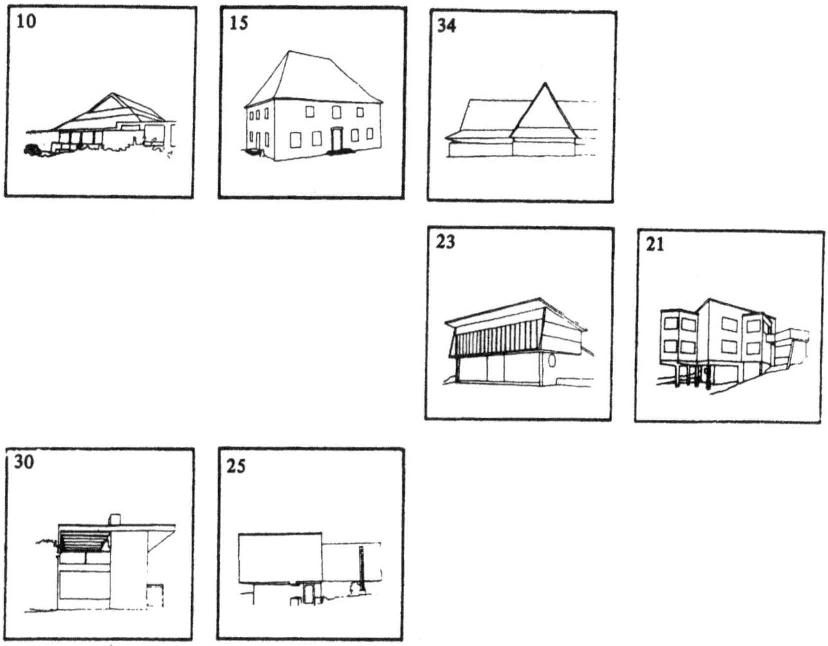

Figure 4.28. Substudy 2: the seven building pictures in one of the small clusters. These buildings are probably houses, although some of them could equally be churches.

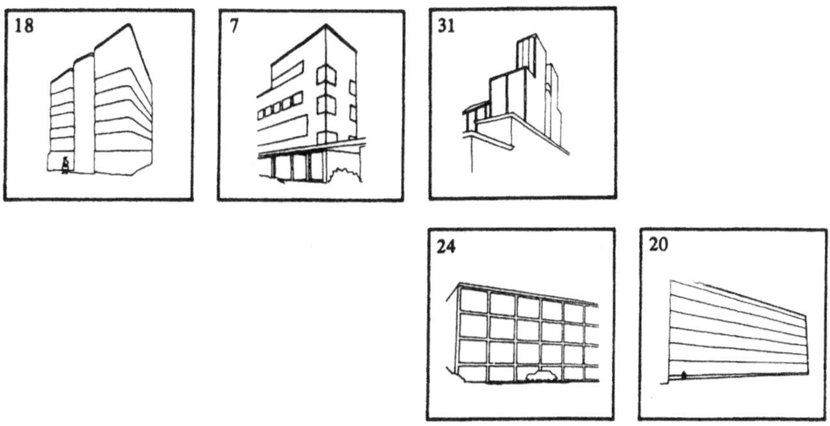

Figure 4.29. Substudy 2: the five building pictures in the other small cluster. These pictures could represent offices, but some perhaps tenement houses.

Figure 4.30. Substudy 2: one picture of a building with a dome. This picture was recognized as a church by the majority of the subjects, as in substudy 1.

Table 4.7. Substudy 2: seventeen properties assigned to the five subjective clusters for their reconstruction by means of the re-cognition model.

		A	B	C	a	b	c	d	e	f	g	h	i	j	k	l	m	n
Larger buildings	1	+	−	−	−	−	−	−	−	−	+	−	−	−	−	−	−	−
	3	+	−	−	−	−	−	−	−	−	−	−	−	−	+	−	−	−
	5	+	−	−	−	−	−	−	−	−	+	−	−	−	−	−	−	−
	6	+	−	−	−	−	−	−	−	−	−	−	−	−	+	−	−	−
	9	+	−	−	−	−	+	−	−	−	−	−	−	−	−	−	−	−
	11	+	−	−	−	−	−	−	−	−	−	−	−	−	+	−	−	−
	16	+	−	−	−	−	−	−	−	−	−	−	−	−	+	−	−	−
	19	+	−	−	−	−	−	+	−	−	−	−	−	−	−	−	−	−
	26	+	−	−	−	−	+	−	−	−	−	−	−	−	−	−	−	−
	32	+	−	−	−	−	+	−	−	−	−	−	−	−	−	−	−	−
	33	+	−	−	−	−	−	+	−	−	−	−	−	−	−	−	−	−
Smaller buildings	2	−	+	−	−	−	−	−	−	+	−	+	−	−	−	−	−	−
	4	−	+	−	−	−	−	−	−	+	−	−	+	−	−	−	−	−
	8	−	+	−	−	−	−	−	−	+	−	−	+	−	−	−	−	−
	12	−	+	−	−	−	−	−	−	+	−	−	+	−	−	−	−	−
	13	−	+	−	−	−	−	−	−	−	−	−	−	−	−	+	−	−
	14	−	+	−	−	−	−	−	−	−	−	−	−	−	−	−	−	−
	17	−	+	−	−	−	−	−	−	+	−	+	−	−	−	−	−	−
	22	−	+	−	−	−	−	−	−	−	−	−	−	−	−	+	−	−
	27	−	+	−	−	−	−	−	−	+	−	−	−	+	−	−	−	−
	28	−	+	−	−	−	−	−	−	+	−	+	−	−	−	+	−	−
	29	−	+	−	−	−	−	−	−	−	−	−	−	−	−	+	−	−
Office	7	−	−	−	−	+	−	−	+	−	−	−	−	−	−	−	−	−
	18	−	−	−	−	+	−	−	+	−	−	−	−	−	−	−	−	−
	20	−	−	−	+	+	−	−	−	−	−	−	−	−	−	−	−	−
	24	−	−	−	−	+	−	−	−	−	−	−	−	−	−	−	−	−
	31	−	−	−	+	+	−	−	+	−	−	−	−	−	−	−	−	−
Individual home	10	−	−	+	−	−	−	−	−	−	−	−	−	−	−	−	+	−
	15	−	−	+	−	−	−	−	−	−	−	−	−	−	−	−	+	−
	21	−	−	+	−	−	−	−	−	−	−	−	−	−	−	−	−	−
	23	−	−	+	−	−	−	−	−	−	−	−	−	−	−	−	+	−
	25	−	−	+	−	−	−	−	−	−	−	−	−	−	−	−	−	−
	30	−	−	+	−	−	−	−	−	−	−	−	−	−	−	−	−	−
	34	−	−	+	−	−	−	−	−	−	−	−	−	−	−	−	+	−
Church	35	−	−	−	−	−	−	−	−	−	−	−	−	−	−	−	−	+

A large (many storeys)
B medium (some storeys)
C small (few storeys)
a very large (very many storeys)

b horizontal versus vertical
c short horizontal, many storeys
d long horizontal, no storeys
e vertical volumes

f multiplication of volume
g duplication of volume
h addition of volumes
i row of small volumes
j cluster of small volumes
k thin vertical volume added

l large entrance

m particular form of the roof
n cupola

Reconstruction by means of the re-cognition model of the classification of the second level. As in substudy 1 the two larger clusters were differentiated from the three smaller by a common notative property A (large'), and from each other by the presence in the second cluster of a second notative property B (possibly 'fewer storeys'). The cluster of one-storey buildings was characterized as having the notative property C ('small'). A number of connotative features were again postulated as being characteristic of the buildings in the different clusters, the second-smallest cluster being described in terms of the connotative property 'horizontal versus vertical' and the round building in terms of the presence of the dome.

These notative and connotative properties are summarized in table 4.7. The results of the SSAP-I analysis conducted on this data are given in figures 4.31 and 4.32 and the results of the Johnson analysis in figure 4.33.

A comparison of the results of these analyses and those obtained from analysis of the subjects' data yields the same result as in substudy 1: although the model reproduced the original clusters, the density of their contents was much higher than in the clusters produced by the subjects.

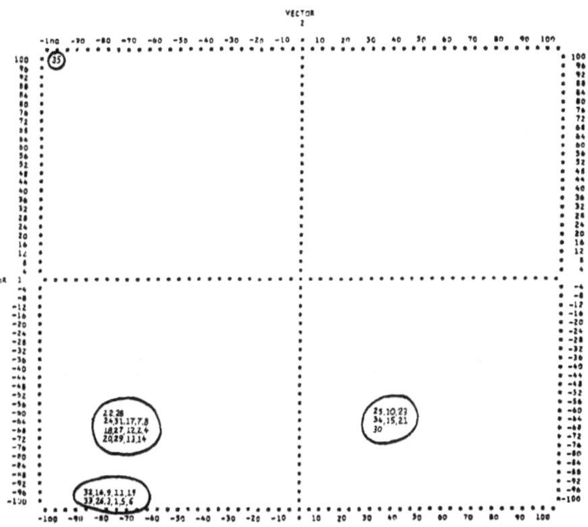

Figure 4.31. Reconstruction of the classifications in substudy 2 by means of the re-cognition model, smallest-space analysis 1. This figure shows the similarities between the thirty-five building pictures, resulting from properties assigned to the subjective clusters in keeping with the model of architectural re-cognition. The boundaries around the clusters are added by using the results of the 'maximum solution' of the Johnson hierarchical cluster analysis shown in figure 4.33(b).

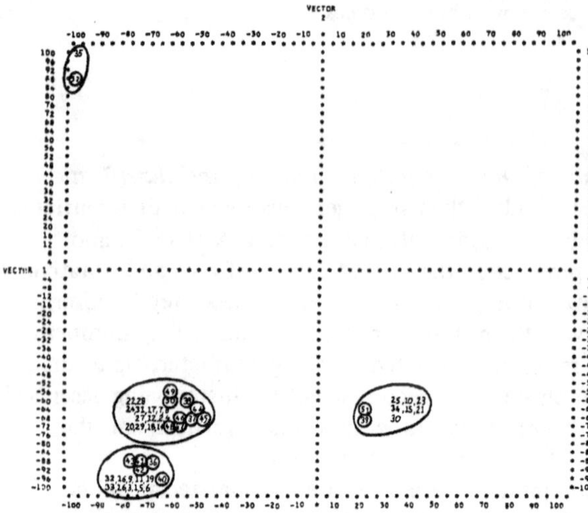

Figure 4.32. Reconstruction of the classifications in substudy 2 by means of the recognition model, smallest-space analysis 2. The thirty-five building pictures (1–35) and the seventeen properties assigned to them (36–52) are shown in a joint two-dimensional space. The seventeen properties and their distribution over the various buildings are shown in table 4.7.

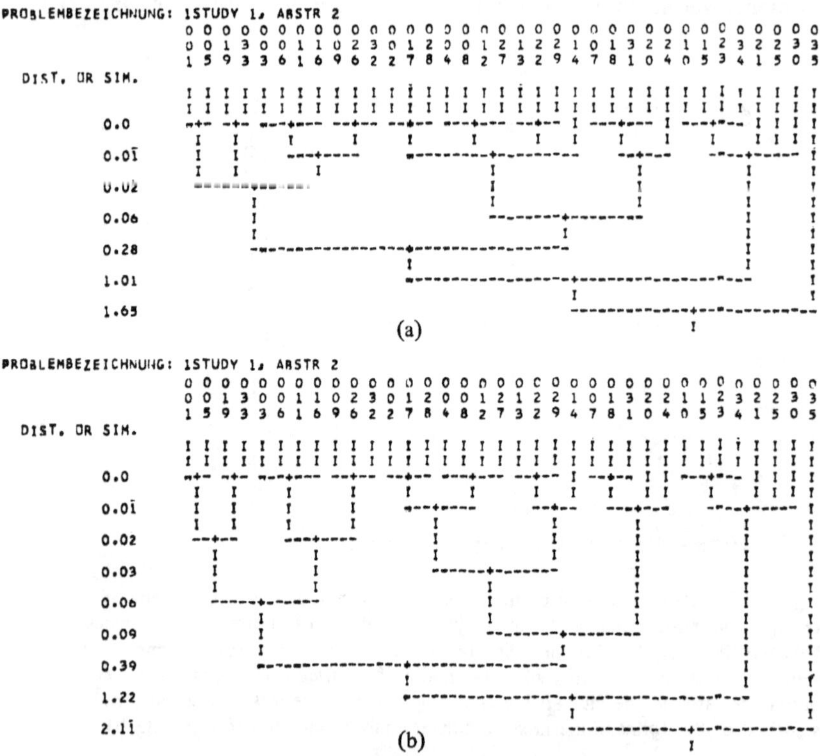

Figure 4.33. Reconstruction of the classifications in substudy 2 by means of the re-cognition model: (a) 'minimum solution' of the Johnson hierarchical cluster analysis; (b) 'maximum solution' of the Johnson hierarchical cluster analysis. This solution was chosen to draw the boundaries in figures 4.31 and 4.32.

Results of substudy 3: the subjects' classification of the pictures for the third level of detail, where the number, form, and distribution of windows was added. The results of the smallest-space analysis for this level, together with cluster boundaries, are given in figures 4.34 and 4.35. The results of the Johnson analysis are given in figures 4.36 and 4.37 and the pictures representing the clusters in figures 4.38–4.42.

At this level clusters were produced by both the minimum and the maximum solutions of the Johnson method. The maximum solution was nevertheless chosen again, as its hierarchy was better articulated. The results of this solution were two main clusters, each of which consisted of two subclusters divided in turn into further lower-level clusters.

The first half of the tree seems to contain houses, including row housing, buildings of up to two storeys, and churches with or without steeples. The second half contains what would most likely be factories, multistorey office buildings, and multistorey tenement houses. It is not clear from this data where schools would be located, but the vector plot of the categories (figure 4.35) shows them as situated between the house and tenement house categories. The additional information about the windows was therefore sufficient to enable the subjects to distinguish between five distinct categories, but still insufficient for identification of the school category.

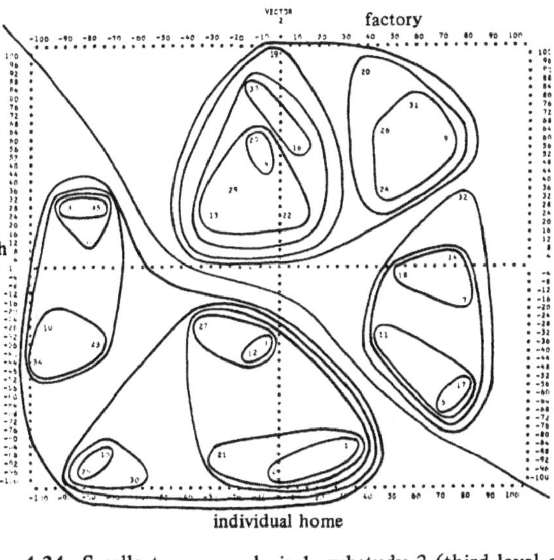

Figure 4.34. Smallest-space analysis 1, substudy 3 (third level of detail). The similarities between the thirty-five building pictures are shown, reflecting the subjects' attempts to sort them into six building categories. The boundaries around the clusters are derived from the results of the 'maximum solution' of the Johnson hierarchical cluster analysis shown in figure 4.37.

Some of the features relevant to the classification process at this level were probably:

the average number of storeys,

the average number of windows,

the relationship between the number of windows and the number of storeys.

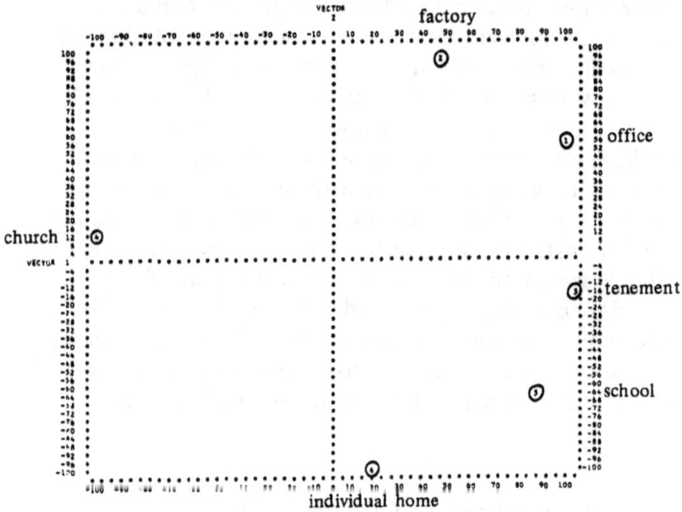

Figure 4.35. Smallest-space analysis 2, substudy 3. The similarities between the six building categories are shown. The details of form, size, and distribution of windows helped the subjects to distinguish between the six categories of buildings, although the distance between tenement houses and schools (3 and 5) is not great enough to put the corresponding pictures into separate clusters.

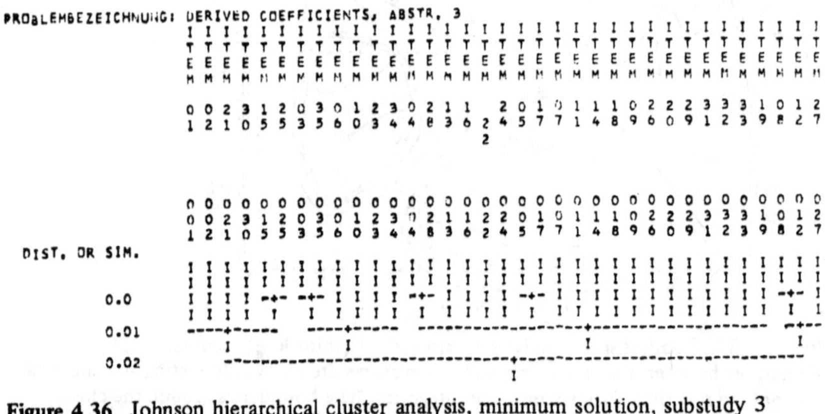

Figure 4.36. Johnson hierarchical cluster analysis, minimum solution, substudy 3 (third level of detail). In the minimum solution of the analysis a hierarchy of three levels was formed. Since the maximum solution yielded more levels, the minimum solution was not adopted.

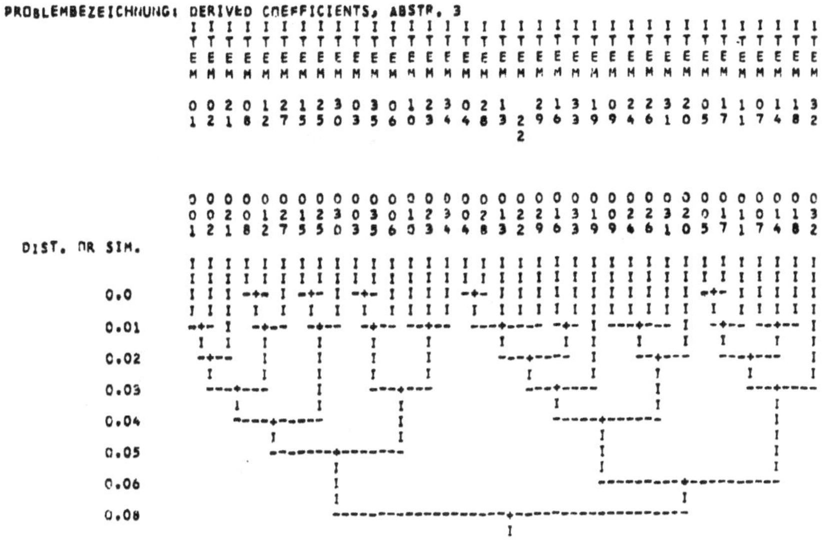

Figure 4.37. Johnson hierarchical cluster analysis, maximum solution, substudy 3. A well-articulated hierarchical structure of eight levels is shown. The groupings indicated by this solution were used to draw the boundaries in figure 4.34.

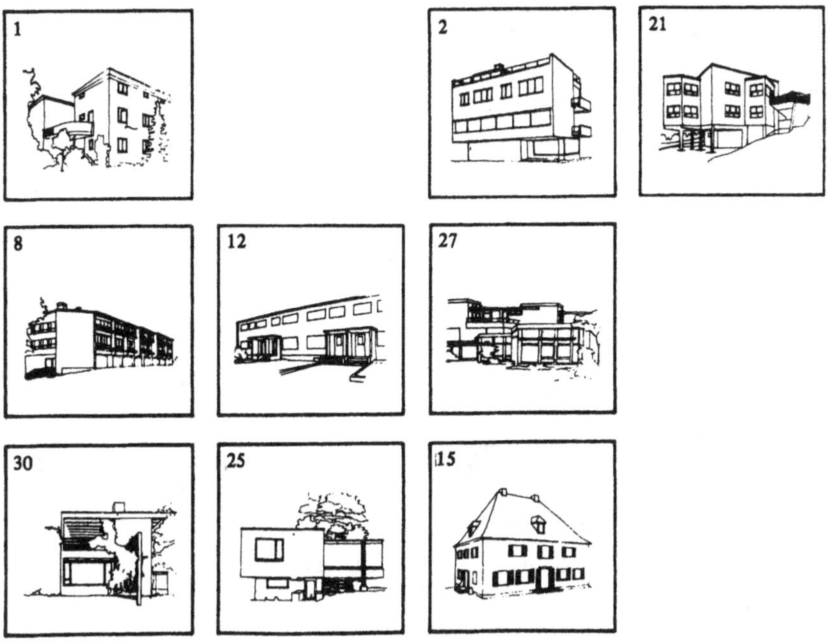

Figure 4.38. Substudy 3: the nine building pictures in the largest cluster. This cluster represents the category of individual houses.

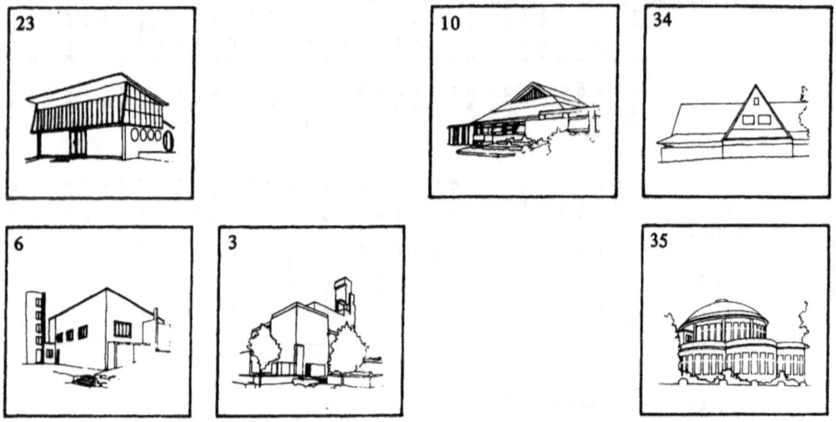

Figure 4.39. Substudy 3: the six building pictures in the cluster adjacent to the cluster of houses. This cluster represents the category of churches.

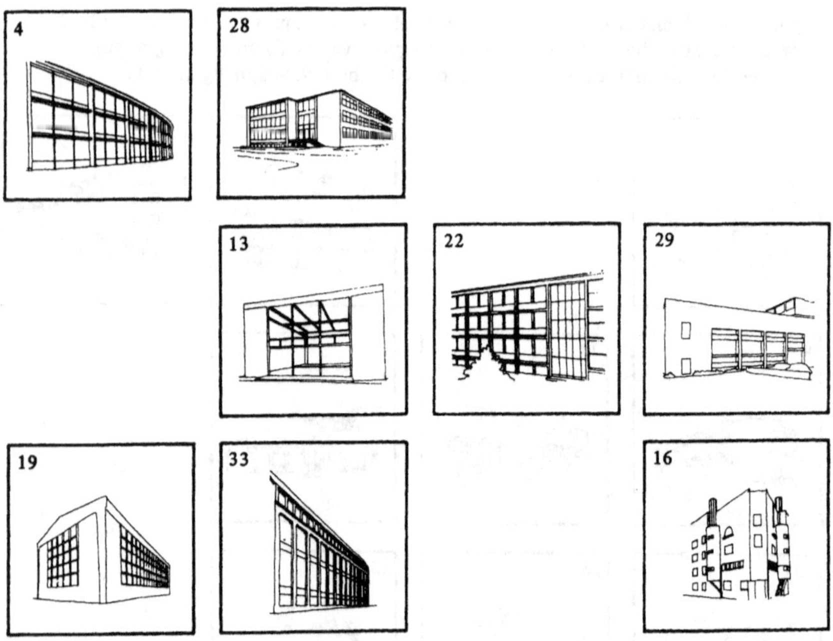

Figure 4.40. Substudy 3: the eight building pictures in the second-largest cluster. This cluster represents the category of factories.

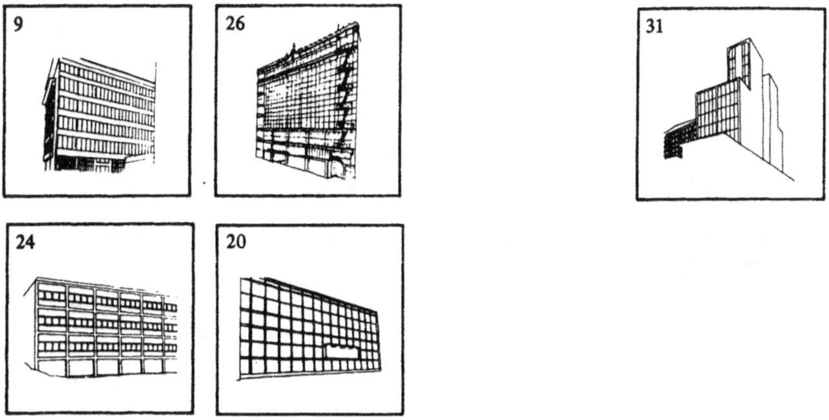

Figure 4.41. Substudy 3: the five building pictures in the cluster adjacent to the cluster of factories. This cluster represents the category of offices.

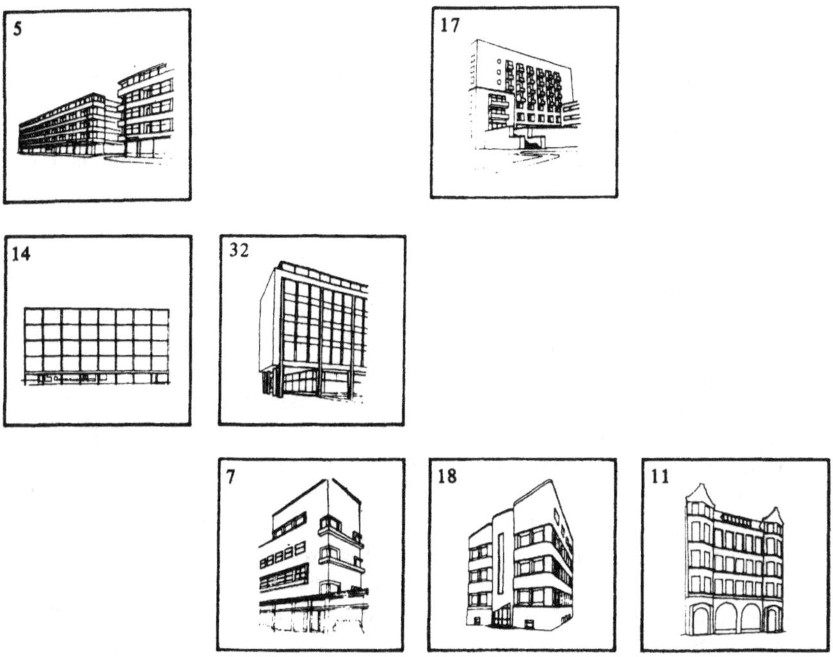

Figure 4.42. Substudy 3: the seven building pictures in the third-largest cluster. This cluster represents the category of tenement houses but could also contain schools.

The numerical value of these indices is given in table 4.8. It can be seen from this table that the perception of a building's size would depend in part on the number of windows distributed over the different storeys. Where there are few storeys and few windows, the building would most likely be perceived as small. Where there are few storeys and a large number of windows, or a large number of windows and a large number of storeys, the building would be perceived as large.

Table 4.8. Quantitative comparison of three distinctive features of five building categories.

	Average number of storeys	Average number of windows		Window/storey ratio	
		minimum	maximum	minimum	maximum
Small buildings					
individual homes	1·5	6·4	15·8	4·2	10·5
churches	0·3	6·3	8·8	19·1	29·3
Not-small buildings					
factories	1·4	24·4	33·9	17·4	24·2
office buildings	4·5	100·0	108·3	22·2	24·1
tenement houses	3·7	29·3	40·1	7·9	10·8

Reconstruction of the subjects' classification by means of the model
Table 4.8 lists some of the properties, probably relevant in distinguishing the buildings in the different clusters. The first half of the tree was distinguished from the rest by attributing to it the notative property A (small). It was then subdivided by attributing to its subclusters connotative properties such as 'one or two storeys high', 'enlarged by secondary volume', etc. These subclusters were in turn subdivided in terms of further connotative characteristics chosen from the pictures by the experimenter.

The second half of the tree was distinguished from the first by the notative property D (large number of windows per storey). It was then subdivided by attributing to the branch with offices and factories the notative property C (large). The factories were in turn distinguished from the offices by the connotative property 'low storey average'. The tenement houses were differentiated from the offices and factories by the notative property B (medium size). All the categories were then further subdivided in terms by a number of connotative properties.

These connotative and notative properties are given in table 4.9. The results of the smallest-space analysis conducted on this data are given together with cluster boundaries in figures 4.43 and 4.44 and the results of the Johnson analysis in figure 4.45.

A comparative analysis of the subjects' data and that of the model produced the same results as in the previous two levels. The clusters produced by the model were again identical to those of the subjects, but

Table 4.9. Substudy 3: twenty-four properties assigned to the five subjective clusters for their reconstruction by means of the re-cognition model.

	A	B	C	D	a	b	c	d	e	f	g	h	i	j	k	l	m	n	o	p	q	r	s	t
Individual home																								
1	+	-	-	-	-	-	-	-	+	-	-	-	-	-	-	-	-	-	+	-	-	-	-	-
2	+	-	-	-	-	-	-	-	+	-	-	-	-	-	-	-	-	-	+	-	-	-	-	-
21	+	-	-	-	-	-	-	-	+	-	-	-	-	-	-	-	-	-	+	-	-	-	-	-
8	+	-	-	-	-	-	-	-	+	-	-	-	-	-	-	-	-	-	-	-	-	-	-	-
12	+	-	-	-	-	-	-	-	+	-	-	-	-	-	-	-	-	-	-	-	-	-	-	-
27	+	-	-	-	-	-	-	-	+	-	-	-	-	-	-	-	-	-	+	-	-	-	-	-
15	+	-	-	-	-	-	-	+	+	-	-	-	-	-	-	-	-	-	-	-	-	-	+	-
25	+	-	-	-	-	-	-	+	+	-	-	-	-	-	-	-	-	-	+	-	-	-	-	-
30	+	-	-	-	-	-	-	+	+	-	-	-	-	-	-	-	-	-	+	-	-	-	-	-
Church																								
3	+	-	-	-	-	-	-	-	-	-	-	-	-	-	-	-	+	+	-	-	-	-	-	-
35	+	-	-	-	-	-	-	-	-	-	-	-	-	-	-	-	-	+	-	-	-	+	-	+
6	+	-	-	-	-	-	-	-	-	-	-	-	-	-	-	-	+	+	-	-	-	+	-	-
10	+	-	-	-	-	-	-	-	-	-	-	-	-	-	-	-	-	+	-	-	-	+	-	-
23	+	-	-	-	-	-	-	-	-	-	-	-	-	-	-	-	-	+	-	-	-	+	-	-
34	+	-	-	-	-	-	-	-	-	-	-	-	-	-	-	-	-	+	-	-	-	+	-	-
Factory																								
4	-	-	+	+	-	-	-	-	-	-	+	+	-	-	-	-	-	-	-	-	-	-	-	-
28	-	-	+	+	-	-	-	-	-	-	+	+	-	-	-	-	-	-	-	-	-	-	-	-
13	-	-	+	-	-	+	-	-	-	+	+	-	-	-	-	-	-	-	-	-	-	-	-	-
22	-	-	+	+	-	-	-	-	-	-	+	+	-	-	-	-	-	-	-	-	-	-	-	-
29	-	-	+	-	-	+	-	-	-	+	+	-	-	-	-	-	-	-	-	-	-	-	-	-
16	-	-	+	+	-	-	-	-	-	-	+	-	-	-	-	-	-	-	-	-	+	-	-	-
33	-	-	+	+	-	+	-	-	-	+	+	-	-	-	-	-	-	-	-	-	-	-	-	-
19	-	-	+	+	-	-	-	-	-	+	+	-	-	-	-	-	-	-	-	-	-	-	-	-
Office																								
9	-	-	+	+	+	-	-	-	-	-	-	-	+	-	-	-	-	-	-	-	-	-	-	-
24	-	-	+	+	+	-	+	-	-	-	-	-	-	-	-	-	-	-	-	-	-	-	-	-
26	-	-	+	+	+	-	-	-	-	-	-	-	+	-	-	-	-	-	-	-	-	-	-	-
31	-	-	+	+	+	-	-	-	-	-	-	-	+	-	-	-	+	-	-	-	-	-	-	-
20	-	-	+	+	+	-	-	-	-	-	-	-	+	-	-	-	-	-	-	-	-	-	-	-
Tenement house																								
5	-	+	-	+	-	-	-	-	-	-	-	-	-	+	-	-	-	-	-	+	-	-	-	-
17	-	+	-	+	-	-	-	-	-	-	-	-	-	-	-	-	-	-	-	+	-	-	-	-
11	-	+	-	+	-	-	-	-	-	-	-	-	-	+	+	-	-	-	-	-	-	-	-	-
7	-	+	-	+	-	-	-	-	-	-	-	-	-	-	+	-	-	-	-	-	-	-	-	-
14	-	+	-	+	-	-	-	-	-	-	-	-	-	+	-	-	-	-	-	-	-	-	-	-
18	-	+	-	+	-	-	-	-	-	-	-	-	-	+	+	-	-	-	-	-	-	-	-	-
32	-	+	-	+	-	-	-	-	-	-	-	-	-	+	-	-	-	-	-	-	-	-	-	-

A small
B medium
C large

D many windows per storey
a small subdivision of glass surface
b vertical articulation of large glass surface
c ground floor without windows

d very small, 1 storey
e small, 1-2 storeys
f large, no storeys
g large, low storey average
h large, 2 storeys
i large, very many storeys
j medium, 4 storeys

k horizontal addition of vertical volumes
l vertical addition of vertical volumes
m thin vertical volume added
n addition of large secondary volume

o 1-2 small secondary volumes (balcony, etc)
p many small secondary volumes (balconies, etc)
q smokestacks

r particular form of the roof
s gabled roof
t cupola

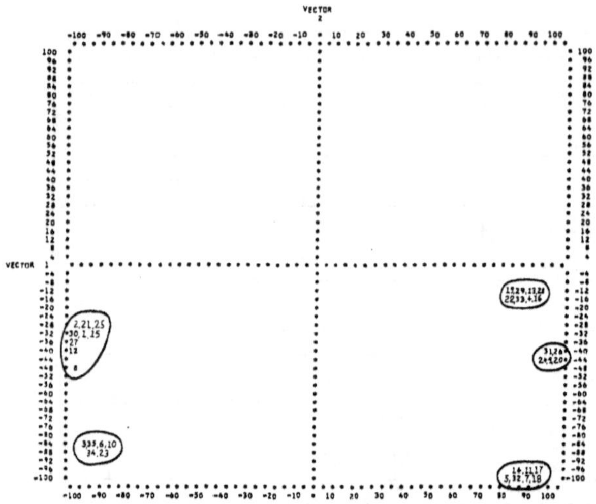

Figure 4.43. Reconstruction of the classifications in substudy 3 by means of the re-cognition model, smallest-space analysis 1, showing the similarities between the thirty-five building pictures. The boundaries around the clusters are added by using the results of the 'maximum solution' of the Johnson hierarchical cluster analysis shown in figure 4.45(b).

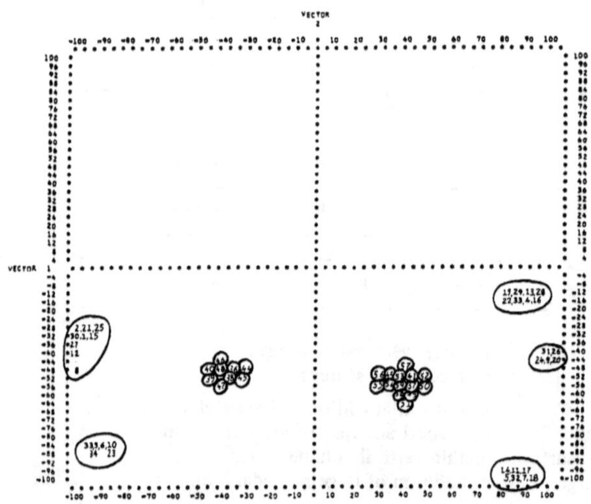

Figure 4.44. Reconstruction of the classifications in substudy 3 by means of the re-cognition model, smallest-space analysis 2. The thirty-five building pictures (1–35) and the twenty-four properties assigned to them (36–59) are shown in a joint two-dimensional space. The twenty-four properties and their distribution over the various buildings are shown in table 4.9.

the density of content within the model's clusters was higher and the tree structure less elaborate.

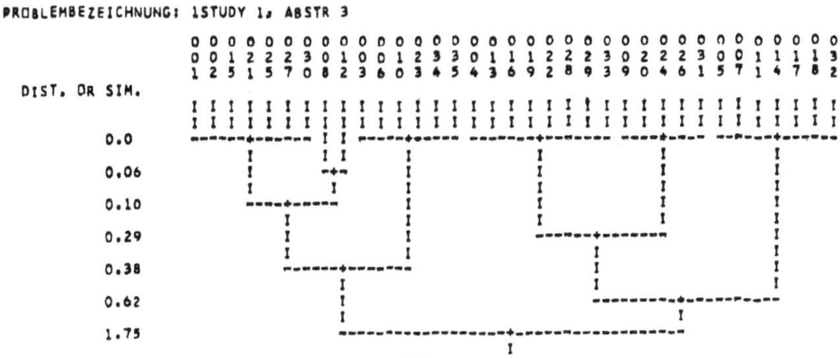

(a)

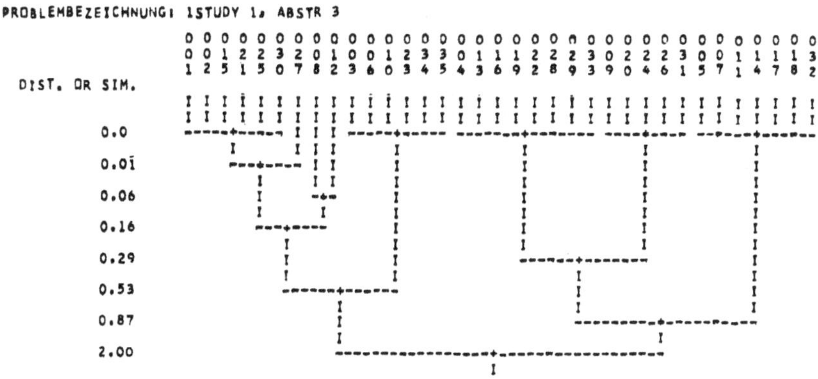

(b)

Figure 4.45. Reconstruction of the classifications in substudy 3 by means of the re-cognition model: (a) 'minimum solution' of the Johnson hierarchical cluster analysis; (b) 'maximum solution' of the Johnson hierarchical cluster analysis. This solution was chosen to draw the boundaries in figures 4.43 and 4.44.

Results of substudy 4: the subjects' classification of the original photographs. The results of the smallest-space analysis, including cluster boundaries, are given in figures 4.46 and 4.47. The results of the minimum and maximum solutions of the Johnson analysis are given in figures 4.48 and 4.49 and the pictures grouped in their clusters in figures 4.50–4.55.

It will be recalled that the approximation of the SSAP-I output to the precise relationships existing between the elements of the input matrix can be indexed by the Guttman–Lingoes 'coefficient of alienation'. According to Laumann and House (1970), this coefficient, calculated for each iteration, becomes acceptable at values below 0·15. In the present two-dimensional study a coefficient of 0·15734 was arrived at after seventy-five iterations.

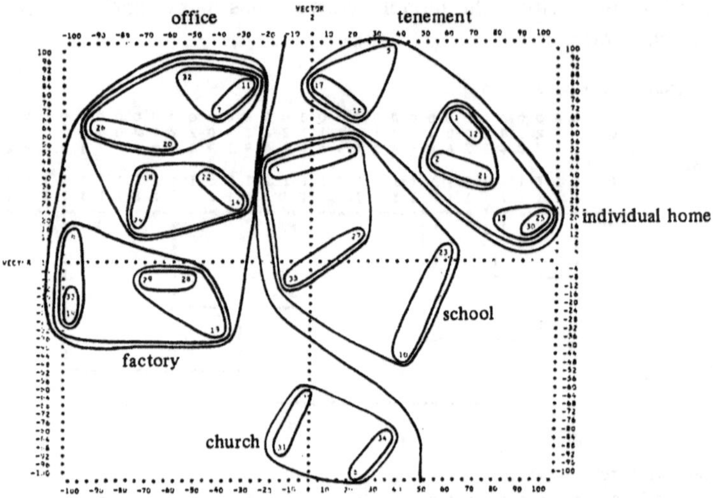

Figure 4.46. Smallest-space analysis, substudy 4, two-dimensional solution (photographs): the similarities between the thirty-five building photographs as they were sorted by the subjects into six building categories. The boundaries around the clusters are derived from the results of the 'maximum solution' of the Johnson hierarchical cluster analysis shown in figure 4.49.

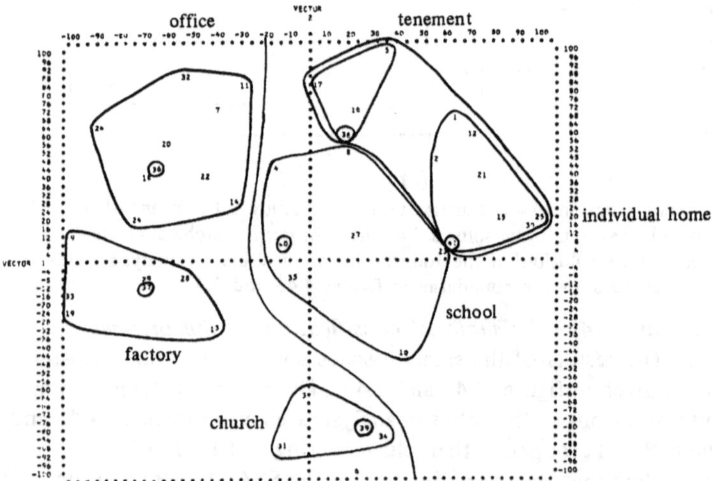

Figure 4.47. Smallest-space analysis 2, substudy 4, two-dimensional solution: the similarities between the thirty-five building photographs (1–35) and between the six building categories (36–41) are shown in a joint two-dimensional space. The photographic detail enabled the subjects to distinguish between the six categories of buildings. The boundaries around the clusters are derived from the Johnson hierarchical cluster analysis in figure 4.49.

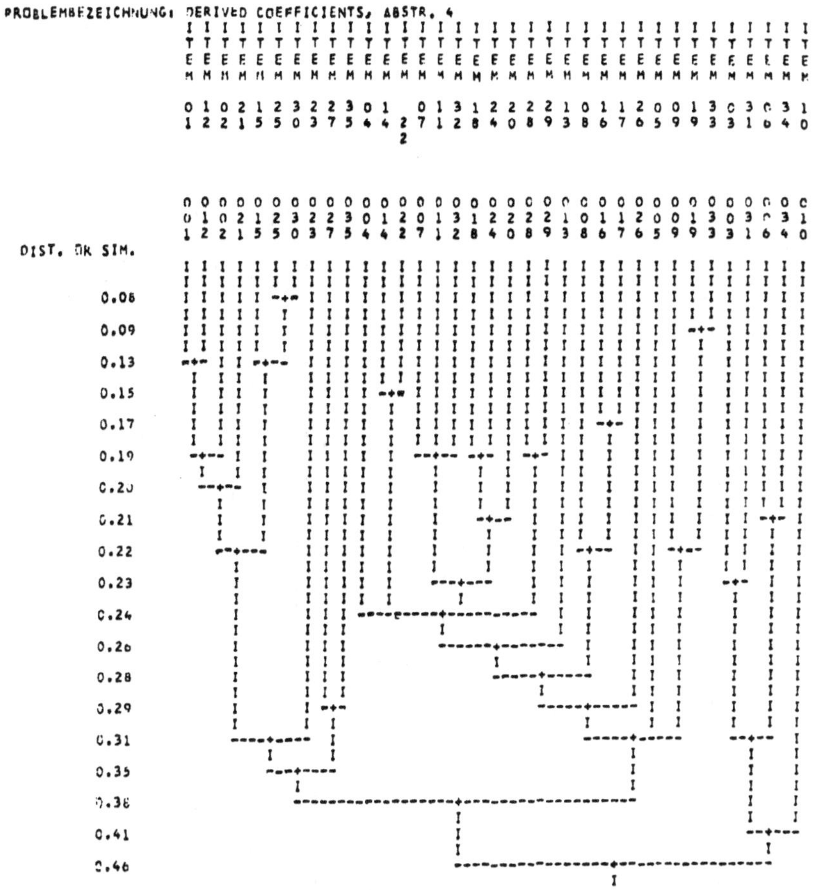

Figure 4.48. Johnson hierarchical cluster analysis, minimum solution, substudy 4 (photographs). In the minimum solution a well-articulated hierarchy was formed; however, the maximum solution was even more articulated.

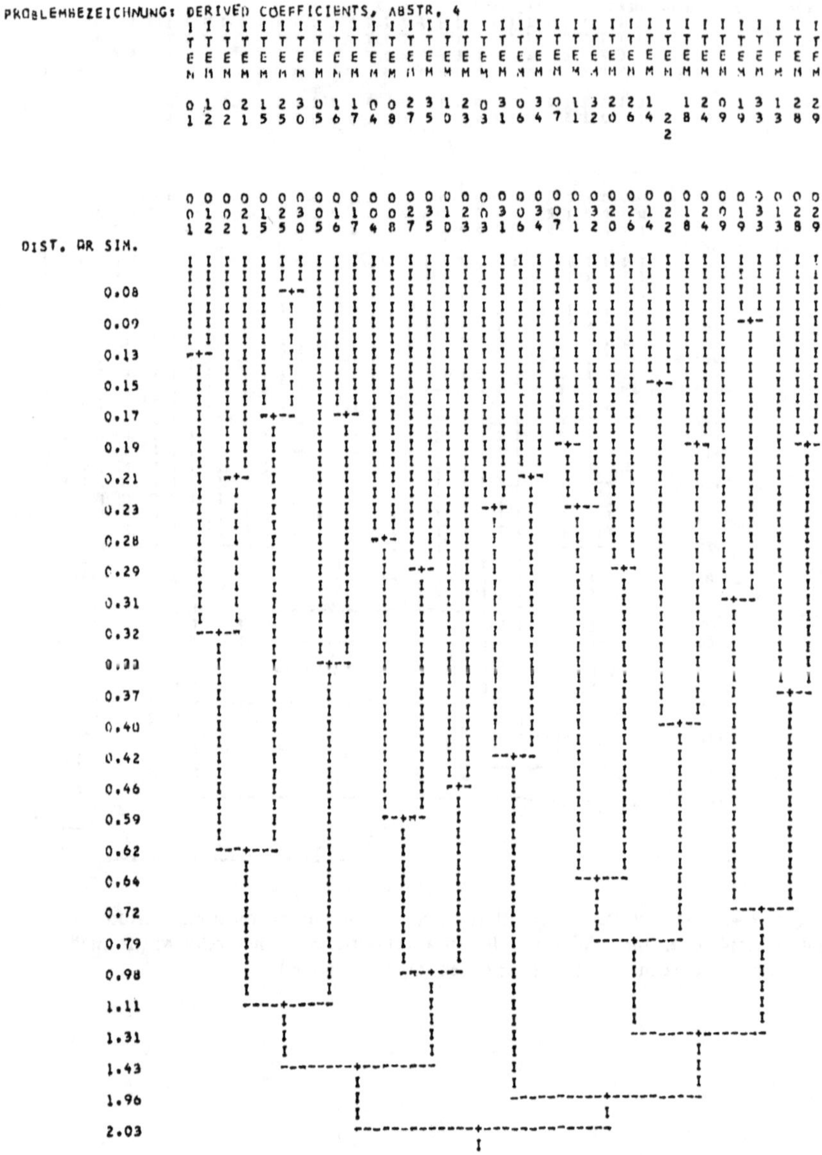

Figure 4.49. Johnson hierarchical cluster analysis, maximum solution, substudy 4. A very detailed articulation is shown. Its groupings were used to draw the boundaries in figures 4.46 and 4.47.

As this value was just barely acceptable, a three-dimensional SSAP-I solution was calculated in order to discover whether an added dimension would improve the value and at the same time the conditions for interpretation of the results. This time a coefficient of 0·100 was arrived at after one hundred iterations.

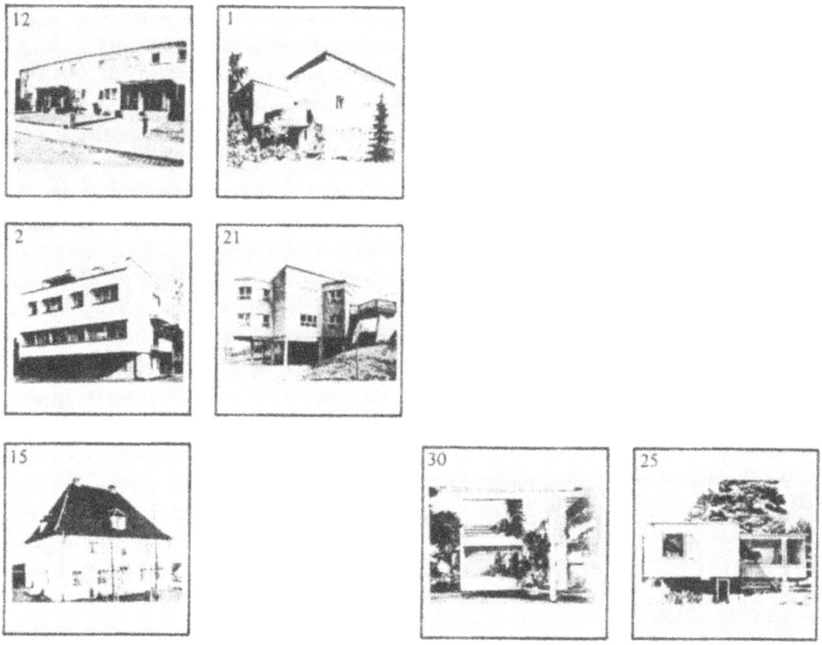

Figure 4.50. Substudy 4: the seven photographs that form the cluster of large and small houses.

Figure 4.51. Substudy 4: the three photographs that form the cluster of tenement houses.

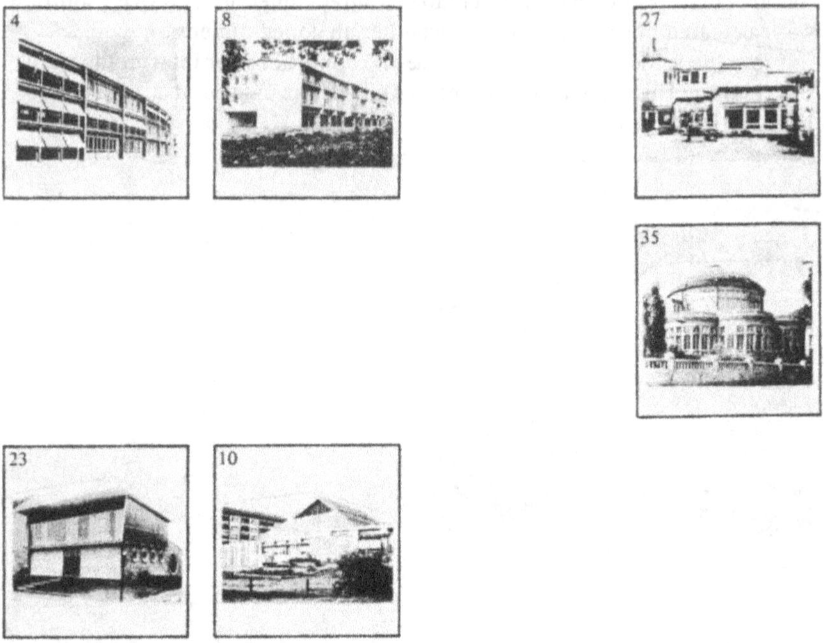

Figure 4.52. Substudy 4: the six photographs that form the cluster of schools.

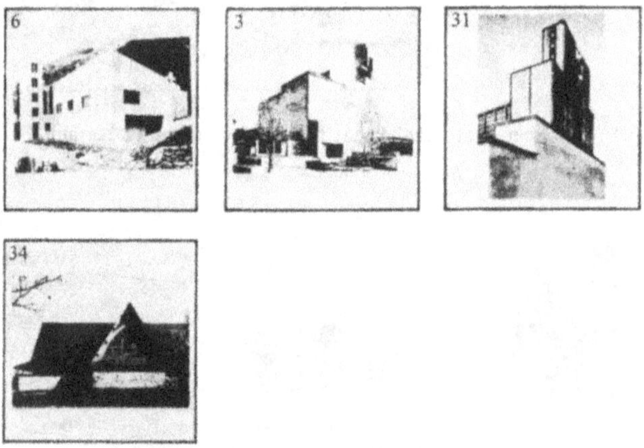

Figure 4.53. Substudy 4: the four photographs that form the cluster of churches.

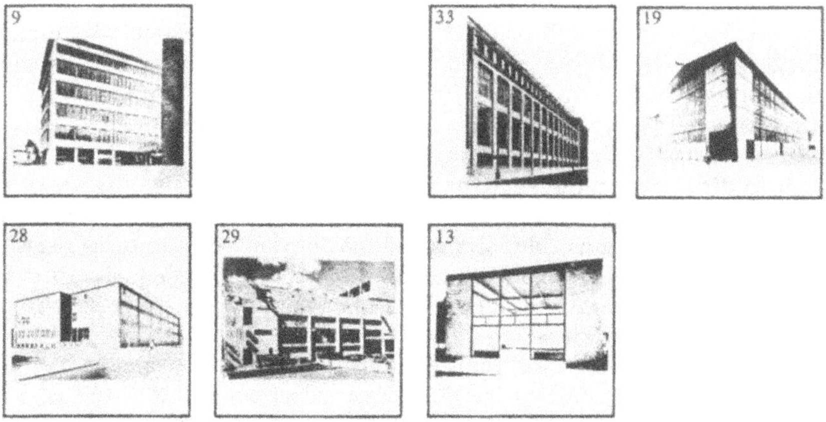

Figure 4.54. Substudy 4: the six photographs that form the cluster of factories.

Figure 4.55. Substudy 4: the nine photographs that form the cluster of office buildings.

Two vector plots, one with boundaries derived from the minimum solution and another with boundaries derived from the maximum solution, are given in figures 4.56 and 4.57. The results of the Johnson analysis are given in figure 4.58. In figure 4.59 the SSAP-I data are converted into perspective. The maximum results of the Johnson analysis produced six distinct clusters, each corresponding to one of the original categories. Some difficulties as regards interpretation were, however, raised by the presence of an additional cluster located between the factory and school clusters. The minimum solution made no distinction between offices and schools and failed to categorize two pictures. Apart from these drawbacks comparison of this SSAP-I solution with the two-dimensional one showed a fair degree of coincidence.

The concept of 'fuzzy set' (cf pages 121, 122) is well illustrated in the perspective representation of the three-dimensional output. A fuzzy set is characterized by the fact that each element in the set possesses its own degree of membership, indexed by a value between 0 and 1. The closer the value of an element is to one, the higher its degree of membership will be. In the perspective drawing (figure 4.59) the degree of membership of the pictures in their various categories is indicated by the different lengths of the connecting bars between the spheres representing the pictures and the spheres representing the categories. Some of these picture-spheres are completely unconnected from any category, although the three-dimensional solution indicates that they do in fact belong to a definite cluster. This is a particular feature of the perspective program used here, which isolates a picture if its degree of membership in a cluster is too low (cf page 113).

The comparison of the two-dimensional and three-dimensional solutions nevertheless indicated no specific advantage in the latter, so the two-dimensional solution was used. The maximum solution of the Johnson analysis for the two-dimensional solution was chosen as before, since it again produced a better articulated tree. It distinguished between two clusters, each divided again into two subclusters, which were in turn subdivided into lower-level clusters.

The first half of the tree groups houses and tenement houses in one cluster and schools in another. The housing cluster is then divided into a cluster of large and small houses and a cluster of tenement houses (characterized by balconies). The schools are divided into multiple and single buildings. The second half of the tree groups churches in one cluster and factories and offices in another. The cluster of factories seems to consist of two subclusters, one of old-styled and the other of modern buildings, whereas the cluster of office buildings seems subdivided into two further subclusters, one with buildings containing offices and flats, the other only offices.

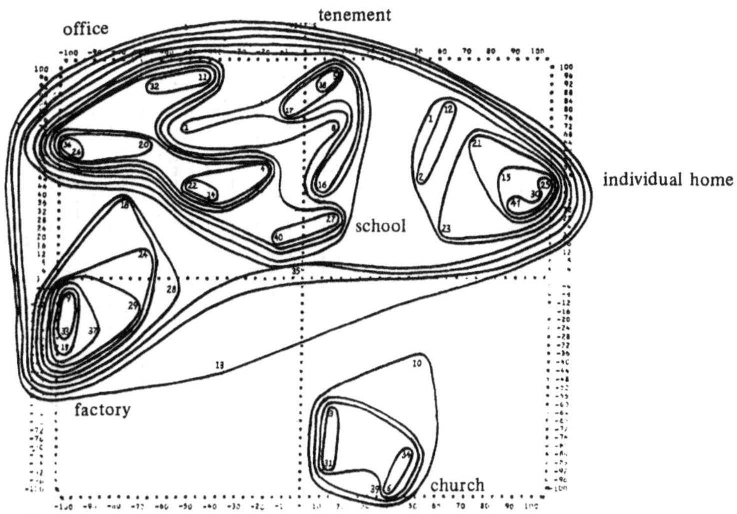

Figure 4.56. Smallest-space analysis 1, substudy 4, three-dimensional solution (photographs). The similarities between building pictures (1-35) and between the six building categories (36-41) are shown in one aspect of a three-dimensional space. The boundaries around the clusters were derived from the 'minimum solution' of the Johnson hierarchical cluster analysis shown in figure 4.58(a).

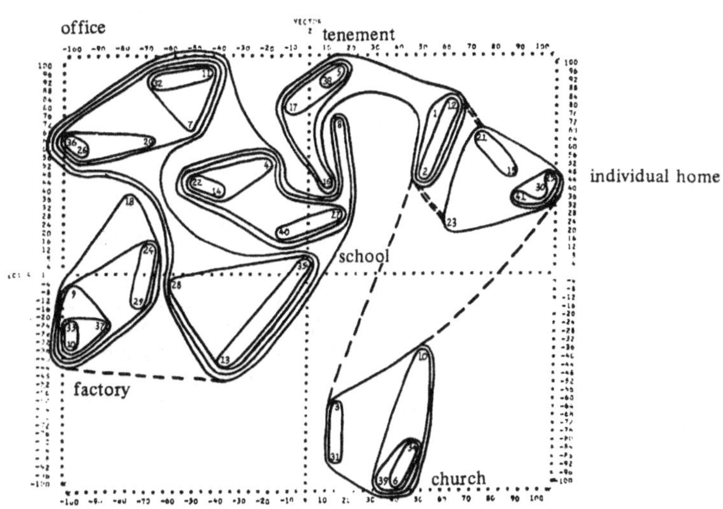

Figure 4.57. Smallest-space analysis 2, substudy 4, three-dimensional solution (photographs). The similarities between the building pictures and between the six building categories are shown in one aspect of a three-dimensional space when the boundaries are derived from the 'maximum solution' of the Johnson hierarchical cluster analysis shown in figure 4.58(b).

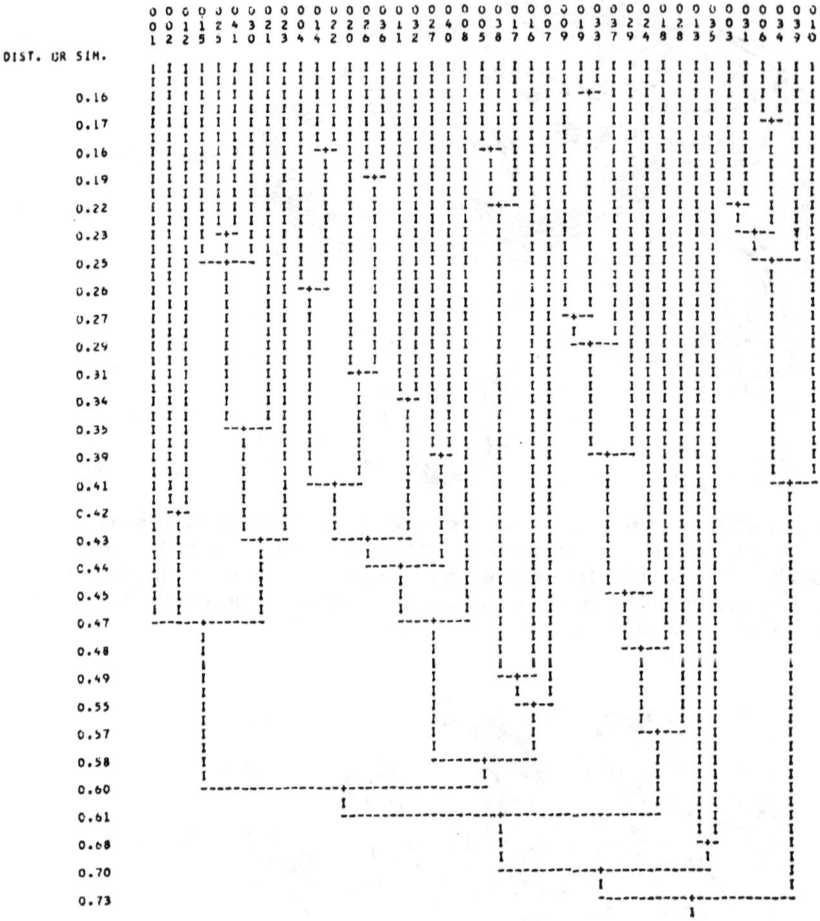

(a)

Figure 4.58. Johnson hierarchical cluster analysis, substudy 4 (photographs), resulting from an input of coefficients of similarity from the three-dimensional solution of the smallest-space analysis shown in figures 4.56 and 4.57. (a) Minimum solution: the groupings of this hierarchy were used to draw the boundaries in figure 4.56.
(b) Maximum solution: the groupings of this hierarchy were used to draw the boundaries in figure 4.57.

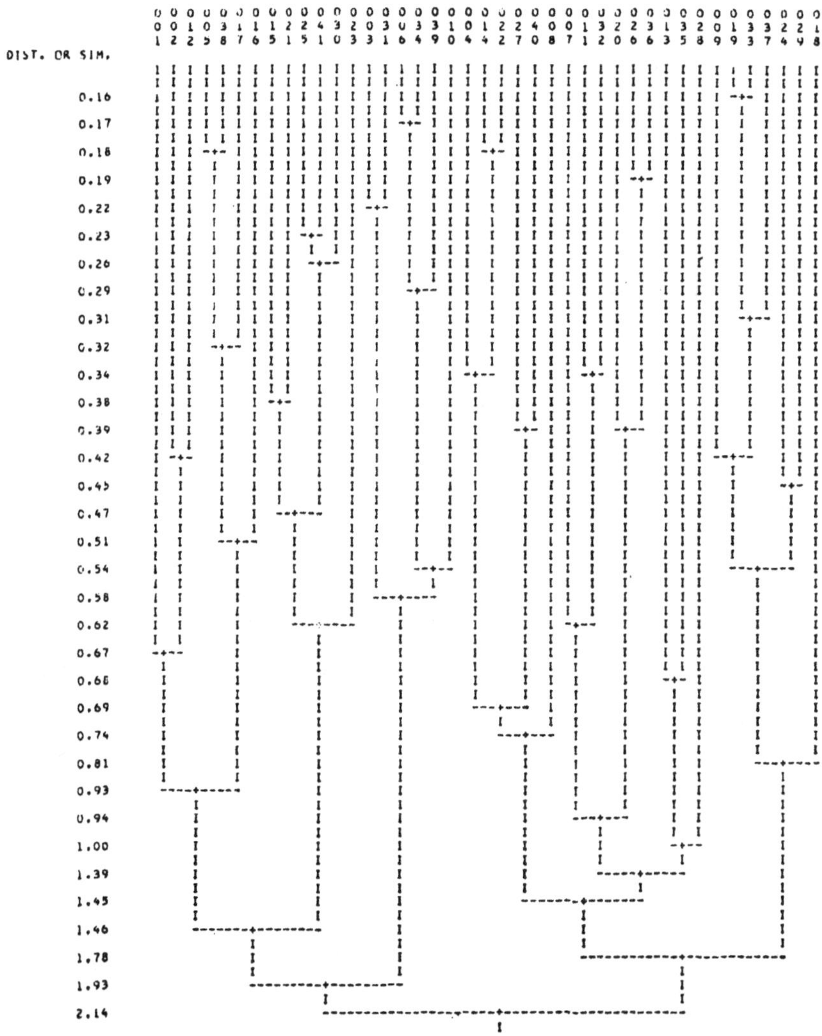

(b)

Figure 4.58. (continued).

Examination of the pictures within the various categories suggests that size, the complexity of building outline, and the number, size, form, and distribution of the windows are the features by means of which the subjects classified the photographs. These features are grouped in their respective categories in table 4.10.

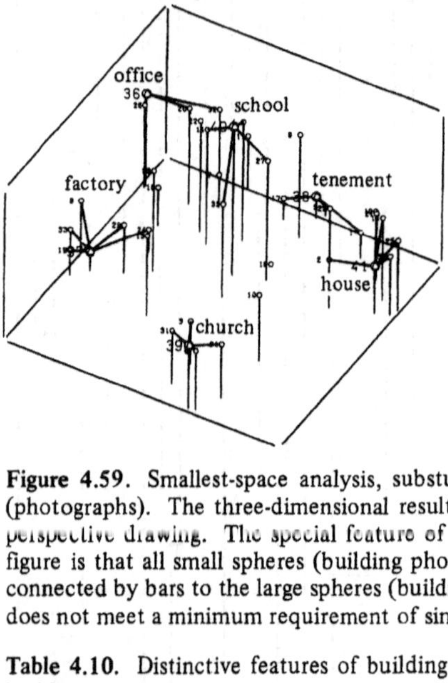

Figure 4.59. Smallest-space analysis, substudy 4, three-dimensional solution (photographs). The three-dimensional results are shown in a computer-generated perspective drawing. The special feature of the perspective program used to draw this figure is that all small spheres (building photographs, 1-35) which are very similar are connected by bars to the large spheres (building categories, 36-41). Any picture which does not meet a minimum requirement of similarity remains unconnected in space.

Table 4.10. Distinctive features of building shape and windows applicable to six building categories.

Size	Shape	Windows
Large buildings		
office buildings	simple box-shape volume	many windows
factories	simple box-shape volume	horizontally continuous large windows
churches	complex shape by addition of vertical volume	
tenement houses	complex shape by addition and repetition of volume	variation of window form
Not-large buildings		
individual homes	complex shape by addition of volumes	variation of window form
schools	complex shape by repetition or addition of volumes	large uniform windows

Reconstruction of the subjects' classification of the fourth level by means of the model. According to table 4.10 the three principle variables in terms of which the buildings in the different clusters may be described are size, shape and the windows. These variables may alter in character. Size is dichotomous: large or small. Shape is likewise dichotomous: simple or complex. Complexity can be a result of different operations such as simple repetition of one volume or addition of other volumes. The variable windows is multidimensional: number, size, distribution, form. Combinations of these variables should constitute the distinctive features of the various clusters.

For this final level the notative property A (large) was assigned to that half of the hierarchical tree containing churches, factories, and offices. This same property also had to be assigned to the cluster containing tenement houses in the other half of the tree. Offices and factories were further distinguished by the notative property B (unity of volume), and the offices were distinguished from all the others by the notative property D (many windows). Factories were characterized by the notative property F (horizontal orientation of glass surface). The large buildings—offices, factories and churches—were further subdivided by means of connotative properties.

In the other half of the tree the houses and tenement houses were described in terms of a notative property C (complexity) and were distinguished from schools by the notative property E (variation of window form). The distinctive notative property of the schools was G (large modular windows). These buildings were again all further differentiated by a number of connotative properties.

All the notative and connotative features used to distinguish the buildings in the clusters are given in table 4.11. The results of the SSAP-I analysis on this data are given in figures 4.60 and 4.61, together with the boundaries derived later from the Johnson analysis. The minimum and maximum solutions of the latter are given in figure 4.62.

As before, the model reproduced the same clusters that the subjects had originally produced and, although the density of the points in the vector plots was still comparatively greater than that of the subjects' data, the discrepancy at this level was much reduced. Moreover, the positions of the clusters relative to each other in two-dimensional space resembled more closely those of the subjects' data. This indicates a lesser degree of randomness in the subjects' data and a closer correlation between the features chosen by the experimenter and those used by the subjects. Similar effects are apparent on the hierarchical trees. Although the subjects' data again produced a more complex tree (explained by the greater simplicity of the modelling process) the difference was less than in the first two studies.

One might require of the experimenter that the modelling process be conducted 'blind', that is, without prior knowledge of the subjects'

Table 4.11. Substudy 4: twenty-four properties assigned to the six subjective clusters for their reconstruction by means of the re-cognition model.

	A	B	C	D	E	F	G	a	b	c	d	e	f	g	h	i	j	k	l	m	n	o	p	q
Home																								
1	−	−	+	−	+	−	−	−	−	−	+	−	−	+	+	−	−	−	+	−	−	−	−	−
12	−	−	+	−	+	−	−	−	−	−	+	−	−	−	+	−	−	−	+	−	−	−	−	+
2	−	−	+	−	+	−	−	−	−	−	+	−	−	+	−	−	−	−	+	−	−	−	−	+
21	−	−	+	−	+	−	−	−	−	−	+	−	−	+	−	−	−	−	+	−	−	−	−	−
15	−	−	−	−	+	−	−	−	−	−	+	−	−	−	−	−	−	−	−	−	+	−	−	+
25	−	−	+	−	+	−	−	−	−	−	+	−	−	−	−	−	−	−	+	−	−	−	−	−
30	−	−	+	−	+	−	−	−	−	−	+	−	−	−	−	−	−	−	+	−	−	−	−	+
5	+	−	+	−	+	−	−	−	−	−	−	−	+	−	−	−	−	−	+	+	−	−	−	+
16	+	−	+	−	+	−	−	−	−	−	−	−	−	−	−	−	−	−	+	+	−	−	−	−
17	+	−	+	−	+	−	−	−	−	−	−	−	−	−	−	−	−	−	+	+	−	−	−	−
School																								
4	−	−	+	−	−	−	+	−	−	−	−	−	−	−	−	−	−	−	−	−	−	−	−	−
8	−	−	+	−	−	−	+	−	−	−	−	−	−	−	−	−	−	−	−	−	−	−	−	−
27	−	−	+	−	−	−	+	−	−	−	−	−	−	−	−	+	−	−	−	−	−	−	−	−
35	−	−	+	−	−	−	+	−	−	−	−	−	−	−	−	+	−	−	−	−	−	+	−	−
10	−	−	+	−	−	−	+	−	−	−	−	−	−	−	−	−	+	−	−	−	−	−	−	−
23	−	−	+	−	−	−	+	−	−	−	−	+	−	−	−	−	−	−	−	−	−	−	−	−
Church																								
3	+	−	−	−	−	−	−	+	−	−	−	−	−	−	−	−	−	+	−	−	−	−	−	−
31	+	−	−	−	−	−	−	+	−	−	−	−	−	−	−	−	−	+	−	−	−	−	−	−
6	+	−	−	−	−	−	−	+	−	−	−	−	−	−	−	−	−	+	−	−	−	−	−	−
34	+	−	−	−	−	−	−	+	−	−	−	−	−	−	−	−	−	+	−	−	−	−	−	−
Office																								
7	+	+	−	+	−	−	−	−	−	−	−	−	+	−	−	−	−	−	−	−	−	−	−	−
11	+	+	−	+	−	−	−	−	−	−	−	−	+	−	−	−	−	−	−	−	−	−	+	−
32	+	+	−	+	−	−	−	−	−	−	−	−	−	−	−	−	−	−	−	−	−	−	+	−
20	+	+	−	+	−	−	−	−	−	−	−	−	−	−	−	−	−	−	−	−	−	−	−	−
26	[illegible]	[illegible]		+	−	−	−	−	−	−	−	−	+	−	−	−	−	−	−	−	−	−	−	−
14	+	+	−	+	−	−	−	−	−	−	−	−	−	−	−	−	−	−	−	−	−	−	−	−
22	+	+	−	+	−	−	−	−	−	−	−	−	−	−	−	−	−	−	−	−	−	−	−	−
18	+	+	−	+	−	−	−	−	−	−	−	−	−	−	−	−	−	−	−	−	−	−	−	−
24	+	+	−	+	−	−	−	−	−	−	−	−	−	−	−	−	−	−	−	−	−	−	−	−
Factory																								
9	+	+	−	−	−	+	−	−	−	+	−	−	−	−	−	−	−	−	−	−	−	−	−	−
19	+	+	−	−	−	+	−	−	−	+	−	−	−	−	−	−	−	−	−	−	−	−	−	−
33	+	+	−	−	−	+	−	−	+	+	−	−	−	−	−	−	−	−	−	−	−	−	−	−
13	+	+	−	−	−	+	−	−	+	−	−	−	−	−	−	−	−	−	−	−	−	−	−	−
28	+	+	−	−	−	+	−	−	−	+	−	−	−	−	−	−	−	−	−	−	−	−	−	−
29	+	+	−	−	−	+	−	−	+	−	−	−	−	−	−	−	−	−	−	−	−	−	−	−

Key:

A large
B unity of volume
C complexity of building
D many small windows
E variation of window form
F horizontal orientation of glass surface
G large modular windows
a opaque windows
b vertical articulation of horizontal glass surface
c small subdivision of glass surface
d few varied windows
e round windows
f shop windows
g second storey
h repetition of small volumes
i cluster of small volumes
j lateral addition
k vertical volume added
l small secondary volumes (balconies, etc)
m many small secondary volumes (balconies, etc)
n gabled roof
o cupola
p uniform curtains
q individual curtains

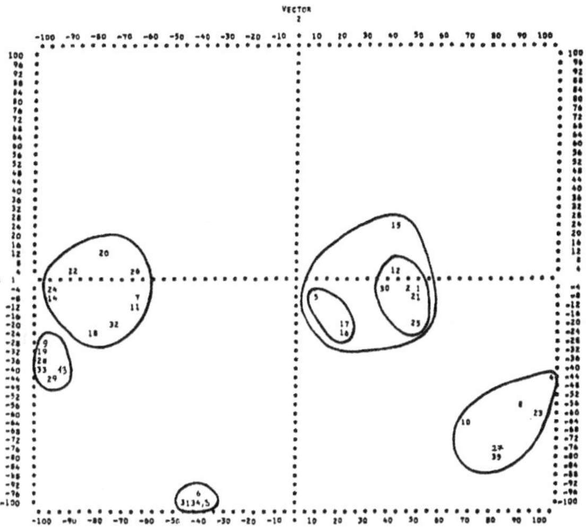

Figure 4.60. Reconstruction of the classifications in substudy 4 by means of the re-cognition model, smallest-space analysis 1, showing the similarities between the thirty-five building photographs, that result from the properties assigned to the subjects' clusters. The boundaries are derived from the 'maximum solution' of the Johnson hierarchical cluster analysis shown in figure 4.62(b).

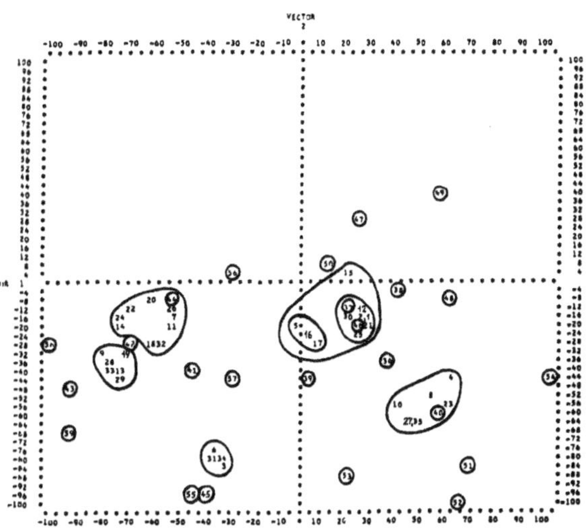

Figure 4.61. Reconstruction of the classifications in substudy 4 by means of the re-cognition model, smallest-space analysis 2. The thirty-five building pictures (1–35) and the twenty-four properties assigned to them (36–59) are shown in a joint two-dimensional space. The twenty-four properties and their distribution over the various buildings are shown in table 4.11.

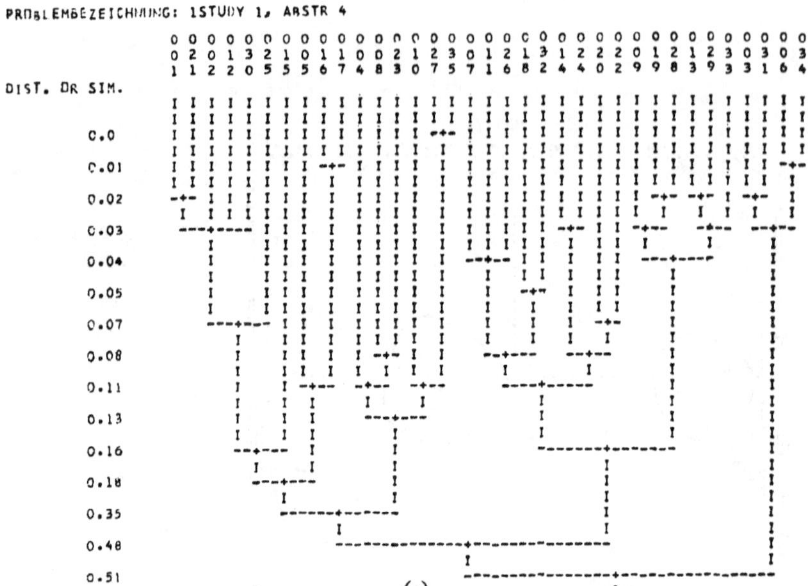

(a)

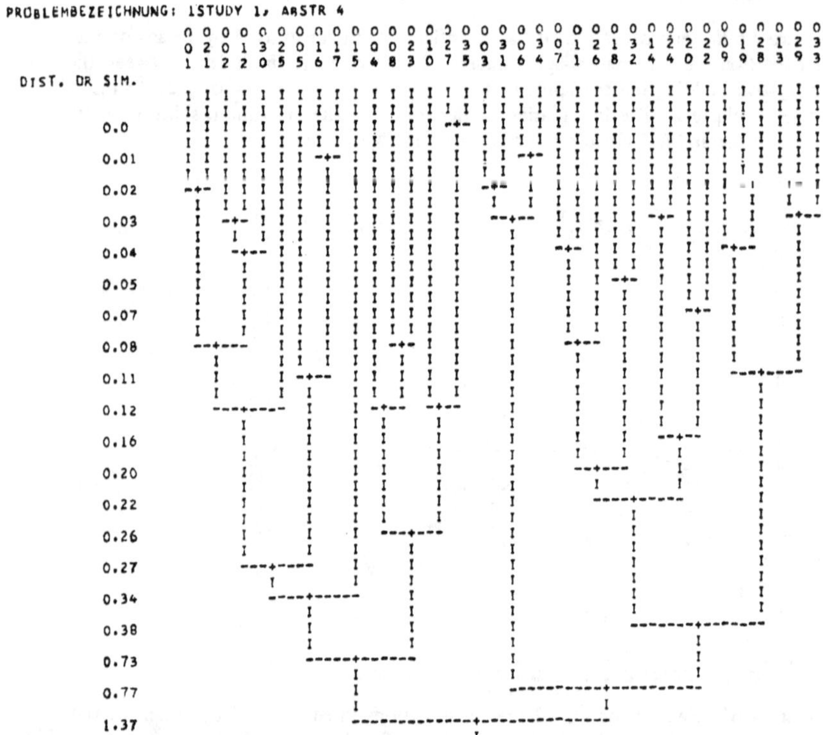

(b)

Figure 4.62. Reconstruction of the classifications in substudy 4 by means of the re-cognition model. (a) Minimum solution of the Johnson hierarchical cluster analysis. (b) Maximum solution of the Johnson hierarchical cluster analysis. This solution was chosen to draw the boundaries in figures 4.60 and 4.61.

performance, but this would not in fact achieve the desired sense of the modelling experiment. This was to establish whether there is any coherent set of architectural features capable of reproducing the subjects' performance. (Whether these features were actually used by the subjects themselves is not the issue.) Of course no such set was apparent in the first two substudies. There was probably one in the third substudy and definitely one in the fourth. All that can be required of the experimenter is a guess at the particular features the subjects had used. The subjects' comments were simply an aid. Demanding more would amount to demanding an experiment with a one-person sample, in which case it would be preferable to have thirty-five or more 'blind' experimenters.

4.2.4 Discussion of the study as a whole

Some basic findings of the study

The results of the study give some insight into the probable nature of the semiotic structure which regulates the re-cognition of different building functions. This had been assumed to be an internalized coordination of a universe of a signifier (a classification system of architectural design features) with a universe of the signified (a classification system of building functions, or their characteristics). It is apparent that both of these universes have been only indirectly identified.

The first level of detail provided only sufficient information for the subjects to identify houses (by the notative cue of their small size) and one church (by the connotative cue of its dome). Additional information at the second level about the number of storeys made no appreciable difference. At the third level the design feature which enabled the subjects to provide five of the six categories was the number, size, and distribution of the windows. Only at the fourth level was complete classification possible. The presence of both notative and connotative detail seemed essential for successful classification of a building with a basic volume in the form of a simple rectangular box without addition of secondary volumes (for example, a steeple), a specific roof formation, or other protrusions (for example, balconies). This lends some support to the hypothesis that the simple 'functionalist' architectural (or object) design is 'semantically nonfunctional' because it is polysemic.

Another finding was that architectural syntagms, that is, complex architectural forms, are formed by either horizontal or vertical addition of volumes, the basic functional description of the unit volume being one storey. Where ground rent is high, vertical addition of volume can be expected; where it is low, horizontal sprawl of architectural syntagms is possible. Transinformation about the degree of complexity or the particular nature of the function of a one-storey unit can thus be 'read' off the horizontal or vertical chains of these units.

There was also evidence in the study of the hierarchical nature of the process of architectural re-cognition. It was possible to reconstruct the subjects' performance by feeding design features which were possibly used by the subjects into a hierarchically structured process model. The resulting topological similarity of the subjects' clusters and those produced by the model did in fact leave much to be desired, particularly in the first two substudies, but this was explained by the difference between the logical application of the model and the individual nature of the subjects' choices.

The problem of structuring building functions

One remaining problem is the ad hoc definition of the term 'building function' by simply using labels such as office, factory, tenement house, etc. Such definition lacks systematic structure, as it is not made in terms of a system of sets of functional characteristics, nor is it based on any theory capable of explaining the origins and evolution of systems of building functions. (Such a theory should also be capable of explaining the different building types which are determined by these systems of building functions.)

An attempt will now be made to sketch such a system. As was pointed out earlier (pages 60–62), the guiding anthropological axiom of the theory underlying this research is that human beings create the conditions of their survival by collective work. Social functions would, according to this premise, be regarded as specific types of collective (that is, social) activities implementing specific historical forms of collective production (and maintenance) of the conditions of human life. It is assumed here that the form these activities will take will depend on the development of the 'forces of production' (the means of production and human working capabilities) and the correlated development of the 'social conditions of production' (ownership of the means of production, the type of division of labour, and forms of cooperation and leadership, etc). The implementation of joint production by means of particular historical forms of social activity produces and maintains other social activities called institutions, which are in turn geared to maintain the original activities upon which they depend. Historical dialectical changes in the relationship between the forces of production and the social conditions of production tend to occur whenever the particular framework of social conditions is no longer capable of containing the forces of production. These historical changes are dialectically related to changes in the corresponding institutions. The development of general descriptive, explanatory, and predictive theories concerning such frameworks is the task of political economy. The effects these frameworks have on the structure of a given society are described, explained and predicted by sociology. The mechanics by which the objective conditions of a society are transmitted to and experienced by the individual is studied by social psychology, while individual psychology is concerned with the native conditions of this individual experience and personality differences.

The purpose of this cursory examination of the politico-economic, sociological, sociopsychological, and psychological aspects of the production and maintenance of human life is to sketch the framework within which a system of building functions can be developed. One of the reasons for the lack of faith on the part of architects and urban planners in the ability of sociology to contribute to this field is that sociology is chiefly interested in the objective and global aspects of social structures and social functions. Social psychology, on the other hand, studies the circulation of psychological determinants between social groups and the individual, concentrating for the most part on two issues which are of direct relevance to those concerned with architecture:

1 how the individual is incorporated into a group or class, that is, how a group maintains itself—the question of socialization;
2 how the circulation of psychological determinants operates during group interaction—the question of cooperation.

From the perspective of social psychology it can be said that social functions in architecture are generally small group functions determined by particular tasks. The term 'small group' should be operationalized here by indicating a high degree of acquaintance between the members of the group. This is the case, for example, within a particular working group in a factory, office, classroom, or family. The opposite holds for 'large groups'—for example, all the employees of a factory, all the members of a religious community and all the pupils of a high school.

It is from the point of view of small-group processes that the application of social psychology to the urban environment seems more productive than that of sociology. While there is as yet no satisfactory classification of small groups, the variables determining the composition of these groups can be more or less completely defined. The most important dimension of group composition is the group task. A task may be operationally defined by the following characteristics and their variables:

the number of operations necessary to complete the task
(few, many),
the degree of difference between the operations
(homogeneity, heterogeneity),
the spatial coordination of the operations
(common space, spatial separation),
the temporal coordination of the operations
(simultaneous, consecutive),
the network of communication necessary for coordination
(unilateral, reciprocal).

Every operation will also involve the necessary equipment to carry it out. If this framework is applied to a family's task of reproduction and socialization, for example, the tasks may be defined as consisting of partial operations which: are large in number (as compared to those involved in office work), are heterogeneous, are spatially separated, are simultaneous, involve reciprocal communication.

Differences between various types of production would depend on the complexity of the task, on the constraints of the spatiotemporal coordination of the operations, and on the communication network necessary to maintain this coordination. The multiplication of small groups along assembly lines is a usual form of industrial cooperation, while the flow of message production in administrative work tends to be spatially separated (although there is an increasing tendency to coordinate the small groups involved in the paperwork within a common layout).

In summary, the following levels of classification should be considered:

production	versus	maintenance (especially institutional),
large groups	versus	subdivision into small groups,
complexity	versus	simplicity of task,
coordination	versus	independence of group operations,
coordination	versus	independence of individual operations.

If these levels of classification are applied to the social functions dealt with in the preceding study, the matrix given in table 4.12 results. Each of the factors in the matrix should have architectural consequences, in the layout of storeys as well as in the transinformation visible from the outside and deriving from the arrangement of groups and objects in the different storeys of a building.

Because of the criterion of maximum profitability, buildings used in production would generally be larger and simpler in volume. Offices tend, moreover, to be located in city centres where high ground rent is compensated for by vertical extension of the buildings.

The subdivision of large groups into small groups of similar size should result in a regular modular subdivision of storey layout visible on the façade. Classrooms have to provide space for small groups of up to forty pupils and thus exhibit an extremely large modular subdivision. Where large groups have to be assembled under one roof, a hall with a particularly large span would result.

If the task of a small group is particularly complex and if there is need at the same time for spatial and temporal separation of the individuals, as is the case with a family, a complex layout of compartments of different sizes results. This would be visible on the outside of the building by the

Table 4.12. Matrix of group functions applicable to social activities.

	Production	Subdivision of large group	Complexity of small-group task	Spatiotemporal coordination	
				among individuals	among small groups
Family	–	–	+	–	–
School	–	+	–	+	–
Church	–	–	–	+	–
Factory	+	+	+	+	+
Office	(+)	+	–	+	–

complex subdivision of the façade and the variations in window size. If spatiotemporal coordination of the individuals within a small group is necessary to execute a certain task, they will generally be placed together into one modular unit. Where a task requires few group members and is relatively homogeneous, a subdivision into small regular modular units usually results, as is the case in the average office building.

If spatiotemporal coordination is necessary among small groups or individuals, as is the case with an assembly line, there is either no subdivision of the storey layout, or separation of space into very large sections. If heavy equipment or large products are involved, the coordination of the small working groups in an assembly task requires an elongated hall.

It should be kept in mind that these considerations apply only at a given historical moment and at a given point in the development of the means of production and in the corresponding social conditions of production.

The problem of classifying design features

The proposed contingencies for architecture of cooperation within and between small groups has led to a tentative classification system of characteristics of social functions as they are carried out at a certain moment in history and as they are applicable to buildings. If there had been available at this point a stringent classification system of distinctive architectural features which could be coordinated with the system of characteristics of social functions, a semiotic structure would have been established. As has been seen, however, the modelling experiment yielded what was at best only a tentative classification system of design features. It proved to be difficult to establish exactly which distinctive features the subjects had in fact used. It is intuitively apparent that they could only have used those features visible in the pictures. But different subjects could possibly have used different features or even entirely different classification systems. This difficulty of establishing the exact features used in classification points to the necessity of further research.

One direction such research could take, could be the ontogenetic study of architectural re-cognition. It would be interesting to know more about at what age children re-cognize which building functions by which design features, how they label them verbally, and how they learn to distinguish the design features.

A hypothetical classification system of design features could be verified by having subjects sort the pictures from which the features are derived into different classes labelled by appropriate combinations of design features. An alternative method would be to have the subjects select combinations of design features from a list of such combinations and attach them 'correctly' to the appropriate pictures. A further experiment of the same nature could involve two lists, one of combinations of design features, the other of combinations of building-function features. Subjects could then be asked to attach matched couples of design-feature combinations and characteristics of building functions to pictures of buildings.

A more stringent formulation of the hypothesis

The preceding discussion can be summarized by a more precise formulation of the hypothesis used in the main study. The hypothesis concerns the relation between the perception (classification) of an oppositional structure of *architectural design features* and the perception of an oppositional structure of *characteristics of social activities.*

The term 'architectural design features' is defined by dimensions such as size and uniformity of building volume, the number of storeys, and distribution and form of windows in each storey. These variables should be observable in pictures of actual buildings.

The term 'social activity' is defined in terms of tasks of varying complexity occurring in the sectors of production and maintenance and carried out by large groups which may or may not be subdivided into small groups. These small groups may or may not be coordinated and will consist of numbers who likewise may or may not be coordinated in space and time.

The term 'perception' is defined as the operation of classifying something on the basis of an oppositional structure.

The hypothesis is a statement about the relation between design features and social activities as they exist in space and time. For this to be verified, the null hypothesis that there is no such relationship would have to be negated. Once this is done, the hypothesis would figure as an empirical statement, comprising the first step in the following schema (Hempel and Oppenheim, 1948).

1 Empirical statement: *There is a relationship* between a perceived system of design features and a system of social functions to which buildings are attributed by means of certain labels.
2 Law-like statement: *If* a system of object features is correlated in memory with a system of function features, *then* the function of an object can be recognized by recognition of its objective features.
3 Explanandum: *Therefore* the function of an architectural object can be re-cognized on the basis of the correlation between that function and a particular set of architectural design features.

The semiotic structure given in figure 4.63 can thus finally be hypothesized as existing between certain small-group and large-group activities and their variables and certain building design features and their variables. It is very important to remember at this point, that the study almost entirely dispensed with the role of context in the recognition of building functions. On the importance of context this book has left no doubt (cf page 66). In that sense the results of this study could be considered somewhat artificial. However, they could also be considered all the more interesting. The subjects had to work without the help of contextual cues (or to supply them by memory). If they came up with positive results these should therefore constitute more 'conservative' estimates of the functioning of architectural re-cognition since they resulted from a more difficult task.

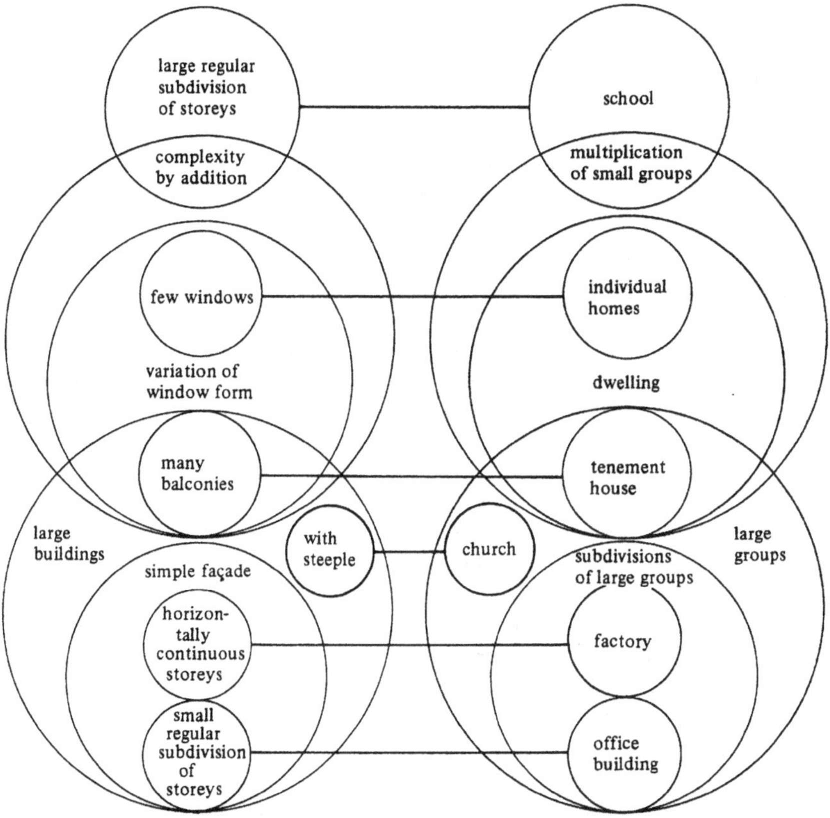

Figure 4.63. The hypothetical semiotic structure between the level of the signified, namely small-group and large-group activities defining the social functions of buildings (on the right) and the level of the signified, namely building design features (on the left).

4.2.5 Summary

Thirty-five students from the School of Architecture at the University of Geneva were asked to sort pictures of buildings into the six categories of office building, factory, tenement house, church, school, and house. Three sets of pictures were traced off the original set of thirty-five photographs, each set containing a different degree of detail. The first set consisted only of building outlines and certain cues acting as scale references. The second set gave indications of the number of storeys, the third added windows, and the fourth set consisted of the original photographs. Each subject sorted one set per session into the six categories, each of the four sessions being placed at least one week apart to minimize learning effects.

The subjects were then asked to choose the most representative picture for each category and to comment on the design features which had led to this choice.

The preliminary evaluation of the data involved simply identifying the pictures chosen by the majority of the subjects as representative of the six categories for each of the levels. These pictures were then examined in relation to the subjects' comments.

In the second phase of analysis the initial data matrix of 6 (categories) x 35 (pictures) was converted into a square matrix of similarity coefficients. This was then submitted to a distance analysis which yielded two-dimensional vector plots. The boundaries of the clusters in these vector plots were derived by feeding the distance measures into a hierarchical cluster analysis.

It became apparent from this analysis that the information provided by the first two levels of detail was insufficient to enable the subjects to produce the required six categories. The third level was sufficiently detailed for the subjects to form five of the six categories. Only on the last level, where the original photographs were presented, were all six categories produced.

The pictures comprising the different categories were then examined in terms of their distinctive design features according to a hypothetical, hierarchically structured model of the recognition process. This model regards re-cognition as a hierarchical process of classification. Re-cognition of an architectural object thus involves classifying its visual design details on increasingly differentiated hierarchical levels. A classification system of design features was chosen from the pictures by the experimenter simply on the basis of his own reactions to the pictures. The property-by-object matrix formed by these features was then subjected to the same cluster analyses as the subjects' data had been. The resulting clusters were identical in content to the subjects' clusters, although the density of content in each cluster was considerably higher, presumably because of the biased and simplified nature of the model.

The basic finding of the results was that the transinformational cues on the outside of a building, which convey information about the building's function, are largely size (defined by the number, size, and distribution of windows in each storey and the horizontal or vertical extensions of the building), homogeneity of building volume and form, and various stylistic features. A new definition of the concept 'building function' was then derived from the axiom that man produces the conditions of human life through cooperative work. The cooperative nature of work implies that productive labour is performed in groups which may or may not be coordinated in space and time.

Finally a hypothetical semiotic structure involving architectural re-cognition was proposed. In this semiotic structure a system of design features forming the universe of the signifier is correlated with a system of social activities forming the universe of the signified.

5

Empirical studies of connotation in architecture

5.1 Connotation in architecture

It is clear from the results of the first empirical study that one cannot conceptualize (de)notative aspects of architecture without simultaneously conceptualizing connotative aspects. Prieto's (1975b) hypothesis that a process of double classification holds for the conceptualization and recognition of any object has in fact been demonstrated to hold for the case under study. The subjects classified pictures of buildings according to an internalized semiotic structure between building functions and design features, while at the same time recognizing differences in 'style' within a cluster of pictures of one specific function. They had thus succeeded not only in identifying '*what*' the members of the clusters were, but also '*how*' they had been variously designed. This suggests that one cannot recognize the style of a design without first knowing what has been designed. (This is not unlike the linguistic phenomenon that one cannot decide *how* one is going to say something before knowing *what* one intends to say.) In the same way one cannot realize a design without at the same time realizing a style.

Because of this close correlation, the study of the different styles of a particular building function becomes a possible, and indeed legitimate, area of research. In other words, once the function of an architectural object has been ascertained, an independent study of its style becomes possible.

This question of the dependence of connotative systems on logically prior notative systems has raised a fair amount of difficulty for many semiologists. In Scalvini's recent book on architectural semiotics (1975), for example, the question is treated as if there were instances when 'architecture' (as the connotative level of different styles of architectural conception) will be superior to 'tectonics' (as the notative level of mere construction). What this disregards is the fact that choosing to design a building as 'mere construction' is just as much a style as any other. That this particular style will be in opposition to more deliberately 'artistic' styles is a foregone conclusion, but does not alter the fact that 'architectural' and 'tectonic' styles are both connotative with respect to a specific set of properties of building functions. The fact that one cannot conceptualize a way of doing something, for example a style, without first conceptualizing what one intends to do, seems to have become confused with the fact that a particular style of doing something can be transfunctionalized to serve an additional purpose, such as artistic or political communication. An architect may, for example, transform the occasion of the design of a building into an occasion for artistic communication. Similarly, the design of an official government building by a Nazi architect may become an instrument of political rhetoric, including perhaps such rhetorical devices as metaphors ('the rhythmical march of the columns') and hyperboles (the exaggeration of empty space, repetition, and size to provoke a sense of being lost as an individual). In both cases the style becomes a 'pretext for communication' and thus belongs in the semiology of communication (cf Markuzon, 1972). Scalvini's criticism of her use in the past of linguistic analogy in the semiology of architecture also remains incomplete as long as no attention is paid to Eco's (1971) warning against the 'aesthetic fallacy' of the semiology of architecture, namely where preoccupation with 'architecture' becomes analogous to literary criticism, while the environment becomes filled with 'tectonics'.

Once the fallacy of regarding "Architecture as a connotative semiotic" (Scalvini, 1975) has been exposed, attention can then be focused on internalized semiotic structures in which certain design features correspond to properties of different conceptual styles of the *same* building function. Systems of such properties are connotative with respect to the previous conceptualization of a notative system of properties of different building functions. The study of the recognition of different ways of conceptualizing buildings will now proceed on the basis of selecting from such a system one set pertaining to one particular building function.

This study must proceed along lines essentially similar to those of the preceding study. An internalized semiotic structure has to be exposed. The substance of the universe of the signifier of this structure consists again of visual images. This may activate a universe of the signified, the substance of which is cognitive. It is, however, equally possible that the inverse may occur, where the cognitive substance of the universe of the signified activates certain visual images in the universe of the signifier.

In the following studies the cognitive substance will be established by activating it by means of pictures of two (historically) different styles of tenement houses. The resulting cognitive substance should be a system of two mutually exclusive sets of properties, indicating recognition of different styles of tenement houses and corresponding to the system of two mutually exclusive sets of design features exhibited by the pictures. This will be followed by an analysis of the design features of the latter system, after which both systems can be put into correspondence by correlation in order to demonstrate the existence of an internalized semiotic structure.

Subjective reactions toward two architectural styles

5.2 The method of the semantic differential and its applications in environmental studies

5.2.1 The functioning of the semantic differential

To accomplish the first step securing the system of two mutually exclusive sets composing the cognitive substance, an instrument called the semantic differential (SD) will be used. This was first proposed by Osgood et al (1957) and has been widely applied in what its authors called 'the measurement of meaning', particularly in the field of psycholinguistics. On the other hand some more or less severe criticisms have been advanced against the SD, particularly by linguists, and these should be discussed before proceeding to its application in the 'measurement of architectural meaning'. The SD consists of seven-point scales located between pairs of polar adjectives such as 'good–bad', 'hot–cold', etc. Subjects are confronted with stimulus material, which can be either verbal or nonverbal, and mark a particular scale between a polar pair, presumably in accordance with some kind of internal reaction towards the stimulus material.

An individual subject's resulting scores can be directly compared by correlation with those of other subjects. It is more often the case, however, that the scores of all subjects involved in a particular experiment are averaged, a mean is obtained for the particular verbal or object stimulus used, and a 'profile' is derived by connecting the average scale points over the set of scales. The average profiles of different concepts or objects are frequently then correlated and the resulting matrices of correlation measures subjected to factor analysis. For readers unfamiliar with factor analysis it is sufficient to understand that it consists of a sequence of (statistical) calculations applied to a matrix of similarity measures or correlation coefficients (which are simply numerical indices of the degree of similarity holding between two series of measurements of the same object, person, or scale). Wells (1957) has written a readable account of factor analysis as applied to SD data. The scope of factor analysis is to describe in a representative manner this matrix of a large number of indices in terms of a limited number of dimensions or factors. Two kinds of factor analyses are generally applied to SD data. One is based on correlations between concept or object profiles (*Q* analysis) and the other on correlations between different scales (*R* analysis). The latter is used to establish groups of scales which are similar in 'meaning'.

A subject's marks on an SD scale are an indirect measurement of internal processes which cannot be directly observed. Such measurement would have been rejected by traditional Watsonian behaviourists, but through the introduction of such theoretical constructs as 'representational mediation processes' neobehaviourists like Osgood have been able to broaden the scope of behaviourist psychology. It is not surprising in this context that the core of the controversy about the SD is essentially the question of 'what the SD really measures', that is, what these theoretical constructs really are.

It must be stated from the beginning that Osgood never pretended to be measuring meaning in its totality. As far as he was concerned, the SD simply tapped one aspect of a complex internal process. This aspect he called 'connotative'. Osgood's use of this term is not to be confused with its use in this book. Connotative meaning is for Osgood the affective, emotion-related aspect of meaning as distinct from lexical meaning, which is the denotative meaning, the use of the latter term being likewise at variance with its use in this book. No one has as yet raised serious doubts that the three dimensions of Osgood's connotative meaning—evaluation (scales like 'good–bad', 'beautiful–ugly'), potency ('strong–weak', 'large–small') and activity ('active–passive')—do have a certain face validity. These three dimensions have turned up in numerous factor analyses, including cross-cultural ones. There has been, however, an increasing tendency recently away from Osgood's classical SD scales, which invariably yielded these dimensions. The main controversy regards the 'meaning of Osgood's connotative meaning'. Some of the positions in this controversy will be reviewed in the following pages in chronological order.

5.2.2 Survey of various positions regarding measurement of the SD

Before the appearance of *The Measurement of Meaning* (Osgood et al, 1957), Hofstaetter (1955) in Germany had described what he called the 'polarity profile' method of Osgood, which he applied to measure 'similarity' of reactions to different stimuli. He used verbal concepts as stimulus material and subjected correlations between SD profiles of different concepts to factor analysis. He held that two factors, potency and activity, accounted best for the distances between the concepts measured by the SD. Weinreich (1958) levelled a lengthy criticism against *The Measurement of Meaning*, which largely took a lexicographer's point of view and which to a certain extent misunderstood the problem of the book in presupposing that a similar lexicographer's intention lay behind it. Weinreich was not able to remove himself sufficiently from the lexicographer's approach to be able to distinguish between his conception of meaning and what Osgood calls connotative meaning. He did not at the time foresee that at least three 'connotative' lexicographical works would appear, one by Jenkins et al (1958) using a corpus of 360 English words, another by Heise (1965) using a corpus of the 1000 most frequent English words, and a third published in an appendix to Snider and Osgood (1969), which studied 550 English words. In addition to criticizing (with apparently insufficient knowledge of experimental design) the limited number of word samples used in the factor analyses and the small number of dimensions yielded by those analyses, Weinreich claimed that the SD did not measure meaning, but only the "emotive influence of words" (page 347), the "affect" or "attitude" of the subjects (page 360). He suggested instead that certain technical characteristics of the SD (such as the rating-scale technique and the sampling of a given language community, which replaces the traditional introspective approach of the lexicographer) be applied to a transformed version of the "game of twenty questions" by which "the meaning Weinreich means" is guessed by successive approximations forming a tree structure such as pictured in figure 5.1.

Osgood's (1959) reply showed quite adequately that in experimental strategy the adequacy of a sample "depends upon the variability in the dependant variable across the instances" (page 197) one is sampling. If that variability is small, a small sample will be adequate. Osgood showed that the variability was in this case indeed small, by repeatedly obtaining the same small number of dimensions from his factor analyses. He justified this small number of dimensions generated by pointing out the necessity for a compromise between computer capacity and the attempt to demonstrate at least a few fundamental and dominant dimensions rather than a mass of unimportant ones. He then gave examples that there is not necessarily a coincidence in communication between the 'descriptive', lexicographic ('denotative') meaning of terms and the other type of meaning he chose to call 'connotative'. In a conversation the following situation could arise:

Denotative plus connotative agreement between the partners.

Denotative agreement without connotative agreement: "I know what you mean, but I disagree with you completely".

Connotative agreement with denotative disagreement given the same verbal stimuli or responses to stimuli: a Soviet Russian and an American saying, "Democracy is good".

Connotative agreement with denotative disagreement given different verbal stimuli or responses. In the sentence "The gourmet approaches his food as a lover approaches his mistress" the gourmet and the lover are in connotative agreement about the denotatively distinct objects of their desires.

Neither connotative nor denotative agreement: the differences between Weinreich and Osgood on the meaning of 'meaning'.

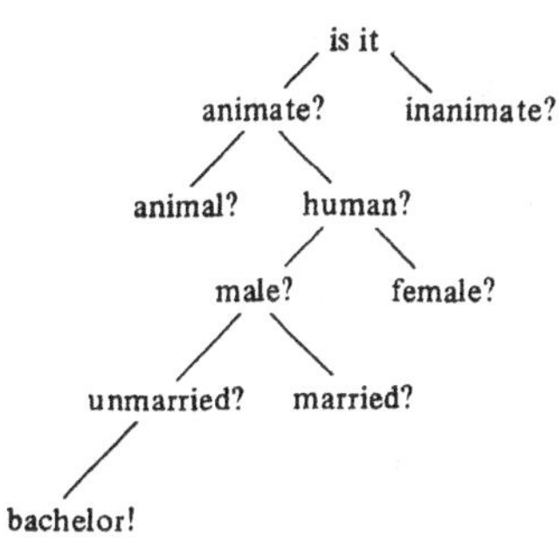

Figure 5.1. A tree structure by which the lexical meaning of 'bachelor' is defined in successive questions and answers.

Weinreich's (1959) final rejoinder was very brief and added no new insights or arguments. In it he largely reiterated his initial position—which would seem to lend a certain credibility to Osgood's argument that he and Weinreich are in neither connotative nor denotative agreement on the 'meaning' of 'meaning'.

Carroll's (1959) criticism of *The Measurement of Meaning* is clearly more competent in dealing with factor analysis techniques. He was at least aware of Jenkins et al's (1958) study of the semantic profiles of 360 English words. Carroll is, however, in accordance with Weinreich in suggesting that Osgood did not adequately sample the 'semantic space': for Carroll neither the number of adjective scales nor the number of concepts was large enough. With large enough samples, he maintained, dimensions could be established almost at will, taking into account the limitations placed on this by the "actualities of human perception and of the phenomenal world. One is unlikely to find a dimension which has not a counterpart somewhere in the real world or in the distinctions which the human race has learned to value" (page 71). Carroll's criticism boils down in the end to the statement that Osgood "has not yet reported research in which he constructs hypotheses concerning what semantic dimensions (beyond the usual ones) might actually be expected to exist". One would imagine that Osgood's answer to this would be similar to the one he gave Weinreich: it would be a legitimate compromise to sacrifice, in the early stages of the research at least, many unimportant dimensions in order to obtain a few fundamental and dominant ones. Carroll maintained finally that in filling out an SD form subjects probably judge on the basis of their own experiences the plausibility of the connection between a concept ('lady') and the scale on which it is to be rated ('rough-smooth'). For this reason he suggested that the SD should rather be called the "Experiential Differential". Thus, according to Carroll, what the SD really measures is 'experience' with objects or stimuli.

Ullmann (1962) regarded the SD in a relatively optimistic light as a tool serving in "experimental semantics" to measure an important component of meaning which is "feeling tone" or "feeling connotation". Ullmann insists that language is not only a means of communication but is also a means of expressing emotions, a function which Jakobson (1963) attributes to communication itself and which Guiraud (1973) calls the "affective and subjective" function of language as distinct from the "cognitive and objective" one. According to Ullmann the sources of the feeling content of linguistic meaning are sound (onomatopoeic) factors, context, catch phrases and slogans, certain word endings (diminutives, perjoratives, etc), incorporated value judgements ('messy'), expression of feelings ('funny' or 'horrible'), feeling associations (foreign words, technical terms or archaic expressions), etc. It would be interesting to draw parallels between sources of emotional content in linguistic meaning and those in 'visual meaning' (sound is material, context is visual context, slogan is imitation of known formal repertoires, etc). Ullmann also refers to certain linguistic devices which produce emotional tones, such as phonetic accentuation and emphasis, lexical metaphors, hyperboles, as well as syntactical factors such as word position (for example, chiasm). He also stipulates the conditions for change or dissipation of feeling tone in words: changing historical conditions; conditions under which word origin is forgotten; extensive repetition ('the law of diminishing returns'); and transition of the style of a specific social group into common use. It is thus, according to Ullmann, the feeling-tone component of meaning, described in terms of its sources, means and changes, which Osgood's SD measures.

Seven years after the appearance of *The Measurement of Meaning* the first comprehensive survey of its impact appeared. Diebold (1965), in a survey of psycholinguistic research covering the decade from 1954–1964, mentioned that Osgood et al's book had acquired wide currency and that the SD had within a relatively short period been adopted in an impressive array of studies dealing with such diverse subjects as attitude change, psychotherapy, cross-cultural semantic factors, aesthetics, advertising, mass media, bilingualism.

Following a series of cross-cultural applications (Osgood, 1960) of the SD with, among others, Japanese, Korean (Kumata, 1957; Kumata and Schramm, 1956), and Greek (Triandis and Osgood, 1958) subjects, Ertel in Germany (1964; 1965a; 1965b) confirmed the validity for German-speaking subjects of the dimensions of evaluation, potency and excitation (activity) and proposed a set of standardized scales for each of the dimensions. He also suggested that what the SD measures is an emotional semantic dimension ('emotion' could be substituted by 'affective' or 'feeling'). Since SD ratings had by then been applied not only to linguistic stimuli but also to art objects, noises, colours, pictures, etc, Ertel held that the term 'semantic' in the designation of the instrument should be replaced by 'impression' (Eindrucksdifferential = impression differential).

One of the most intriguing contributions to the question of what the SD actually measures was made by Orlik (1965; 1967), who showed that the SD can be regarded as a special version of the classical psychophysical scales. He describes a hypothetical experiment in which a number of subjects are confronted with a stimulus in the form of a scale of seven shades of grey merging at the extremes into black and white. They are given various pairs of bipolar adjectives and are required to rate their impressions on separate SD scales. Different average scores should emerge for different adjective pairs. The following are four possible results.

1 It can be expected that in response to the adjective pair 'light–dark' the mean of the subjects' responses will be regularly distributed over the scale in correspondence with the different degrees of brightness of the stimulus. The result will be a correspondence in the scaling between subjective experience and objective reality.

2 In response to adjectives, such as 'heavy–light', which have little relation to the stimuli, the subjects would perhaps interpret their task as nonsensical, and the mean results would be expected to concentrate round the centre of the SD scale—as the closest to a neutral tone. No differentiation between stimulus objects would therefore be registered on the scale.

3 It would, however, be possible for the subjects to interpret the adjectives 'heavy–light' in a metaphorical sense, attributing to the darker areas of the stimulus scale values closer to 'heavy' and to the lighter areas scale values closer to 'light'. In this case there should be a relatively high degree of correspondence between the stimulus scale and the SD scale, but there would be a greater tendency towards the centre than in the first example because of an element of uncertainty on the part of the subjects as to the acceptibility of metaphorical interpretation. There would in this case be a fair degree of correspondence between the objective and the subjective scaling, but the variability between the individual scales would be high.

4 The situation could arise where only one half of the bipolar scale is relevant to the stimulus. If, for example, the scale of varying grey shades was presented as square chips and had to be judged in terms of the adjectives 'square–round', the mean results would be located only in that half of the scale corresponding to 'square'. The same result would occur if the stimulus was babies and the adjectives were 'sweet–sour'.

Orlik concluded that a one-dimensional stimulus variation would yield only one factor through factor analysis. A two-dimensional stimulus, such as square chips of different colours and different sizes, would under factor analysis yield a two-dimensional result. Orlik summarized these conclusions in a mathematical model in which the different sources of variance in SD measurement are accounted for: the different dimensions of the stimulus, the relation of the stimulus to scale relevance, and biased as well as random error. He subsequently demonstrated empirically that correlation coefficients taken as similarity measures between SD profiles do not fulfill the conditions of the model. When a factor analysis based on correlation coefficients was run to assess profile similarity in an experiment with a one-dimensional series, an additional factor appeared which had to be artificial, since there was no additional source of variance to explain it. Again in an experiment dealing with two-dimensional stimuli, a third factor appeared. Orlik then devised a method whereby these superfluous factors could be eliminated: each mean score is balanced by subtracting from it the average mean score of the whole profile, then the sums of the balanced cross products are used as the similarity measures. Experiment shows that artificial factors disappear under these conditions. There is thus a good chance that many SD experiments performed so far that are based on factor analysis of correlation coefficients, are actually affected by additional artificial factors.

This should be taken into account in any application of this principle to questions of architectural connotation (style). It would be expected according to Orlik's model, that a factor analysis performed on building profiles of two different architectural styles would yield two factors, one for each style. Orlik's studies suggest that one should aim at composing an SD with bipolar items which are relevant to the investigation at hand, although nonrelevant items should do little harm, as they would in effect be reduced to a mean value of zero.

There are further reasons for using Orlik's method. Three types of information are contained in profiles: information about level (whether the profile is situated high or low on the scales), information about shape (the particular zigzag form of the profile), and information about dispersion (how the raw scores are distributed around the mean of scales). If a rank correlation coefficient is used to compare two profiles only the information on shape difference is used, while the information on level and dispersion gets lost in the standardization procedure connected with the computation of correlation coefficients. Many authors (for example, Osgood and Suci, 1952; Cronbach and Gleser, 1953; Nunally, 1962) have therefore recommended the use of the *d* measure (*d* being the square root of the sum of squared differences between profile points). Osgood et al (1957) derived a method for factoring matrices of *d* measures. But a factor analysis of *d* is actually only a special case of raw-score factor analysis. Nunally (1962) points out that it is actually better to use a function of *d* than *d* itself in analyzing profiles. Profiles may be considered as points in a Euclidean space. Studying relationships among profiles can then be pictured as measuring interpoint distances in Euclidean space. If one wants to find out how many types of profiles there are, similar kinds of profiles would have to be clustered in that space and each cluster would be indicative of a type. Thus Nunally (1962) advocates factor analysis of the sum of raw-score cross products between the scales of profiles, done in the same way as for correlation coefficients, rather than *d* measures. But since Orlik has shown that the raw-score cross products still yield additional artificial variance in a way similar to that in which the artificial factor is produced in the case of correlation coefficients, it is only when the individual scores are balanced by subtracting from them the average score of the whole profile that the sum of their cross products between profiles does not produce any artificial variance.

Despite Orlik's call for relevance of scaling adjectives, there is a great deal of evidence in factor analyses of cross-cultural samples of as many as twenty-five languages, to show that the 'evaluation', 'potency', and 'activity' dimensions are actually universal (Osgood, 1971). This is generally true for factor analyses of the mean results of a number of subjects. If, however, the profiles are based on individual scores, dimensions in addition to these three often emerge. One example is where the evaluation factor is divided into subdimensions, sometimes three (for example, functional, hedonic, and ethical value), sometimes only two (Komorita and Bass, 1967; Smith, 1961). This is also confirmed in a study by Miller (1967) which used another method. Miller put through the Johnson hierarchical cluster analysis (described earlier in this book) one hundred polar adjective pairs which had been used by Osgood et al and had been sorted by twenty subjects according to similarity of meaning. He found that the evaluation dimension was divided into three clusters expressing three types of evaluation which he called moral ('good-bad', 'fair-unfair'), intellectual ('complete-incomplete', 'true-false'), and aesthetic ('beautiful-ugly', 'graceful-awkward'). There is nevertheless ample evidence (for example Heise, 1969) that the three dimensions of evaluation, potency, and activity generally emerge if more than twenty stimuli and mean scale values on a large number of scales are used. This is important if one wants to study the meaning of objects rather than individual differences in meaning.

In 1969 a source book on semantic differential technique appeared (Snider and Osgood, 1969) which contained the major contributions to SD research that had appeared up to that time. It also contained a bibliography of 1066 titles of literature relevant to SD technique and presented a 'semantic atlas' of the evaluation, potency and activity scores of 550 English stimuli.

This impressive manifestation of the technique's prestige was followed shortly afterwards by a critical survey of the SD by Heise (1969). He reviewed studies dealing with the metric properties, the sources of variance and the factor structure of SD ratings. The results of the studies of the metric properties are summarized as follows:

> "... the basic metric assumptions of the SD are not quite accurate, but... violations of the assumptions are not serious enough to interfere with many present applications of the SD".

As far as the sources of rating variance are concerned, Heise deals with two forms of biased error. One concerns the error of social desirability, where subjects distort their ratings in the direction of a socially desirable response. The other concerns scale-marking styles, which result from the tendency of some subjects to exaggerate their impressions by consistently marking the extreme ends of the scale, while others tend to attenuate. Although the effect of the first bias must, according to Heise, await confirmation in further studies, it is possible to counteract the second "by random assignment of subjects into experimental and control groups or by attempting to measure the effect in order to control for it statistically". Heise in addition gives a rough estimate of the approximate proportions according to which the rating variance can be split up:

> "One tenth... due to differences between subjects in the use of scales, one quarter due to bias and/or deviations of subjects' true scores from the population true scores, one quarter due to momentary deviations of subjects from their own true scores and two-fifths due to random error."

As regards the structure of SD ratings Heise writes:

> "In general, the findings and arguments reviewed... lend strong support to the proposition that E[valuation], P[otency], and A[ctivity] are the most significant factors underlying averaged SD ratings... the situation is more complicated when one deals with the structure of individual judgements rather than group means."

As regards concept-scale interaction (discussed above in Orlik's imaginary experiment, of which Heise apparently had no knowledge) Heise draws particular attention to the situation where the scale is irrelevant to a stimulus concept or object. He proposes that for this to be avoided "means that an SD ideally should be validated and adjusted for every new stimulus class with which it is used". Heise suggests that in constructing an SD one should obtain a large number of contrasting adjectives from the literature dealing with the issue and from the subject population of interest. From these only those which most clearly and relevantly represent the full structure of evaluation, potency, and activity (including any other relevant dimensions) should be retained. One could possibly end up with thirty new scales, to which perhaps three of Osgood's basic reference scales for each of the three dimensions should be added, giving a total of thirty-nine scales. Heise concludes his review by quoting Carroll's (1959) earlier remark on *The Measurement of Meaning*:

> "Nevertheless, the reviewer is inclined to characterize the book by asserting: it is good, it is active, it is potent."

He adds:

> "The 'successful' profile for the SD still seems warranted after more than 10 years of additional studies and applications."

The person perhaps best qualified to judge the powers and limitations of semantic differential technique should be Osgood himself. In his article "Explorations in semantic space: a personal diary" (Osgood, 1971) he discussed the advantages and disadvantages of his "first vehicle for exploring semantic space". He drew attention to four main advantages:

1 The SD is highly adaptable to the procedures of multivariate statistics (that is, factor analysis).

2 The SD as a componential model describes the meaning of a large number of concepts or objects in terms of a relatively small number of distinguishing features, which are continuous rather than discrete in coding and which exhibit no logical priority of certain features over others.

3 The SD allows for systematic sampling of the distribution of concept usage instead of argument by examples as is customary in linguistic and philosophical semantics.

4 When applied cross-linguistically and cross-culturally the SD yields strong evidence for the universality of evaluation, potency, and activity as affective features of meaning.

Osgood finally asks why the SD yields these massive affective features instead of 'denotative' features such as concrete/abstract, animate/inanimate, human/nonhuman, etc. The answer he gives is

> "that the SD technique literally forces the metaphorical usage of adjectival scale terms, and shared affective features appear to be the primary basis for metaphorical extension. If I were to predict what little essence of my life's work might go down in history... I would say it would be the demonstration of this simple fact, that shared affective meaning of concepts is the common coin of metaphor."

He continues that the reason for the presence of evaluation, potency, and activity dimensions is at the same time the reason for its insufficiency in discovering all the features of semantic space. Since all concepts or objects must be rated against all scales, metaphorical usage of the scales is in many instances unavoidable.

> "This pressure toward metaphorical usage of scale terms means that most scales used with most concepts must rotate in the semantic space toward the affective feature on which they have their dominant loading —sweet–sour toward E, hard–soft toward P, hot–cold toward A. Since in factor analysis the major dimensions are mathematically inserted through the largest clusters of variables, this means that the shared affective features, E, P, and A will be amplified and the many subtler denotative features of meaning will be damped. This is why the SD does not provide a sufficient characterization of meaning."

(Osgood's proposition to remedy this, called the semantic interaction technique, cannot be dealt with here. By this method the minimum or critical set of semantic features of a word necessary for its correct usage is inferred from a systematic variation of context combinations).

A monograph on the SD has recently appeared in the German language (Bergler, 1975). After carefully reviewing the literature dealing with all the issues involved in the SD technique and background theory, the authors come to the conclusion that the SD is still an acceptable measuring instrument, provided that the adjective scales are relevant to the stimuli, that these scales are a representative sample of the adjectives generally used by the subject population concerned to describe the stimuli in question, that they are dimensionally differentiated so as to take account of the multidimensionality of concepts and objects, that they discriminate effectively between concepts and objects, and that they are geared to the particular subject population.

In summary, the following definitions of what the SD actually measures have emerged:

connotative meaning (Osgood et al, 1957),
the similarity of concepts (Hofstaetter, 1955),
the emotive influence of words, affect, attitude (Weinreich, 1958),
the plausibility of the connection between scales and concepts on the basis of experience (Carroll, 1959),
the emotive component of meaning (Ullmann, 1962),
an emotional semantic dimension, affect, feeling (Ertel, 1964; 1965a; 1965b).

Three different names have been proposed for the instrument itself: 'polarity profile' (Hofstaetter), 'experiential differential' (Carroll), and 'impression differential' (Ertel).

5.2.3 The SD and related techniques in the study of meaning changes in visual objects

There is a technique inspired by the SD which has closer application to the issues in this study. It was used by Lindekens (1971) in empirical studies of type faces, trademarks, and photographs, in support of the construction of a semiological theory of iconic objects. Since the true level of the iconic object is by definition nonverbal, primary responses to it can only be visual, that is, nonverbal. For this reason the type of verbal response demanded by rating scales would be able to give only indirect indications of shifts in iconic 'meaning', as would occur for example in commutation tests of iconic distinctive features. As a method of indirectly capturing the visual effect in the verbal medium, Lindekens proposes the compilation of a number of small bipolar lexicons containing words which would in some cases be indirectly relevant to the particular iconic object under study (be it in a descriptive 'denotative', or affective 'connotative' manner), and in others wholly irrelevant. Rating scales would then be applied to these words. Thus, if the effect of the iconic distinctive features 'contrasted' versus 'shaded' is studied by presenting a photograph of a human eye in several phases from shaded to hard black-and-white contrast, the effect would be indirectly registered in the mean scale values attached to those 'lexicons' which are indirectly relevant to the iconic object. Such a lexicon might contain symbolic expressions connected to the eye, for example 'love–hate', 'consciousness–unconsciousness', 'God–devil'; expressions which refer to the immediate perceptual impression of the photograph, for example, 'clarity–darkness', 'photograph–drawing', 'natural–cultural', 'reality–unreality'; or which refer to physical characteristics of the eye, for example, 'mobility–immobility', 'fragility–resistance', 'transparency–opacity'. The effect of changes in the iconic distinctive features would not be registered by lexicons containing expressions which are only metaphorically relevant to the photograph of the eye, for example, 'eagle', 'cat', 'bull', or by lexicons in which connections would actually be absurd, for example, 'ashtray', 'electron', 'city'.

Lindekens compares the indirect repercussions in the verbal medium of changes in the iconic medium with the circular wave trains set up by the impact of a pebble thrown into water. A more scientific analogy would be the Laue diagram, in which atoms are made indirectly visible by the patterns of x-ray waves with which they interfere. In the Laue diagram something invisible is rendered visible, whereas with the SD or some form of bipolar lexicon with rating scales the effect of changes in the nonverbal iconic medium is verbalized. This has the advantage of being able to avoid the question of whether the type of meaning exposed is denotative or connotative as well as the problem of whether it is the affective component of meaning which is measured. Lexicons used in this type of measurement are never definitive—one set of words could be replaced by another, as long as they are lexicons of the same kind (indirectly relevant to the iconic object or not). Since in this study what is measured is the effect of different design styles, it is in the sense of the operational definition given above that the SD will be used.

5.2.4 The SD and measurement of subjective responses to the environment

The SD has been applied to the study of paintings (for example Choynowsky, 1969; Simmat, 1969), of product design (for example Doblin, 1965), of theatre (Smith, 1961) and even of such remote problems as acoustics in architecture (Reinecke, 1966a; 1966b), to name but a few studies. But it has been particularly used in architectural research.

The increasing interest researchers are taking in people's responses to the urban environment can be seen, particularly in the USA, in the increasing number of publications on this topic since the early 1960s. This could be explained in part by the increased awareness of urban blight (especially since black uprisings in US ghettos such as Watts) in contrast to the overall economic potential of what Galbraith (1958) called the 'affluent society'. Since the start of this trend, the SD has played an important role among the instruments used to measure subjective response to environmental stimuli.

One such study in which the SD was used was concerned with the terminology of architecture and urban design. Sommer (1965) used Osgood's typical 'good–bad', 'beautiful–ugly', 'valuable–worthless' scales for the evaluation dimension; 'strong–weak', 'large–small', 'potent–impotent' scales for the potency dimension; and the 'fast–slow', 'energetic–placid', 'active–passive' scales for the activity dimension in measuring the affective effects of the terms 'room', 'space', 'town (7000 inhabitants)', and 'city (1 0000000 inhabitants)'. The subjects were forty-nine students of psychology from a small-town campus in California and 125 students of architecture from a Detroit university. The hypothesis was that the concept 'space' would have a stronger affective meaning for architectural students than for psychology students. The hypothesis was supported by the results: except for the scale 'large–small' the average ratings of the architectural students were higher than those of the psychology students.

The architectural students registered 'space' as better, more valuable, more beautiful, stronger, more potent, faster, more energetic, and more active than the psychology students. In a subsequent study dealing with the exception of the 'large–small' scale it was discovered that the psychology students had frequently interpreted 'space' as 'outer space' and had therefore given it a higher rating on this scale than had the architectural students. 'City' was judged by both groups as being very 'active' and very 'potent'. In the evaluation dimension the architectural students found 'city' to be better, more valuable, and more beautiful than did the psychology students. Both groups had largely the same responses to 'town', finding it less potent, energetic, fast, and active but at the same time more beautiful and better, albeit not more valuable than the city. They had similar responses to the term 'room', rating it as 'good', 'valuable', 'beautiful', and low in activity but high in potency.

Holschneider (1969), using an SD of forty-one point scales, investigated the concepts 'space', 'structure', 'system', 'element', 'proportion', 'rhythm', and 'visual sequence' for their effects on 153 architects and architectural students. He hypothesized that these terms were often used by architects as pseudoscientific catch phreases. The SD profiles turned out to be very largely the same, an indication that the subjects did not affectively differentiate between them to any significant extent. In a second stage of the study, the same profiles were subjected to factor analysis together with profiles of concepts such as 'male', 'female', 'success', 'progress', 'intelligence', for which data were available in Hofstaetter's previous investigations. The aim of the method was to analyze affinities between the two sets of concepts. The result was that the greatest affinity occurred in a segment of the two-dimensional factor space which Hofstaetter called 'elitist meaning'. This segment contained the concepts 'success', 'intelligence', 'personality', 'well-to-do', and 'male', when the two coordinates were rotated so that the y axis passed through 'male' and the x axis through 'female'. Holschneider concluded that architects use the first seven concepts as prestige formulae, to protect their profession against criticism, to advertise themselves and their work, and to upgrade their reputation.

Another type of SD study of architecture is directly concerned with the measurement of the effect buildings have on subjects. In what is perhaps the earliest study of this kind, Berger and Good (1963) postulated the use of the SD in research leading to better psychiatric hospital designs. The SD was frequently employed later to compare the impressions different groups of subjects had of various buildings. In 1966, Krampen (1971), for example, investigated the effect on twenty-three architectural students and twenty-two urban and regional planning students of three buildings in Toronto which had been the subject of much discussion in architectural circles. These buildings (figures 5.2–5.4) were the New City Hall (two semicircular, multistorey buildings facing each other, completed in 1965), the Old City Hall (a classical building from the end of the last century, which was scheduled to be demolished), and a modern psychiatric clinic on the campus of the University of Toronto. The same scales were used as had been used by Sommer (1965). The planning students gave the New City Hall higher average ratings on all three dimensions (evaluation, potency, and activity) than did the architectural students, who in turn gave higher ratings to the Old City Hall than did the planning students. This was, however, simply a difference in degree, as both architectural and planning students rated the New City Hall higher than the Old City Hall.

Figure 5.2. New City Hall, Toronto, Ontario, Canada; begun 1961; opened 1965; architects: Viljo Revell and Associates.

The planners rated the psychiatric clinic higher on evaluation and potency and lower on activity. The architectural students ranked the three buildings in turn: New City Hall, Old City Hall, psychiatric clinic. The planners gave almost the same ranking to the Old City Hall and the clinic, both being ranked below the New City Hall.

The utility of the SD for semantically differentiating buildings of the same function but different style was also demonstrated in another study (Krampen et al, 1971). Two theatre buildings, one old, one modern, and located side by side in Stuttgart (FRG) were rated on a fourteen-scale SD as to their effect on sixty-eight subjects. Half of the subjects rated one building, the other half the other building. The subjects were addressed in a park in front of and at approximately the same distance from the two buildings (figures 5.5 and 5.6). The SD consisted of four bipolar adjective scales for each of the three dimensions, translated from Osgood's Thesaurus study (Osgood et al, 1957) and selected on the basis of their relevance to the study of the two buildings. Two further relevant adjectives were added. For the evaluation dimension the following pairs were used: 'beautiful-ugly', 'functional-unfunctional', 'expressive-inexpressive', 'timeless-fashionable'.

Figure 5.3. Old City Hall, Toronto; begun 1891; opened 1899; architect: Edward James Lennox.

For the potency dimension the following pairs were used: 'festive–everyday', 'heavy–light', 'large–small', 'coarse–fine'. The following pairs were used for the activity dimension: 'moving–matter-of-fact', 'multiple–simple', 'calm–moved'. The added pairs were 'interesting–boring', 'inviting–repellent'. On the evaluation dimension the new theatre was rated higher than the old one, whereas the latter had a significantly 'heavier' effect than the former. Women and elderly subjects tended to rate the traditional building higher than did the men and middle-aged subjects.

At the School of Architecture in Geneva, Krampen (1969) had architectural students rate four Geneva bank buildings: Comptoir Bancaire et Financier, Rue d'Italie; Banque de Dépôts, Rue du Rhône; Banque d'Investissements Privés, Rue de l'Université; First National City Bank, Lake of Geneva

Figure 5.4. Clarke Institute of Psychiatry, Toronto; begun 1964; opened 1966; architects: John Parkin and Associates.

(figures 5.7–5.10). The subjects registered their impressions of photographs of the buildings, although they all knew the actual buildings well. The SD scales were the same as the ones used in the study of the theatres, but were translated into French. The buildings were easily differentiated by the subjects. In the evaluation dimension the ranking was Banque d'Investissements Privés, Comptoir Bancaire et Financier, Banque de

Figure 5.5. Theatre building, 'Old House', Stuttgart.

Figure 5.6. Theatre building, 'New House', Stuttgart.

Dépôts, First National City Bank. In the potency dimension the First National City Bank ranked first, followed by Banque de Dépôts (the effect of potency here is conceivably influenced by knowledge of the 'financial potency' of the banks). In the activity dimension the rank order was First National City Bank, Comptoir Bancaire et Financier, Banque d'Investissements Privés, Banque de Dépôts. These three studies give some

Figure 5.7. Comptoir Bancaire et Financier, Rue d'Italie, Geneva.

indication of the utility of the SD in measuring differences in architectural style as perceived by different groups of subjects—a utility which is to be elaborated on in the main study.

In another SD study Sanoff (1970) investigated the effects that different house forms had on American Negroes of low income and on Caucasians of middle income. The stimuli were photographs of low-income housing.

Figure 5.8. Banque de Dépôts, Rue du Rhône, Geneva.

In addition to this the subjects rated their affective impression of the concept 'ideal house' on the same SD. Both groups coincided to a large extent on the 'ideal house' profile, but reacted quite differently to the pictures. Sanoff concluded that sociocultural backgrounds on the one hand and life experiences on the other tended to influence the impressions within and between the two groups (cf Sanoff, 1973). Joedicke et al (1975) studied people's reaction to complex space-frame structures with the SD.

Figure 5.9. Banque d'Investissements Privés, Rue de l'Université, Geneva.

They correlated these 'subjective' measures with expert ratings on the 'objective' features of the buildings. The results were used to better adapt the form of space-frame structures to the feeling of the people.

The SD study of different styles of housing façades constitutes a major part of the investigation of connotation in architecture in this book. The most comprehensive study in this area (Bortz, 1972) will therefore be reviewed in detail in the main study.

Figure 5.10. First National City Bank, situated on the Lake of Geneva.

The SD has not only been applied to the study of architectural objects and individual buildings, but also to entire city districts as well. Franke (1969) and Franke and Bortz (1972) investigated the experience of city districts in West Berlin. They used an SD with forty-one bipolar adjective pairs which had been chosen simply on the basis of relevance. Groups of twenty subjects, each group taken from one of ten homogeneous districts of West Berlin, rated the districts they lived in. According to the investigators, correlations between the profiles of subjects within a district ranged between 0·81 and 0·96. These high correlations indicate that even with samples as small as twenty subjects the profiles within a group showed a high degree of stability. Of course the profiles varied considerably between the districts. The data were subjected to factor analysis to ascertain whether there were common dimensions underlying the impressions the subjects had of their life in the city. Three dimensions resulted. The first was a dimension of attractiveness, represented by such adjectives as 'colourful', 'attractive', 'familiar', 'free', 'beautiful', 'varied', 'pleasant', 'diverse'. This dimension reflected the impressions of subjects living in districts with a great deal of greenery and little traffic. The second dimension reflected the impressions of subjects living in 'old' and somewhat dilapidated areas. Typical adjectives in this case were 'old-fashioned', 'familiar', 'lively', 'dense', 'cheap', 'natural', 'dirty', 'shabby', 'poor'. These subjects apparently registered the poverty of their districts but seemed nevertheless to identify with them. [The same phenomenon was found in studies of slums in the USA (cf Fried, 1963)]. The third dimension was characterized by 'functionality' with some negative undertones. Typical adjectives were 'matter-of-fact', 'plain', 'modern', 'airy', 'uniform', 'boring', 'impersonal', 'clear', 'grey', 'articulated', 'familiar', 'sober', 'cheap', 'meaningless'. The first factor apparently corresponds to Osgood's evaluation dimension, the second to the potency dimension, the third seems to correspond to some sort of functional evaluation, while the activity dimension does not seem to appear at all (perhaps because it was represented by too few or no adjectives).

In the earlier phases of the SD research on architecture the three dimensions of connotative meaning, evaluation, potency, and activity were used in the interpretation of the data. Later, at the end of the 1960s, there was a tendency to try to discover meaning dimensions specific to architectural environments by applying factor analysis to the SD scales rather than to objects or subjects selected on the basis of architectural relevance. Work in this direction by Vielhauer, Canter, Craik, Collins, Brittel, and Hershberger was reviewed by Collins and Seaton (1970) and Hershberger (1972). A set of semantic scales to measure the meaning of architectural environments (Hershberger, 1972) emerged.

In other parlance, there seemed to be emerging a set of lexicons especially suited to register verbally variations in the visual medium of architecture. Hershberger (1972) has listed a total of twenty such lexicons shown with examples of polar adjective pairs in table 5.1.

Recently, a certain disenchantment with the SD can be observed among architectural researchers. There is to be sure an almost incredible stability of SD ratings across persons. Within-group correlation coefficients between 0·60 and 0·95 are frequently obtained. Consequently, increasing the sample of subjects per object to be rated beyond $N = 20$ seldom causes any change in the resulting profiles (for example Franke and Bortz, 1972). On the other hand, the new cluster analysis techniques based on the classification of stimuli by subjects (for example Johnson, 1967; Guttman, 1968; Lingoes, 1973) seem to fit semantic and architectural data particularly well. Nevertheless, disenchantment with the SD seems to be motivated more by the curiosity of researchers about more sophisticated methodological gadgetery than by any operational or theoretical weakness in the SD. The instrument, it seems, is too simple and functions too well; it cannot therefore hold the interest of sophisticated researchers any longer.

The following studies on architectural connotation do not subscribe to this fashionable point of view. It can be seen in the first study in this book that cluster techniques function well when many buildings of different styles have to be classified according to function. On the other hand, when subjects are required to distinguish between different styles of one functional category, the SD technique is better suited. Whereas the first method simply produces clusters of objects, the SD reveals in addition to this the adjectives by which the subjects qualify the objects in the clusters. The reader must, however, be warned again that the term 'connotation' is used in this book as synonymous with style and not in the sense that Osgood uses it in reference to the affective component of meaning.

Table 5.1. Dimensions of architectural meaning with their typical scales (lexicons) according to Hershberger (1972).

Dimension	Scale	Dimension	Scale
Aesthetics	unique-common	Lighting	light-dark
Friendliness	friendly-hostile	Privacy	private-public
Organization	ordered-chaotic	Shape	angled-curved
Potency	rugged-delicate	Ventilation	drafty-stuffy
Space	loose-compact	Noise	quiet-noisy
Ornateness	ornate-plain	Rigidity	rigid-flexible
Colouring	colourful-subdued	Formality	formal-casual
Neatness	clean-dirty	Texture	rough-smooth
Size	large-small	Age	old-new
Temperature	warm-cool	Utility	useful-useless

5.2.5 Some general misunderstandings about the SD

A general criticism of the SD is that it (allegedly) claims to measure an absolute set of meanings in its realm of application. This confuses the persistent and cross-cultural emergence of Osgood's three dimensions with absolute meaning. It is true that Osgood and his colleagues found a surprising cross-cultural stability in the dimensions he had originally discovered when using American subjects. This means, however, only that although the scales may be applicable to fundamental feeling dimensions, the application of the scales to a given object yields different results for different individuals, groups, or cultures. Thus, the SD does not measure absolute meaning but can be used to test the null hypothesis that there is no difference in the 'impression' objects make on individuals or groups. The result is certainly no 'socially neutral knowledge'. Even if it were true true that Osgood's evaluation dimension, optimally expressed by scales such as 'good–bad', 'beautiful–ugly' is cross-culturally stable, these scales would register a considerable difference between a group of communists and a group of middle-class Americans if applied to the 'hammer and sickle' symbol. It has, however, been shown that the cross-cultural stability of Osgood's dimensions is not a necessary factor in being able to register variations in the visual medium on a verbal scale.

Another criticism frequently levelled against the SD is that the experimentor's bias in the choice of adjectives limits the amount of information one can obtain about the effect a stimulus has on subjects. This can, however, actually be interpreted as an advantage in that the subjects' ratings constitute a valid test of the accuracy of the experimenter's hypothesis concerning the subjects' feeling responses to an object and should thus reveal the degree of his bias.

A further argument is that factor analysis, which is often applied to SD data, "does not reveal more than what has been put in". This is in fact true and is one of the reasons why any statistical measure, even a simple average, is calculated. It is useful to reduce a large number of correlation coefficients to a few meaningful dimensions and a large number of individual scores to an average scale. Besides which, it is factor analysis which can help eliminate irrelevant adjectives and focus attention on the relevant dimensions of the experimenter's hypothesis.

Beneath these arguments, which are, incidentally, more or less applicable to all psychological measuring instruments (including the nondirective interview with its 'biased' leading questions), is the assumption that positivist empirical data gives but a flashlight view of a small and subjectively chosen segment of an ever-evolving historical reality. This half-truth is the result of a subjectivist and idealist reduction of human knowledge of the material world to cognition. There is in fact no thought possible without initial sensory input, which is subsequently elaborated and rendered 'concrete' (applicable to more than the immediate input) by cognition. The results of cognitive elaboration in turn direct and sharpen the sensory analysis of material reality. It is therefore a principal epistemological error to dissolve the dialectic unity of sensory and cognitive acquisition of knowledge of the world in favour of either its sensory or its cognitive phase. Since scientific procedure is nothing but a systematic and methodological application of the principles of everyday acquisition of knowledge by perception and reasoning, the principle of dialectic unity applies equally to the unity of the empirical and the theoretical phase in the acquisition of scientific (systematic) knowledge. There is no theory without a basis in empirical reality (unless one wants to grant theoretical status to theological constructs), and empirical data remain abstractions removed from the complex flux of reality if they are not rendered 'concrete' in their applicability to a complex and multilaterally connected context by theoretical elaboration. This method has been described as "the ascent from the abstract to the concrete".

5.3 Study 1: reactions of subjects with higher educational level toward two different styles of tenement house façades

5.3.1 Introduction

Problem

The purpose of this study was to find out whether subjects recognize two differently conceptualized styles of façade found on buildings with the same social function—in this case that of 'tenement house'. Although every tenement house is to some degree different from every other, there are certain differences in the design of their façades which reflect particularly well differences of 'style'. Thus there might be façades which differ mainly by the presence of decorative properties absent in other merely functional designs. This difference should show up clearly in an experiment comparing the highly decorated façades of buildings constructed before 1900 in Germany with the typical façades of buildings constructed there after the Second World War.

Background: an SD study of subjective reactions towards façades

There exists one large study of subjective façade impressions which bears some resemblance to the one to be presented in this book. The goal of this broadly conceived pilot study by Bortz (1972) was to "to detect the design parameters of façades which determine the impressions had during their perception". The SD was used to quantify subjective reactions; the stimulus material in turn was quantified 'objectively' by information measurement. Twenty-six tenement house façades were judged by fifty subjects on twenty-five SD scales. The twenty-five scales were, with some exceptions, the ones used by Franke and Bortz (1972) in their study of the environmental experience of inhabitants of eleven Berlin districts. These adjectives are given with their English translation in table 5.2.

Since Bortz was going to use photographs of the façades instead of the façades themselves he first conducted a small pilot experiment: the same subjects rated the façade of a real building as well as a photograph of the same façade on an SD. The resulting profiles differed only very slightly. Bortz admitted that this result could not be generalized, since only one façade was involved but he nevertheless felt that it was indicative enough to justify the use of photographs instead of real buildings.

Since the same use of photographs for real façades will be made in this study, another experiment may be cited, to support this use. The SD reactions of subjects to real settings (including natural landscapes, man-made environments and buildings) were compared with their reactions to colour photographs and slides of the same settings in an experiment carried out by Shafer and Richards (1974) to compare the three forms of presentation. The authors concluded that "when colour-slide or picture presentations adequately depict most of the variation of natural and man-made environments, the adjective-pair measurement of response to the picture presentations agrees favourably with similarly measured on-site responses to the same scenes" (page 79). On average, correspondence to the 'on-site' reaction was closer for colour slides than for colour photographs.

Bortz used black-and-white slide projections in his façade experiment. He subjected the mean ratings of the fifty subjects on the twenty-five SD scales to factor analysis by correlating the scale items over the eighteen façades. He found three factors which he called 'evaluation' (represented by scales such as 'pleasant–unpleasant'), 'structural order' (scales such as 'smooth–broken up') and 'stimulation' (scales such as 'exciting–calming').

Table 5.2. Twenty-five SD scales, with their English translations, used in a study of building façades (Bortz, 1972).

German		English	
anziehend	-abstoßend	attractive	-repulsive
angenehm	-unangenehm	pleasant	-unpleasant
lustig	-traurig	joyous	-sad
häßlich	-schön	ugly	-beautiful
plump	-elegant	clumsy	-elegant
bedrückend	-befreiend	oppressive	-liberating
heiter	-düster	cheerful	-gloomy
hell	-dunkel	light	-dark
gepflegt	-schäbig	well-looked-after	-shabby
muffig	-luftig	close	-breezy
interessant	-uninteressant	interesting	-uninteresting
eindrucksvoll	-nichtssagend	impressive	-insignificant
lebhaft	-ruhig	vivid	-calm
wuchtig	-zart	weighty	-delicate
einheitlich	-zerrissen	smooth	-broken up
schlicht	-verziert	simple	-decorated
übersichtlich	-verwirrend	distinct	-confusing
natürlich	-technisch	natural	-technical
modern	-altmodisch	modern	-old-fashioned
verbaut	-gegliedert	disordered	-structured
eintönig	-vielfältig	monotonous	-diverse
aufregend	-beruhigend	exciting	-calming
aufdringlich	-zurückhaltend	intrusive	-restrained
solide	-überzüchtet	solid	-overdone
fremdartig	-vertraut	strange	-familiar

This result implies that the average façade experience in Bortz's study is made up of three independent aspects. The first and primary aspect is a feeling of subjective value. The second aspect is an experience of orderliness and (stylistic?) unity. The third aspect, least common among the subjects, is a sensation of being stimulated. According to Bortz this stimulation was due to new and uncommon building forms, and depended on the subjects' personal experience of, and interest in, modern architecture. On the basis of factor scores calculated for each of the twenty-six façades Bortz concluded that the subjective evaluation was strongly determined by the perceived age of the buildings. The two most positively evaluated façades were 'designed according to modern ideas', while the two least positively evaluated façades were 'gloomy and unfriendly prewar buildings'. Also, the presence of balconies had in Bortz's judgement to do with positive evaluation, as had the dominance of glass surface and windows on a façade. The feeling of high structural order was apparently determined by geometrical and regular disposition of windows and balconies, whereas low structural order was felt in the presence of many and irregularly displayed columns, arches, bays, half-timbered work, pinnacles, little towers, etc. Bortz doubted whether in every architectural context the aspect of positive evaluation would be independent of structural order. He suggested that in one kind of façade a certain decoration, though diminishing its structural order, might contribute to positive evaluation while in another kind decoration might have the opposite effect. Some façades which seemed to stimulate had also been rated lowly in structural order. Little stimulation was felt, regardless of building age, in the case of normally and solidly constructed houses, furnished with gardens.

Bortz concluded that one could postulate eight combinations of façade experience based on the three bipolar dimensions he found (table 5.3).

Table 5.3. Eight combinations of subjective façade experience based on a three-factor solution in Bortz's (1972) study.

Unpleasant	- orderly	- calming	Unpleasant	- orderly	- exciting
Pleasant	- orderly	- calming	Pleasant	- orderly	- exciting
Unpleasant	- disorderly	- calming	Unpleasant	- disorderly	- exciting
Pleasant	- disorderly	- calming	Pleasant	- disorderly	- exciting

Hypothesis

The present study, unlike the one of Bortz, did not set out to discover by induction dimensions of subjective façade experience. The intention was rather to test a hypothesis deduced from theory. From the theory of connotation developed earlier in this book it may be deduced that the subjects would be able to recognize two different manners of conceptualizing façades in the design features of buildings of the same function constructed before 1900 and after 1945. This hypothesis could be formulated more precisely in terms of the factor analytic technique applied in this study: only two factors should result from the analysis of a matrix of similarity measures calculated between the SD profiles of the two kinds of façades. These two factors should group the façades in two clusters: those constructed before 1900 and those constructed after 1945.

5.3.2 Method

Construction of the SD

In selecting the bipolar couples of adjectives for the SD a suggestion of Hershberger (1972) was followed. Hershberger differentiated two stages of environmental comprehension: the first is the representational stage in which the physical environment serves as a stimulus and "is represented in the human organism as a percept, concept, idea, image or whatever"; the second is the responsive stage which "consists of internal responses to the already internal representations".

A similar distinction was made by Ertel (1964) when he differentiated between SD scale adjectives describing objects and adjectives describing impressions. Ertel suggested that object-describing adjectives tend to produce on many occasions a metaphorical effect, such as when subjects are forced to apply the adjective 'round' to the concept 'sleep'. Given the hypothesis of this study, the bipolar couples of adjectives needed for the SD should not merely induce the subjects to describe the perceptual stage of façade experience: this perceptual experience was already determined by the selection of the two groups of buildings—those constructed before 1900 and those constructed after 1945. Thus scales such as 'decorated - nondecorated', 'old-fashioned - modern' should not be particularly useful for this experiment. Instead, other SD scales are needed which allow the expression of internal subjective feelings in reaction to the perception, that is, to the internal representation of these two kinds of façades.

To select the final 'dictionary' of adjective couples the scales quoted by Hershberger (1972), Franke and Bortz (1972), Bortz (1972), and a collection of 225 scales for SD research prepared by Krampen (1969) were screened. The final selection of adjectives was designed to allow the subjects verbally to express visual impressions evoked by façades of different style. Many of these bipolar couples of adjectives pertain to what has been called the 'friendliness' and the 'aesthetic' dimensions (Hershberger, 1972) of architectural meaning. Others would occasion linguistic reactions to the visual complexity or simplicity of façade design. The German, Italian (an Italian control group was to be used), and English versions of the final choice of scales are given in table 5.4.

To control for possible position effects three different versions of the SD form were prepared: a random sequence, the reverse order of that sequence, and a sequence running from the middle of the original random sequence to its beginning and then from its end back to the middle. There are six

Table 5.4. The twenty-two SD scales used in the studies of subjective reactions toward two different styles of tenement house façades in German, Italian, and English.

German		Italian		English	
warm	-kalt	caldo	-freddo	hot	-cold
natürlich	-technisch	naturale	-tecnico	natural	-technical
bedeutungsreich	-bedeutungsarm	ricco di significato	-privo di significato	meaningful	-meaningless
einladend	-abweisend	invitante	-respingente	inviting	-repelling
unzweckmäßig	-zweckmäßig	afunzionale	-funzionale	nonfunctional	-functional
angenehm	-unangenehm	piacevole	-spiacevole	pleasant	-unpleasant
aufgelockert	-starr	rilassato	-rigido	loose	-inflexible
abwechslungsreich	-langweilig	vario	-noioso	varied	-dull
überladen	-nüchtern	sovraccarico	-sobrio	overdone	-sober
kunstvoll	-kunstlos	aristico	-non artistico	artistic	-unartistic
aufregend	-beruhigend	eccitante	-calmante	exciting	-calming
orginell	-banal	originale	-banale	original	-banal
durchgestaltet	-unförmig	elaborato	-amorfo	well-thought-out	-ill-formed
lebendig	-tot	vivo	-morto	alive	-dead
verschwenderisch	-geizig	generoso	-avaro	generous	-stingy
vielfältig	-eintönig	molteplice	-monotono	diverse	-monotonous
fruchtbar	-steril	fertile	-sterile	fertile	-sterile
verspielt	-sachlich	brioso	-serio	playful	-serious
ungeordnet	-geordnet	disordinato	-ordinato	irregular	-orderly
persönlich	-unpersönlich	personale	-impersonale	personal	-impersonal
ausdrucksvoll	-ausdruckslos	espressivo	-inespressivo	expressive	-inexpressive
freundlich	-unfreundlich	amichevole	-ostile	friendly	-unfriendly

possible combinations of these three different SD forms. Each of these combinations was repeated to make up six different types of eighteen-page questionnaires (since eighteen façades had to be rated).

Stimuli
Photographs of eighteen façades (nine constructed before 1900, nine after 1945) were selected from a larger sample of sixty-nine façade pictures in such a way that the contrast between the design features of the two halves of that selection was maximized. The photographs depicted a series of buildings situated in a marginal zone of the city of Karlsruhe (FRG), which was scheduled for urban renewal. The original series of photographs from which the slides were made were used in an international urban design competition which took place in 1970. From the eighteen selected photographs 35 mm slides were made for projection during the experiment.

In these slides care was taken that the buildings adjacent to the façade to be rated were visible so that the rating would take the façade's context into account. This form of presentation was chosen to approximate a real field situation. The eventual effect of the adjacent buildings on the rating of a façade was to be compensated for by an instruction to pay attention only to the façade indicated by the experimenter. The pictures used in the experiment are shown in figure 5.11.

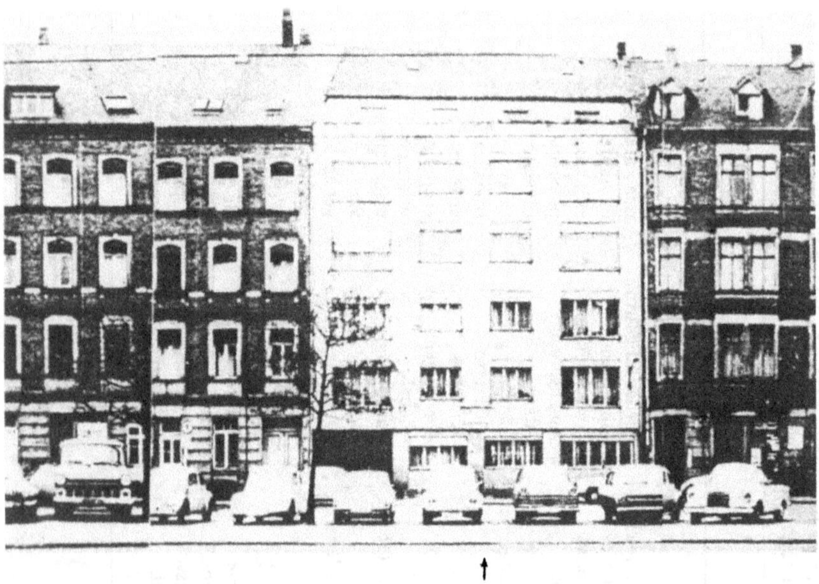

Façade 1 A4 1955

Figure 5.11. Photographs of eighteen façades, nine constructed before 1900, nine after 1945, used as stimuli. These photographs were selected from a larger sample of façades shown in figure 5.23 depicting a series of buildings situated in a marginal zone of the city of Karlsruhe, FRG, which was scheduled for urban renewal.

Façade 2 B9 1883

Façade 3 B4 1948

Figure 5.11. (continued).

Façade 4 B4 1882

Façade 5 B5 1954

Figure 5.11. (continued).

Façade 6 B2 1954

Façade 7 B4 1952

Figure 5.11. (continued).

Façade 8 B3 1883

Façade 9 A6 1892

Figure 5.11. (continued).

↑
Façade 10 A3 1958

↑
Façade 11 B11 1866/1883

Figure 5.11. (continued).

Façade 12 A7 1896

Façade 13 B2 1882

Figure 5.11. (continued).

Façade 14 B7 1894

Façade 15 B5 1955

Figure 5.11. (continued).

Façade 16 B2 1800

Façade 17 B9 1959

Figure 5.11. (continued).

Façade 18 A8 1958

Figure 5.11. (continued).

To offset any effects arising from the order of presentation, four different sequences of slides were prepared—to be given to four different groups of subjects. The first three sequences were obtained in the same way as the orders of the SD adjective couples: a random sequence; a sequence running from the middle to the start, then the end to the middle of the first sequence; and third the reverse of the first sequence. The fourth order of presentation was obtained by alternating slides of old and new façades.

Subjects

A total of seventy-six subjects viewed the slides of the eighteen façades and rated them on the twenty-two SD scales. Sixty subjects were male German nationals of whom at least forty had completed high school education with the appropriate certificate. (In Germany that certificate is called 'Abitur'. It is given after nine years of advanced schooling following four years of obligatory elementary education.) These forty were cadets of the medical corps of the FRG army, while the remaining group of twenty were a mixture of cadets, their instructors and other military personnel. None of the forty cadets were older than twenty-two; the remaining group may have been a little older on the average, but no subject was over thirty.

To obtain some control over the results and to check whether results could be generalized beyond the German language region sixteen Italian male and female nationals, students of the University of Bologna (mostly in the field of theatre science or communications) were added. This group again was young, the oldest of them being twenty-five. The entire sample, not a 'representative' random sample to be sure, was in any event relatively homogeneous with respect to age and level of education.

General procedure

The experimental procedure used was generally the same for all seventy-six subjects. Different groups of them were seated in a room equipped with a projector and a screen. A dummy slide of a façade was projected on the screen to help the experimenter explain the procedure. Two dummy scales, on the cover of the eighteen-page questionnaires enabled the subjects to get practice in rating while viewing the dummy slide. Then questions regarding the procedures were answered by the experimenter. The subjects were instructed to check after each slide of the experimental façade series that they had rated the façade on all twenty-two scales. They were also encouraged to give quick and spontaneous ratings in 'the way they felt about each façade'. In this way it was possible to gather the 18 x 22 = 396 ratings per subject in approximately forty-five minutes. Rating the first slides took the subjects generally longer (about four minutes) but they finally settled around an average time of two and one-half minutes for each slide.

The sixty German nationals were divided into three groups of twenty, each group rating a different order of slide presentation—sequences 1, 2 and 3 respectively—during the same morning in the same room (a classroom in the barracks).

The sixteen Italian students rated the fourth order of slide presentation (in what was announced to them as an experimental session) during a course on 'the effects of mass communications' in which they participated. This session, too, was held during a morning in a classroom and was completed in forty-five minutes.

After each session, the experimenter explained briefly the hypothesis of the study and engaged in a short discussion with the subjects.

Analysis of the data

Each subject's twenty-two ratings of each façade were punched into a separate computer card. Thus there were eighteen cards for each of the seventy-six subjects. Mean ratings and standard deviations for each façade, on each of the twenty-two scales, were computed (a) separately for each of the four groups (different orders of presentation); (b) for all sixty German nationals, and (c) for all seventy-six subjects. The mean ratings of a single façade on the twenty-two scales can be represented as a profile. Likewise, the average profiles of the group of nine 'old' and nine 'new' façades can be derived.

Subsequently, two factor analyses were carried out on the 22 (adjectives) x 18 (façades) matrix mean ratings derived from the individual ratings of the seventy-six subjects. At first, a *Q* analysis was made in order to obtain clusters of similar façade profiles. Each façade profile consisted of the twenty-two mean values calculated from the seventy-six individual scores the subjects gave that façade on each of the twenty-two SD scales. Thus an 18 x 18 matrix of similarity measures resulted when each of the eighteen façade profiles was compared with all others. Second, an *R* analysis was made in order to see which 'dimensions of meaning' the subjects had used to produce the clusters of similar façades. In this case the profiles to be compared consisted of the eighteen mean values calculated from the seventy-six individual scores for each pair of adjectives used in the SD. That comparison resulted in a 22 x 22 matrix of similarity measures when the profile over the eighteen façades of each adjective pair was compared with all others.

Orlik's method of taking as a similarity measure the sum of balanced cross products between the scores of two profiles, was followed in the factor analyses presented here, since it does not have the defect of correlation coefficients (cf pages 185–188).

Therefore, in this study, the similarity measures entering the factor analysis were calculated as follows. The mean scores derived from the individual scores of the subjects were treated as 'raw scores'. From the mean score of each scale in a given profile the average mean score computed over all twenty-two means of that profile was subtracted. Each mean of a scale balanced in this way was then multiplied 'across' by the corresponding mean in another profile (similarly balanced). The resulting cross products of the two profiles were summed, this sum constituting the measure of similarity for those two profiles. This operation was repeated for all possible pairs of profiles entering either the *Q* analysis or the *R* analysis. [The routine for this calculation was programmed by Dr Joachim Kutz from the faculty of Social Sciences at the University of Göttingen. He also carried out the factor analyses using the principal axis method as contained in the "Statistical Package for the Social Sciences" (SPSS) on the Univac 1100 of the Gesellschaft für Datenverarbeitung at the University of Göttingen (FRG).]

5.3.3 Results

'Old' and 'new' façade profiles by groups (order of presentation)
The façade profiles based on the mean ratings of the different groups (order of presentation and nationality) are given separately for façades constructed before 1900 and after 1945 in appendix figures A1–A4. A comparison of the 'old' and 'new' profiles shows that the 'old' profiles tend (with some exceptions) to bundle towards higher average scale values (on the left side of the scale) while the 'new' profiles tend (with some exceptions) to bundle towards lower values (on the right side of the scale).

Average 'old' and 'new' façade profiles by groups (order of presentation
The façade profiles based on separate averages for the four different groups of subjects over the nine 'old' and nine 'new' mean profile values are given in figures A5(a)–A5(d). It can be seen that the average 'old' and 'new' façade profiles clearly differ, the 'old' profiles (with a few exceptions) giving a higher average scale value than the 'new' profiles. This detachment of the average profiles of the nine 'old' from the nine 'new' façades suggests that the subjects in the different groups felt the old façades to be

different from and more positive than the 'new' façades. Only in the case of group 3 (the German military group mixed by age and education) did the average profiles overlap—and there for only two SD scales ('pleasant–unpleasant' and 'well-thought-out–ill-formed').

Average 'old' and 'new' façade profiles for all German subjects and for all subjects

The façade profiles based on separate averages over nine 'old' and nine 'new' mean profile values for all German subjects and for all subjects including the Italian control group are given in figures A5(e)–A5(g). In both cases 'old' and 'new' profiles are completely detached from each other, the 'old' profiles showing higher scale values than the new ones. The same result appears even more clearly if the scales are rank-ordered according to the size of the mean values of the 'old' façades. This result is given in figure 5.12. It clearly shows that the two groups of façades had been distinguished by the subjects of this study.

Clusters of profiles for 'old' and 'new' façades

According to the hypothesis (page 207) the factor analysis of the similarity measures between the profiles of the façades should yield only two factors, grouping the 'old' and the 'new' façades into separate clusters. This hypothesis is supported by the results of the factor analysis: the two-factor solution accounts for a total of 94·7% of the variance in the data, factor 1 explains 83·4%, factor 2 explains only 11% of the variance. A further factor would have explained only 2·4% more of the variance.

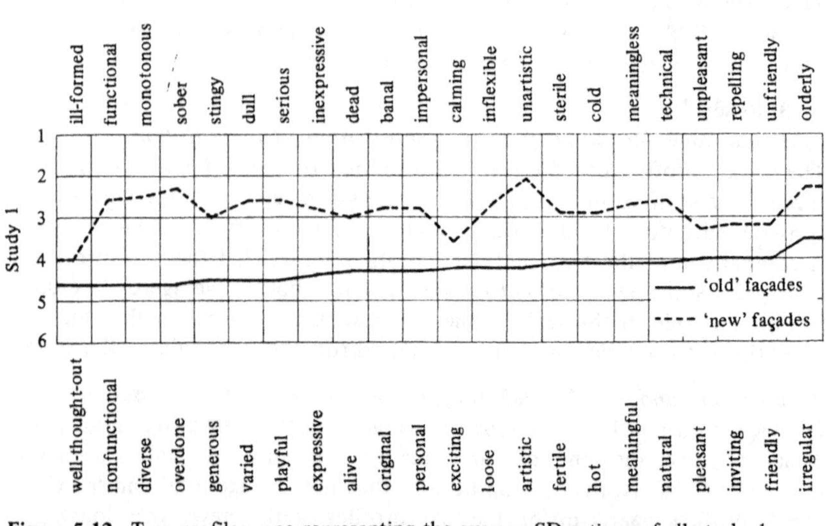

Figure 5.12. Two profiles, one representing the average SD ratings of all study 1 subjects of the 'old' façades, the other representing average ratings of the 'new' façades. The scales of the 'old' façades are rank-ordered according to the size of the mean ratings, the 'new' façade ratings are coordinated accordingly.

The two-factor solution satisfies all criteria of an optimum solution. The factor loadings of the eighteen façades on the two factors are given in table A3. When one plots the two factor loadings for each of the eighteen façades on a graph in which factor 1 is represented by the *x* axis and factor 2 by the *y* axis, two distinct clusters result. The resulting graph is given in figure 5.13. It shows a tight cluster of façade profiles in the lower left-hand quadrant, which comprises seven 'old' façades. The upper right-hand quadrant contains all the 'new' façades and two 'old' ones in a dispersed formation, the two old façades (9 and 11) being located at one end of this cluster. The numbers of the façades in the two clusters along with their years of construction are summarized in table 5.5.

A Fisher exact-probability test (Siegel, 1956) for the independence of two samples was calculated for the two clusters. It is significant well beyond the 0·005 level of probability. The null hypothesis, that there is no difference between the two clusters, can thus be rejected with a high degree of confidence. It can also be stated that the difference between the clusters is due to the average age of the façades. One cluster contains only 'old' façades, and the null hypothesis that the other cluster could be composed of old façades or by a random mixture of old and new façades can be rejected (binominal test, $p = 0 \cdot 033$).

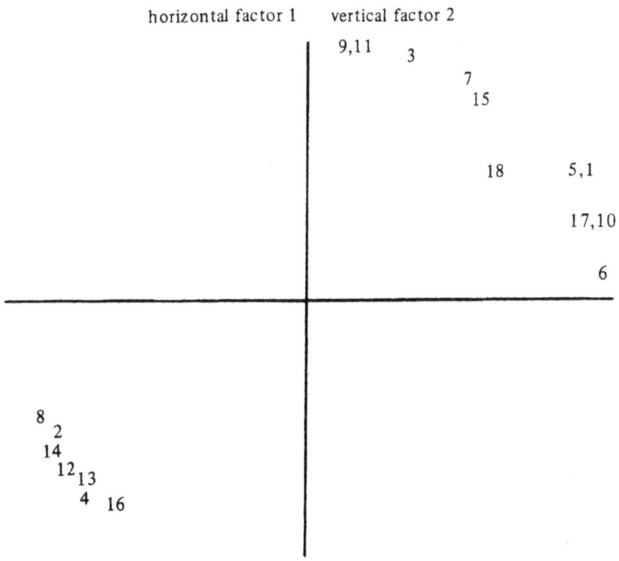

Figure 5.13. Factor analysis, study 1. The scatter plot shows the factor loadings of the eighteen façades located in two quadrants of a coordinate system in which factor 1 is represented by the *x* axis and factor 2 by the *y* axis. The cluster in the lower left-hand quadrant consists of seven 'old' façades. The upper right-hand quadrant contains all the 'new' façades and two 'old' façades (9 and 11), which are located at one end of this cluster.

The results of this factor analysis thus lend support to the hypothesis formulated at the beginning of the study; the subjects of this study apparently recognize the different manners of conceptualizing façades reflected in the different design features of the 'old' buildings and the 'new'.

Table 5.5. The buildings, with their years of construction, found in the two clusters resulting from the factor analysis of the eighteen buildings in study 1 on connotation in architecture.

Cluster 1: 'old' façades							
Façade	13	4	16	12	14	8	2
Year of construction	1882	1882	1880	1896	1894	1883	1883
Cluster 2: 'new' façades							
Façade	9	11		18	3	10	5
Year of construction	1892	1866/1883		1958	1948	1958	1954
Façade	1	7		17		6	15
Year of construction	1955	1952		1959		1954	1955

Dimensions of meaning for 'old' and 'new' façades

Aside from the fact that the subjects differentiated between 'old' and 'new' façades it is of interest to know which SD scales, that is, which 'dimensions of meaning', especially mark this differentiation. To discover these an *R* analysis was run on the 22 x 22 matrix of measures representing the similarity between the eighteen point profiles of the SD scales (over the eighteen façades). There was no particular prediction as to the number of factors which would result. However, something about the number and their character might be deduced, since the adjective pairs had been chosen to help the subjects to articulate their inner reactions towards the façades, to express their feelings about the 'friendliness' and 'aesthetic' dimensions of architectural meaning and their reactions toward (for example) visual complexity or simplicity.

The result of the analysis was a three-factor solution which accounts for a total of 84% of the variance, of which factor 1 explains 36%, factor 2 27%, and factor 3 21%. The factor loadings of the twenty-two SD scales on the three factors are given in table A2. A fourth factor extracted only 4·7% more of the variance—below the current criterion of 5% used to justify the extraction of a further factor. In any case this four-factor solution would not satisfy other criteria of when to stop factoring, and offers no explanatory advantage. Factor 1 of the three factors obtained has the highest positive loadings on the scales 'pleasant–unpleasant', 'friendly–unfriendly', 'inviting–repelling' and the highest negative loadings on 'overdone–sober', 'playful–serious', and 'diverse–monotonous'.

This factor is bipolar, and obviously indicates some semantic connection between 'pleasant friendliness' and 'monotonous sobriety'. Factor 2 has its highest loadings on the scales 'loose–inflexible', 'meaningful–meaningless', 'personal–impersonal', and 'original–banal' and indicates the personal aesthetic meaning of spatial phenomena. The third factor has its highest positive loadings on the scales 'generous–stingy', 'expressive–inexpressive', and has a very high negative loading on 'natural–technical'. This factor, bipolar like the first, reflects the opposition 'expressive generosity' versus 'technicality'. The first factor reflects something of Hershberger's (1972) 'friendliness' and Bortz's (1972) 'evaluation' dimensions. The second factor could have to do with Hershberger's aesthetic (unique–common) and space (loose–compact) dimensions. The third factor has no particular equivalent in either Hershberger's dimensions of architectural meaning or in the results of Bortz's façade study.

These results indicate that the subjects used three dimensions to distinguish between two manners of conceptualizing façades. The old façades were felt to be comparatively pleasant, friendly and inviting as well as overdone, playful, and diverse, while the 'new' façades were felt to be comparatively unpleasant, unfriendly, repelling, sober, serious, and monotonous. Whereas the 'old' façades appeared more loose, meaningful, personal, and original, the 'new' ones were perceived as relatively inflexible, meaningless, impersonal, and banal. Finally, while the 'old' façades gave the impression of being generous, expressive, and natural to the subjects, the new façades were regarded as relatively stingy, inexpressive, and technical.

5.3.4 Discussion

The two 'old' façades in the cluster of 'new' façades

The two-factor solution obtained clearly divided the façades into one 'old' and one 'new' cluster. There were, however, two old façades which fell by their factor loadings at one extreme of the cluster of new façades [see figures A1(a) and A2(a), façades 9 and 11]. It is difficult to account for the deviation of the two exceptions. The two groups of young German cadets clearly distinguished these façades from the remaining old façades, while the third group and the Italians did not [see group façade profiles, figures A3(a) and A4(a)]. Comparing the SD scale ratings of the second German group with those of the Italians for these two façades, shows that the Germans rated façade 9 equal once, and lower five times, out of twenty-two while they rated façade 11 lower twenty times out of twenty-two. The null hypothesis that there is no difference must in each case be rejected, for façade 9 at the 0·026 level of statistical significance and for façade 11 beyond the 0·001 level of significance (one-tailed binomial test). Perhaps the more sophisticated Italian subjects (artistically inclined students) rated more in accordance with the hypothesis than the German medical corps cadets. But this explanation is not satisfactory.

How far can the results of this study be generalized?
This difference between German and Italian subjects raises the question of how far the results of this study can be generalized. The observed difference for façades 9 and 11 is perhaps not due to nationality. It could be an effect of the order of presentation. The German subjects had three orders of presentation, the latter two having been derived from the first. The Italian subjects had a more systematic order of presentation in which 'old' and 'new' façades alternated. This may have allowed them to guess the hypothesis and conform to the experimenter's expectations. Even if this was not the case, the Italian subjects, being students in more art-oriented fields, were probably more sensitive to architectural style than their German counterparts, beginners in a natural science discipline. The difference may therefore be due to training rather than to nationality.

Since such a difference in training probably occurs already within the sample used in this study, there is the possibility that a generalization of the results of this study beyond subjects with a finished high school education would lead to wrong conclusions. There are some further indications that this may be the case. For instance, disregarding the peculiar façades 9 and 11, the bundling of the profiles is less pronounced for German group 3 than for German groups 1 and 2, or the Italian group (figures A3 and A4). It will be remembered that German group 3 was less homogeneous with respect to age and education (page 219). In addition, the standard deviations of the mean ratings per scale of German group 3 tend to be larger than in the other groups. This suggests that these subjects may have had difficulties with the procedure or some of the more sophisticated adjectives used in the SD (tables A4–A8).

Last, we must consider why the subjects seemed to react more favourably to the 'old' façades than to the new ones. Could this have been due to some nostalgia effect? This might have been the case with the more artistically inclined Italian subjects, but it is less probable that the Germans were sophisticated enough in the architectural field to express their feelings in terms of such a preference.

Certainly, the more positive evaluations given to 'old' façades were influenced by historical circumstances. Had the experiment been run at the turn of the century, the evaluations of these 'old' façades would have been different since the subjects would have been faced with completely different comparisons (for example, with earlier buildings of neo-classic style).

However, the question of the relationship between the visual parameters of architectural façade design and subjective reaction towards them remains a question of a more general nature. It may one day be possible to establish some dynamic law for the change of historical architectural styles (cf Kiemle, 1967; Jencks, 1971) on the basis of some sort of satiation with a given manner of conceptualizing architecture, comparable to the 'law of diminishing returns' quoted by Ullmann (1962) as a reason for the disappearance of certain feeling tones in words (cf page 184ff).

5.4 Study 2: reactions of subjects with lower educational level toward two styles of tenement house façades and their difference from those of subjects with higher educational level

5.4.1 Introduction

Problem

Since it was found in the first study that subjects of the same educational level—despite national differences—recognized two different manners of conceptualizing façades of buildings with the same social function (tenement house), the purpose of this study was to find out whether subjects of a different socioeconomic class (indexed here by different educational level) exhibit a different recognition performance when confronted with the same two contrasting groups of façades. One of the differences between social classes might in fact consist in the ability to make a distinction between those façades which contain additional signal properties and those conceptualized on a merely functional basis. Such a difference should have shown up in an experiment comparing the recognition performances of subjects of different educational level (due to different social background) when they viewed the same sample of façade pictures used in the previous study, half of the sample portraying façades constructed before 1900, the other half constructed after 1945.

Background of the problem: cultural and social differences in the perceptual recognition of cultural objects

The following study may provide evidence that different social classes have a different perception of 'style' in architecture. This view is based on a previous theoretical remark on possible incompatibilities between architectural production and consumption (pages 65, 66). At that point some reasons were given for the view that differences between producers and consumers in their conceptual repertoires of architectural products might lead to architectural conceptualization which is not re-cognizable and/or to reconceptualization of original architectural concepts. One of the reasons given was that such differences might result where the perceivers' cultural background involved a different building tradition owing to climate and other factors. Another possibility was that particular properties of architectural products and designs might be conceptualized by the producer, which are not re-cognized by the consumer. It was stated that this danger would presumably be greatest in the case of conceptualization of those excess, but intentional, properties of architectural products and designs which might be called 'artistic' design.

There is striking evidence of the influence of cultural background on perception. Cultural differences in visual perception of physiognomic similarity in photographs, of colour and depth cues (perspective) in pictures and especially of optical illusions were reported by Segall et al (1966). These authors, however, eventually reduced cultural differences to variations "from person to person in accordance with selective forces, both inhibiting and reinforcing, provided by culturally mediated experiences". What they appear to say is that cultural differences in perception are due to differential reinforcement of experiences within or between cultures. A further experiment bears out this reduction. Hudson (1960) studied the use made of different visual depth cues, such as object size, overlapping, and perspective, by eleven samples of South Africans varying in educational background, age, race, and national origin, when they viewed pictures and photographs. He summarized his findings as follows.

> "Outline drawings making use of perspective depth cues are less frequently seen three-dimensionally than those using overlap or size depth cues. This finding holds particularly in the case of white primary school pupils. School-going samples perceive three dimensions in a photograph more readily than in an outline drawing, but this finding does not apply to illiterate samples. (None of the illiterates tested saw the photograph three-dimensionally.) Intelligence and educational level are factors which independently influence dimensional pictorial perception, but they appear to do so only with white school-going samples."

This type of evidence brings Segall et al to the conclusion that "both formal schooling and immersion in a culture containing pictures, books, and magazines are factors that heavily determine the tendency to employ depth cues when viewing two-dimensional space".

This suggests that cultural differences in perception are largely determined by differences in formal and informal education. From this conclusion the next question follows directly: which factors determine differences in formal and informal education within a given culture?

In sociolinguistics during recent years the influence of informal socialization on communication competence, through the mechanisms of family and social class, has been stressed by such authors as Bernstein (1971). Similar mechanisms seem to apply also in visual communication. Bourdieu (1968) has shown that both informal and formal education influence this competence.

Bourdieu argues that there is theoretical and empirical evidence (the latter deriving from his observation of people's behaviour in museums) that 'aesthetic' competence—the ability to recognize artistically conceived properties of designs over and above their 'functional' ones—is confined to the members of a civilized elite within the economically dominating class. The members of the other (economically dominated) classes are generally not educated to re-cognize artistic signals. Bourdieu bases his contention on the assumption that the perception of works of art (and of cultural artefacts generally) is a process of decoding. In order for this process to function, there must be learning of a common code, that is, there must be correlation between the education of the artist and the expertise of the beholder. If there is no common code to begin with, there must be misunderstandings in artistic (or cultural) communication. All alleged 'spontaneous' or 'naive' understanding of objects is nothing but a 'naturalization' of acquired expertise.

Now, the codes of civilization vary in complexity and accordingly so does their accessibility to people. Thus works of art are accessible on two main levels. The more superficial of these penetrates only to a first level of 'meaning'. This level of meaning has been defined by Panofsky (1957) as 'meaning of things and simple emotions'. The more refined perception, resulting in aesthetic experience, employs codes derived from the reading of specialized literature, generally dealing with matters of style. The learning of these codes occurs by means of formal and informal education and is prepared from childhood by a suitable social environment.

Confronted with objects of art or civilization, the beholder, lacking the appropriate codes, finds himself in a similar situation to that of an ethnologist observing a ritual of a newly discovered tribe. The so-called 'connoisseur' on the other hand plays the role of a tribal ethnocentric who regards this cultural ritual to be quite 'natural'.

Those who do not have the appropriate code for works of art and civilization often try, as a consequence, to improve the embarrassing situation by charging their perception with categories and values which are, in fact, inappropriate to aesthetic objects. According to Bourdieu, these categories and values tend to be those which organize everyday perception and orientate practical decisions and judgements.

Works of art or civilization can thus only be works of art or civilization for those equipped to appropriate them intellectually. Aesthetic competence results from the possession of the means of its appropriation. One of the most important means of appropriation is the knowledge of stylistic properties and indicators.

The code of works of visual art (and for that matter objects of civilization at large) consists of an oppositional system of sets of stylistic properties correlated with a corresponding oppositional system of sets of image concepts current within a given society. Aesthetic competence is thus complete only with the knowledge of all possible potential subdivisions of both universes, that of stylization of 'anecdotal content' (different manners of expressing the same content) and that of all the different images in a society.

Some reflection will lead to the conclusion that the motivation for appropriating the 'capital' of civilization increases with the possession of the means of appropriation: each appropriation leads to an augmentation of the means of appropriation. On the other hand, the consciousness of being deprived of these means disappears all the more rapidly the scarcer their supply.

The mechanism of successful appropriation by formal and informal training is the repetition of perception provided through art education and by a social and family environment interested in the arts. To deny this cumulative effect of exposure effectively conceals the social conditions of successfully (or unsuccessfully) appropriating the capital of civilization. The most blatant form of this type of concealment occurs in the charismatic idea of 'natural talent'—the absurd and insipid contrivance of 'natural culture' or 'cultural nature'.

The seemingly symbolic usage of works of art or civilization by connoisseurs—in contrast to the more superficial and practical perception of these works by those unequipped with the appropriate code—does not prevent their refunctionalization in very 'practical' terms: they become a means of social 'distinction'. Symbolic displacement from the economic level (which provides the cultural advantage) to the level of civilization differentiates the 'connoisseurs' from the 'ignorants'. Through this operation one augments the original economical difference (indicated by the possession of material goods) with a difference arising from the possession of symbolical goods (or the competence to use these symbolic goods).

There is evidence in a recent study of Agazzi et al (1977) that the hypothesis of Bourdieu applies also to the perception of architectural objects and urban environments by different social classes. Thus some subjects of higher socioeconomical level (indexed by profession) tended to describe the ancient centres of towns in the Tessin (Switzerland) as places of culture and history. At the same time these subjects declared that they would not want to live there because of the presence of the lower classes in these quarters. In contrast, members of the lower socioeconomic level looked at the historical town centres mainly as places for shopping and habitation, almost the way they looked at the modern quarters of the same towns. They viewed the latter more critically but appreciated them for the inside comfort and equipment of the apartment buildings.

Hypothesis

From the general theoretical framework of this study (cf pages 65, 66, 90) and also from its more detailed elaboration on the basis of Bourdieu's concept of artistic competence it can be deduced that the subjects of lower socioeconomic class indexed by degree of education should less clearly distinguish between façades built before 1900 and those built after 1945. At the same time, factor analysis should still yield two factors, grouping the façades according to design features distinguishing 'old' from 'new'. More precisely, the subjects attending the last years of a trade school should assign lower SD ratings to façades built before 1900 and higher SD ratings to façades built after 1945 than their counterparts with high school degrees, examined in study 1. Thus the mean differences between 'old' and 'new' façades on the various scales of the SD computed over the sample of trade school students should be significantly smaller than the same mean difference computed for the better-educated group.

5.4.2 Method

The SD versions used for comparison

In order to compare the two samples, trade school students on the one hand and subjects with high school degrees on the other, the same SD scales which were used in study 1 had to be adopted. The three versions which the sequence of adjectives took in study 1, were alternated in the eighteen-page questionnaire booklet as before (pages 208, 210).

Stimuli

The eighteen 35 mm slides of façades which were used in study 1 were employed also in this study to guarantee a correct comparison of the two samples. They were projected in the original random sequence (sequence 1) constructed for study 1 (cf page 210). The other slide sequences of study 1 were not used since fatigue effects on the ratings of the last slides shown in each sequence had been negligible.

Subjects

A total of forty male and female students from a trade school in Ulm/Donau (FRG) viewed the eighteen façades in two sessions, one with eighteen, the other with twenty-two subjects and rated them on the twenty-two scales. They were asked to write down the profession of their father or mother and state whether they lived in the rural areas surrounding the city of Ulm or in the city itself. The resulting distribution is given in table 5.6. All subjects frequented the third and fourth classes of trade school. On average they were eighteen years old. This sample was thus homogeneous with respect to age and education.

Table 5.6. Occupation of father (or mother) of the forty trade school students serving as subjects in study 2 on connotation in architecture.

Occupation of father (or mother)	Family living in	
	rural area	city
Farmer	8	-
Skilled worker (for example, automobile mechanic)	6	2
Unskilled worker (for example, janitor)	4	1
Craftsman (for example, carpenter)	7	3
White collar (for example, postal clerk)	3	-
Employed technician (for example, engineer)	2	1
Housewife	1	-
Blank	1	1
	32	8

General procedure

The experimental procedure was the same as in study 1 although the experiment was carried out by another experimenter. He had, however, been carefully prepared and followed written instructions in order to duplicate as closely as possible the experimental situation of study 1. This experimenter reported later that he had noted a drop in attention toward the end of the forty-five minute experiment which he ascribed to fatigue.

Analysis of the data

The analysis of the data was carried out in two phases. The first phase contained the same steps as the analysis of the data in study 1. As in study 1 each subject's ratings of a single façade were punched into a separate computer card. This resulted in eighteen cards for each of the forty subjects. Missing individual ratings (which occurred more frequently in this study than in the previous one) were counted as a rating of ± 0 (as '+4' after transformation of the scale).

Mean ratings and the corresponding standard deviations on each of the twenty-two scales were computed for each single façade and for the combined 'old' and 'new' façades. These mean ratings on the twenty-two scales, which as before were treated as the 'scores' of the façades, were represented as a profile.

Subsequently two factor analyses were computed as in study 1: one *Q* analysis to assess whether the subjects of study 2 did recognize the difference of façade style, and one *R* analysis to obtain information on the semantic dimensions of their ratings.

In the second phase of analysis the results of study 2 were compared with those of study 1. This was begun by calculating, on each of the twenty-two scales, for each of the 116 subjects (seventy-six from study 1 plus forty from study 2) a mean checking score and standard deviation, first for all façades combined, then separately for all 'old' and all 'new' façades. For each scale, this resulted in 116 mean scores and standard deviations each: for all eighteen façades together, for the combined 'old' façades, and for the combined 'new' façades.

Subsequently, for each of the 116 subjects the differences between their mean score for the combined 'old' and for the combined 'new' façades were derived on each of the twenty-two scales.

These mean differences were compared within the groups of study 1 and of study 2 separately by a *t* test. The outcome of this comparison would indicate for each group which scale items yielded significant differences between 'old' and 'new' façades and which items did not. This comparison of the mean differences between 'old' and 'new' façades per single scale gave a different kind of 'item analysis' from that obtained in the factoring (*R* analyses).

Next, the mean difference between 'old' and 'new' façades over all twenty-two scales was calculated and tested by an overall *t* test (Hotelling's T^1). Wherever this difference was significant the group in question could be said to have—over all scales taken together—perceived the 'old' façades to be different from the 'new' ones.

Finally, a discriminant analysis was made to compare the mean values per SD scale (over the old façades) given by the better-educated group with the values given by the trade school students. The aim here was to discover whether the subjects of the two studies perceived the old façades fundamentally differently. The same test was carried out with the nine new façades.

A further discriminant analysis was performed by comparing the two groups' mean scores per SD scale for all eighteen façades together. The purpose of this was to discover whether the perception of façades was generally different for study 1 subjects as compared with study 2 subjects.

As far as the design of this data analysis is concerned it could be objected that one precondition of *t* tests (that the scores are normally

Table 5.7. Design of the analysis comparing study 1 and study 2.

	SD scale 1	SD scale 2	SD scale 3	...	SD scale 22
Experiment 1					
Subject					
1	$\bar{X}_{old\,1,1}\bar{X}_{new\,1,1}$	$\bar{X}_{old\,1,2}\bar{X}_{new\,1,2}$	$\bar{X}_{old\,1,3}\bar{X}_{new\,1,3}$	...	$\bar{X}_{old\,1,22}\bar{X}_{new\,1,22}$
2					
3					
⋮					
40					
41					
42					
⋮					
60					
61					
62					
⋮					
76					
	$\bar{\bar{\bar{X}}}_{old\,1}\bar{\bar{\bar{X}}}_{new\,1}$	$\bar{\bar{\bar{X}}}_{old\,2}\bar{\bar{\bar{X}}}_{new\,2}$	$\bar{\bar{\bar{X}}}_{old\,3}\bar{\bar{\bar{X}}}_{new\,3}$	...	$\bar{\bar{\bar{X}}}_{old\,22}\bar{\bar{\bar{X}}}_{new\,22}$
	1-40 1-40	1-40 1-40	1-40 1-40		1-40 1-40
	1-60 1-60	1-60 1-60	1-60 1-60		1-60 1-60
	1-76 1-76	1-76 1-76	1-76 1-76		1-76 1-76
	t	*t*	*t*		*t*
Experiment 2					
Subject					
1	$\bar{\bar{\bar{X}}}_{old\,1,1}\bar{\bar{\bar{X}}}_{new\,1,1}$	$\bar{\bar{\bar{X}}}_{old\,1,2}\bar{\bar{\bar{X}}}_{new\,1,2}$	$\bar{\bar{\bar{X}}}_{old\,1,3}\bar{\bar{\bar{X}}}_{new\,1,3}$	...	$\bar{\bar{\bar{X}}}_{old\,1,22}\bar{\bar{\bar{X}}}_{new\,1,22}$
2					
3					
⋮					
40					
	$\bar{\bar{\bar{X}}}_{old\,1}\bar{\bar{\bar{X}}}_{new\,1}$	$\bar{\bar{\bar{X}}}_{old\,2}\bar{\bar{\bar{X}}}_{new\,2}$	$\bar{\bar{\bar{X}}}_{old\,3}\bar{\bar{\bar{X}}}_{new\,3}$	...	$\bar{\bar{\bar{X}}}_{old\,22}\bar{\bar{\bar{X}}}_{new\,22}$
	t	*t*	*t*		*t*

distributed) is not met by SD data. This criticism would apply if the raw SD scores themselves were used in the analysis. Instead, not the raw scores but their averages over nine 'old' or nine 'new' façades per scale were processed. It could be expected that these mean values would be normally distributed. The design of the experiment is illustrated in table 5.7. [The author is indebted to M Sund from the Institut für medizinische Datenverarbeitung in Munich for his statistical advice on this problem. He also carried out the processing of the data, adapting the BMD programs 4M and 1D (Health Sciences computing facility, University of California at Los Angeles) to the TR440 computer in the University Computing Centre at the University of Ulm.]

5.4.3 Results: reactions of trade school students to the façades

'Old' and 'new' façade profiles

The façade profiles based on the mean ratings of the trade school students are given separately for façades constructed before 1900 and after 1945 in figures A6(a) and A6(b) (cf table A9). A comparison of the 'old' and 'new' profiles shows that the 'old' profiles (with some exceptions) tend to bundle towards higher scale values (toward the middle of the scale) whereas the 'new' profiles tend (with some exceptions) to bundle toward lower values (on the right side of the scale). As in study 1, the 'old' façade profiles 9 and 11 show exceptionally low scale values. This suggests that most German subjects must have reacted to them in a particular way.

Average 'old' and 'new' façade profiles

The two profiles averaging respectively all 'old' and all 'new' façades for the trade school students are shown in figure A6(c). It can be seen that they are entirely detached, the profile for the average of the 'old' façades holding a higher value on all 22 scales than the profile for the average of the 'new' façades. This detachment shows that the trade school students felt the 'old' façades to be different and somewhat more positive than the 'new' façades. The same result appears even more clearly if the scales are rank ordered according to the size of the mean values of the 'old' façades. This result is given in figure 5.14.

Clusters of profiles for 'old' and 'new' façades

According to the hypothesis the factor analysis of the data in this study should yield two factors, as in study 1, grouping the 'old' and 'new' façades into separate clusters. In fact, the two-factor solution accounts for a total of 86·9% of the variance, factor 1 explaining 71·7%, factor 2 15·2%. This two-factor solution satisfies all criteria for an optimum solution. The factor loadings of the eighteen façades on the two factors are given in table A10. If the loadings of the eighteen façades are plotted into a coordinate system two distinct clusters result, as in study 1. The plot is given in figure 5.15. Again the lower left-hand quadrant shows the same seven 'old' façades as the parallel plot of study 1, although in a

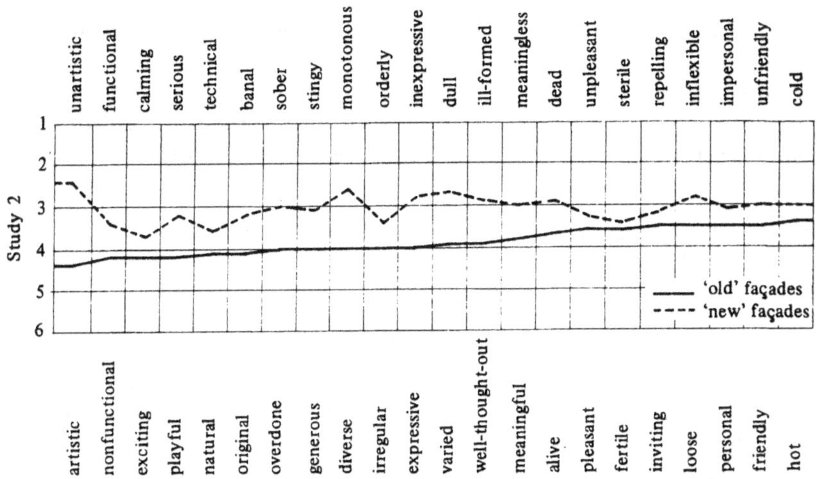

Figure 5.14. Two profiles, one representing the average SD ratings of all study 2 subjects of the 'old' façades, the other the average ratings of the 'new' façades. The scales of the 'old' façades are rank-ordered according to the size of the mean ratings, the 'new' façade ratings are coordinated accordingly.

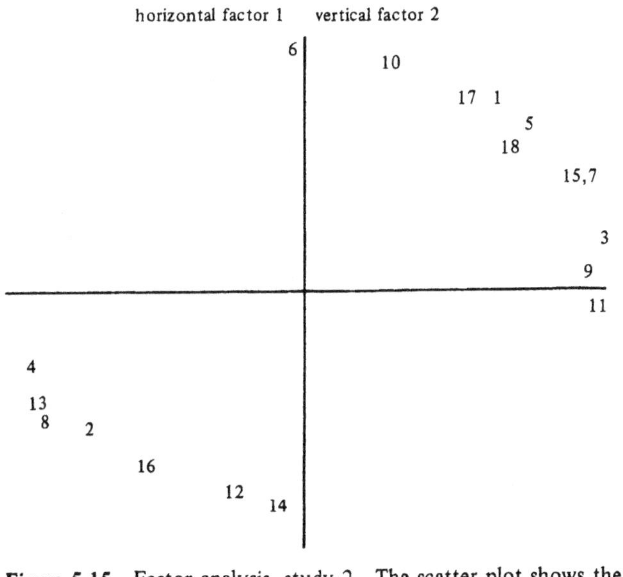

Figure 5.15. Factor analysis, study 2. The scatter plot shows the factor loadings of the eighteen façades located in the quadrants of a coordinate system formed by factor 1 as the x axis and factor 2 as the y axis. The cluster in the lower left-hand quadrant consists of seven 'old' façades. The new façades are dispersed around the upper right-hand quadrant. Two 'old' façades (9 and 11) are also located at one end of that cluster.

different order and much more dispersed. The upper right-hand corner contains all the 'new' façades and in addition the two 'old' façades 9 and 11. The position of the façades in this quadrant is, however, reversed from, and even more dispersed than, in study 1. The clusters are again significantly different and the difference must be due to the age of the façades. Thus the trade school students did recognize the difference between the 'old' façades and the 'new' façades.

Dimensions of meaning for 'old' and 'new' façades

As in study 1 an *R* analysis was run on the 22 x 22 matrix of measures representing the similarity between the profiles of the SD scales. The result of the analysis was a three-factor solution. It accounts for 81·6% of the variance of which factor 1 explains 47·4%, factor 2 22·1%, and factor 3 12·1%. The factor loadings of the twenty-two SD scales on the three factors are given in table A11. A fourth factor would extract a further 5·5% of the variance, but it falls short of other than the 5% criterion and does not offer an explanatory advantage. Factor 1 has its highest positive loadings on the scales 'diverse-monotonous', 'varied-dull', 'artistic-unartistic', 'expressive-inexpressive' and 'well-thought-out' ill-formed', and the highest negative loadings on 'exciting-calming', 'natural-technical', and 'fertile-sterile'. This bipolar factor suggests a semantic connection between 'aesthetic variety' and 'sterile technicality'.

Factor 2 has its highest loadings on 'hot-cold', 'personal-impersonal', 'pleasant-unpleasant', 'friendly-unfriendly', 'loose-inflexible', 'inviting-repelling', and 'alive-dead'. This is obviously a 'friendliness' factor. Factor 3 has its highest loadings on 'overdone-sober', 'playful-serious', and 'irregular-orderly'. This may be a factor of 'overdone playfulness'. The first factor corresponds perhaps to Hershberger's (1972) aesthetic dimension combined with Bortz's stimulation dimension. The second factor reflects Hershberger's friendliness dimension and Bortz's evaluation dimension. The third factor could have to do with Bortz's dimension of structural order. The trade school students felt the 'old' buildings to be relatively varied, artistic, expressive, well-thought-out, exciting, natural, and fertile, while the 'new' buildings were relatively monotonous, dull, unartistic, inexpressive, ill-formed, calming, technical, and sterile. Whereas the old buildings seemed to them personal, pleasant, friendly, loose, inviting, and alive, the new ones were rather impersonal, unpleasant, unfriendly, inflexible, repelling, and dead. The old buildings were thought to be overdone, playful and irregular, the new ones to be more sober, serious, and orderly.

5.4.4 Results: comparisons of study 1 and study 2

The different groups to be compared

For the comparison of the two studies it must be remembered that in study 1 four different groups were tested, differing as to order of slide presentation, education, and nationality:

1 German nationals with high school diploma, receiving the order of slide presentation 1 (subjects 1-20);
2 German nationals with high school diploma, receiving the order of slide presentation 2 (subjects 21-40);
3 German nationals, only in part with high school diploma, receiving the order of slide presentation 3 (subjects 41-60);
4 Italian nationals with high school diploma receiving a different order of slide presentation (subjects 61-76).

The second study had only one group of forty subjects, although these subjects were tested in two different sessions (cf page 231).

Comparison of the average 'old' and 'new' façade profiles of the studies
The façade profiles based on separate averages over nine 'old' and nine 'new' mean profile values resulting from ratings of the subjects in study 1 and study 2 are given in figure A7. These figures show that the subjects in study 1 tended to give both higher ratings to 'old' façades and lower ratings to 'new' façades than did the subjects of study 2.

Comparison of the factor analyses of the studies
A comparison of the Q analyses of the two studies shows no difference in the identity of the façades belonging to the two dimensions. However, in the two plots the façades are differently displayed as far as arrangement and dispersion is concerned. In the study 1 plot, the two clusters are on the whole packed more tightly than in the study 2 plot. Most clearly

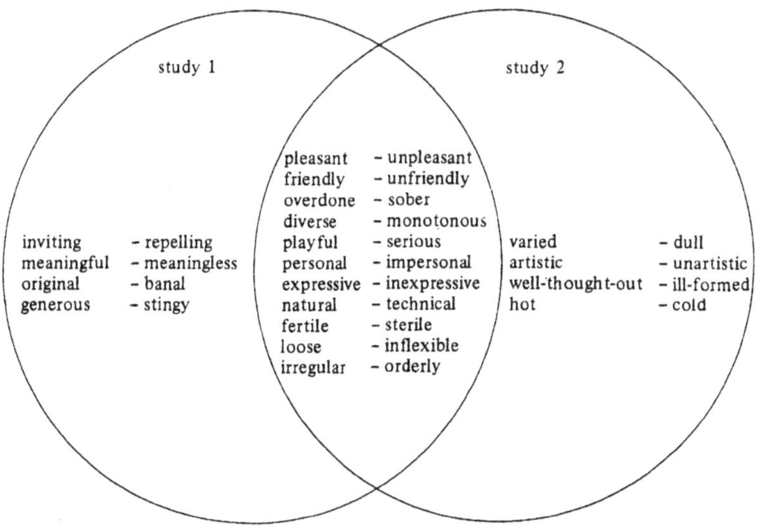

Figure 5.16. Comparison of the dimensions of 'architectural meaning' applied to the façades by the subjects in study 1 and study 2.

in the case of the 'old' façade cluster this indicates that the façades within each of the two clusters were perceived to be more similar by the better-educated group.

The *R* analyses of the studies indicate that the subjects of studies 1 and 2 used somewhat different semantic dimensions to differentiate the same façades. While both groups described the difference between the old and new façades in terms of friendliness and aesthetics, they associated somewhat different affective tones with these dimensions. Nevertheless, there is a considerable overlap in the vocabularies used to differentiate semantically the two kinds of façades. This overlap as well as the differences in the vocabularies of the two groups is summarized in figure 5.16.

Table 5.8. Mean differences between combined 'old' and combined 'new' façades for trade school students (study 2) and forty German subjects with high school diplomas (study 1).

Variable (*V*)		Trade school			High school diploma (subjects 1-40)		
		mean	*s*	*t* test	mean	*s*	*t* test
1 hot	- cold	0·45	1·33	2·13[a]	1·60	1·11	9·13
2 natural	- technical	0·52	1·08	3·01	1·74	1·09	10·10
3 meaningful	- meaningless	0·81	1·07	4·75	1·77	0·99	11·32
4 inviting	- repelling	0·31	1·10	1·74[b]	1·04	1·15	5·71
5 nonfunctional	- functional	0·82	1·15	4·51	2·28	1·24	11·64
6 pleasant	- unpleasant	0·30	1·06	1·76[b]	1·04	1·23	5·34
7 loose	- inflexible	0·78	1·23	4·03	2·14	0·86	15·79
8 varied	- dull	1·22	1·53	5·04	2·30	0·98	14·83
9 overdone	- sober	1·00	1·19	5·32	2·75	0·86	20·13
10 artistic	- unartistic	2·03	1·62	7·93	2·33	1·04	14·24
11 exciting	- calming	0·54	0·87	3·96	0·73	0·86	5·3
12 original	- banal	0·89	1·22	4·61	1·73	1·12	9·17
13 well-thought-out	- ill-formed	1·08	1·27	5·37	0·56	1·16	3·02
14 alive	- dead	0·79	1·47	3·39	1·64	1·11	9·38
15 generous	- stingy	0·91	1·24	4·61	1·85	0·92	12·73
16 diverse	- monotonous	1·40	1·47	6·06	2·45	0·82	18·91
17 fertile	- sterile	0·23	1·11	1·29[b]	1·64	1·10	9·46
18 playful	- serious	0·96	1·22	4·96	2·80	0·92	19·38
19 irregular	- orderly	0·57	1·18	3·07	1·14	0·85	3·46
20 personal	- impersonal	0·46	1·07	2·70	1·94	1·06	11·54
21 expressive	- inexpressive	1·16	1·38	5·32	1·94	1·00	12·28
22 friendly	- unfriendly	0·50	1·18	2·68[a]	1·10	0·97	7·15
Hotelling's T^2		117·17			1129·51		
F value		3·72			23·70		
d.f.		22,18			22,18		

[a]$p < 0{\cdot}05$; [b]not significant; for all other *t* values, $p < 0{\cdot}01$.

Mean differences between combined 'old' and combined 'new' façades for different groups

The mean differences between the averaged 'old' and 'new' façades on each of the twenty-two scales for different groups are given together with their *t* values in tables 5.8 and 5.9. From a comparison of the two groups in these tables the following results were derived.

All groups show a highly significant difference in their ratings of the 'old' façades and the 'new' façades, the 'old' façades generally rating higher on the SD scales than the 'new' ones.

Trade school students show the smallest (although still significant) mean difference between 'old' façades and 'new' façades.

The addition of both the variously educated German group and the Italian nationals group to the forty educationally homogeneous German subjects does not change the general trend of the results.

Table 5.9. Mean differences between combined 'old' and combined 'new' façades for all German, and all German and Italian subjects (study 1).

V	High school diploma (subjects 1-60)			High school diploma (subjects 1-76)		
	mean	*s*	*t* test	mean	*s*	*t* test
1	1·27	1·43	6·86	1·22	1·45	7·35
2	1·62	1·21	10·37	1·48	1·28	10·02
3	1·56	1·13	10·68	1·39	1·20	10·11
4	0·75	1·41	4·12	0·72	1·44	4·38
5	2·24	1·19	14·55	2·02	1·36	12·96
6	0·66	1·48	3·43	0·65	1·57	3·59
7	1·80	1·27	11·04	1·56	1·41	9·65
8	2·06	1·26	12·65	1·88	1·42	11·58
9	2·56	1·09	18·16	2·32	1·26	16·00
10	2·41	1·09	17·06	2·08	1·38	13·11
11	0·73	0·93	6·11	0·63	0·91	6·10
12	1·59	1·17	10·51	1·44	1·27	9·91
13	0·32	1·27	1·99[a]	0·63	1·45	3·78
14	1·43	1·29	8·62	1·28	1·31	8·54
15	1·78	1·03	13·37	1·49	1·14	11·41
16	2·28	1·09	16·21	2·08	1·19	15·26
17	1·46	1·23	9·21	1·25	1·30	8·36
18	2·69	1·08	20·28	2·24	1·40	13·94
19	1·17	0·99	9·16	1·19	1·19	8·74
20	1·54	1·37	8·76	1·49	1·37	9·49
21	1·71	1·26	10·49	1·54	1·31	10·19
22	0·83	1·30	4·93	0·76	1·41	4·67
Hotelling's T^2	1290·26			547·72		
F value	37·77			17·93		
d.f.	22,38			22,54		

[a]Not significant; for all other *t* values, $p < 0{\cdot}01$.

Rank orders of the mean difference values of the SD scales for the different groups of study 1 subjects versus the subjects of study 2 are given in table A12. Inspection of these tables shows that the rank orders of the SD scales according to their capacity of discriminating between 'old' façades and 'new' façades differ only slightly when based on the average performance of the German subjects with high school diploma (study 1, subjects 1–40), the German subjects on the whole (study 1, subjects 1–60), and all subjects of study 1 (subjects 1–76), including Italian nationals). On the other hand, the difference between these three rank orders and the one based on the performance of trade school students is noticeable. This result confirms the trends which were already noticeable in the comparison of the *R* analyses of study 1 and study 2. The subjects in study 1 and study 2 used somewhat different vocabularies to differentiate between old and new façades.

Table 5.10. Discriminant analysis comparing forty subjects with high school diploma (study 1) with forty trade school students (study 2) on 'old', 'new', and all façades.

V	Old façades			New façades			All façades		
	mean 1	mean 2	difference	mean 1	mean 2	difference	mean 1	mean 2	difference
1	4·32	3·42	0·90	2·71	2·96	−0·25	3·51	3·19	0·32
2	4·30	4·11	0·19	2·57	3·60	−1·03	3·44	3·86	−0·42
3	4·17	3·80	0·37	2·40	3·00	−0·60	3·27	3·40	−0·13
4	4·08	3·54	0·54	3·04	3·23	−0·19	3·55	3·38	0·17
5	4·66	4·22	0·44	2·38	3·41	−1·03	3·52	3·81	−0·29
6	4·17	3·55	0·62	3·13	3·26	−0·13	3·65	3·40	0·25
7	4·54	3·54	1·00	2·40	2·75	−0·35	3·47	3·16	0·31
8	4·69	3·88	0·81	2·40	2·66	−0·26	3·55	3·27	0·28
9	4·62	4·02	0·60	1·87	3·02	−1·15	3·25	3·53	−0·28
10	4·29	4·42	−0·13	1·96	2·40	−0·44	3·14	3·41	−0·27
11	4·19	4·20	−0·01	3·46	3·66	−0·20	3·83	3·93	−0·10
12	4·36	4·06	0·30	2·62	3·17	−0·55	3·49	3·62	−0·13
13	4·71	3·94	0·78	4·16	2·86	1·30	4·43	3·39	1·04
14	4·44	3·68	0·77	2·80	2·89	−0·09	3·63	3·29	0·34
15	4·61	4·04	0·57	2·76	3·13	−0·37	3·68	3·60	0·08
16	4·78	4·00	0·78	2·33	2·59	−0·26	3·56	3·28	0·28
17	4·28	3·64	0·64	2·64	3·41	−0·77	3·46	3·52	−0·06
18	4·84	4·15	0·69	2·03	3·19	−1·16	3·43	3·66	−0·23
19	3·34	4·00	−0·66	2·20	3·42	−1·22	2·77	3·71	−0·94
20	4·32	3·54	0·78	2·38	3·08	−0·70	3·35	3·30	0·05
21	4·44	3·94	0·50	2·49	2·78	−0·29	3·46	3·36	0·10
22	4·09	3·52	0·57	2·99	3·02	−0·03	3·53	3·27	0·26
F	7·23			5·65			7·19		
d.f.	22,57			22,57			22,57		
p	<0·001			<0·001			<0·001		

Result of the discriminant analyses

The comparisons of the mean performance per SD scale over all 'old' façades for the various groupings of subjects holding a high school diploma on one hand and trade school students on the other are given in tables 5.10–5.12. The difference in perception of 'old' façades between study 1 and study 2 is highly significant (in all comparisons $p < 0{\cdot}001$), the subjects of study 1 rating 'old' façades higher on almost every SD scale. Similarly, the difference in perception of 'new' façades between study 1 and study 2 is also highly significant (in all comparisons $p < 0{\cdot}001$), the trade school students of study 2 rating the 'new' façades higher on almost all SD scales. Finally the difference in perception of all façades taken together between the two groups is highly significant (in all comparisons $p < 0{\cdot}001$).

Table 5.11. Discriminant analysis comparing all sixty German subjects (study 1) with forty trade school students (study 2) on 'old', 'new', and all façades.

V	Old façades			New façades			All façades		
	mean 1	mean 2	difference	mean 1	mean 2	difference	mean 1	mean 2	difference
1	4·19	3·42	0·77	2·92	2·96	−0·04	3·55	3·19	0·36
2	4·35	4·11	0·24	2·73	3·60	−0·87	3·54	3·86	−0·32
3	4·16	3·81	0·35	2·59	3·00	−0·41	3·36	3·40	−0·04
4	4·04	3·54	0·50	3·30	3·23	0·07	3·66	3·38	0·28
5	4·64	4·22	0·42	2·40	3·41	−1·01	3·52	3·81	−0·29
6	4·05	3·55	0·50	3·39	3·26	0·13	3·72	3·40	0·32
7	4·47	3·54	0·93	2·66	2·75	−0·09	3·57	3·16	0·41
8	4·66	3·88	0·78	2·61	2·66	−0·05	3·64	3·27	0·37
9	4·62	4·02	0·60	2·06	3·03	−0·97	3·34	3·53	−0·19
10	4·41	4·42	−0·01	2·00	2·40	−0·40	3·21	3·41	−0·20
11	4·24	4·20	0·04	2·51	3·66	−0·15	3·88	3·93	−0·05
12	4·41	4·06	0·35	2·82	3·17	−0·35	3·61	3·62	−0·01
13	4·59	3·94	0·65	4·27	2·86	1·41	4·42	3·39	1·03
14	4·43	3·68	0·75	3·00	2·89	0·11	3·73	3·29	0·44
15	4·69	4·04	0·65	2·92	3·13	−0·21	3·81	3·60	0·21
16	4·80	4·00	0·80	2·52	2·59	−0·07	3·66	3·28	0·38
17	4·22	3·64	0·58	2·76	3·41	−0·65	3·49	3·52	−0·03
18	4·82	4·15	0·67	2·13	3·19	−1·06	3·47	3·66	−0·19
19	3·44	4·00	−0·56	2·27	3·42	−1·15	2·85	3·71	−0·86
20	4·26	3·54	0·72	2·72	3·08	−0·36	3·49	3·30	0·19
21	4·46	3·94	0·52	2·75	2·78	−0·03	3·60	3·36	0·24
22	4·04	3·52	0·52	3·22	3·02	0·20	3·62	3·27	0·35
F	3·26			5·84			6·61		
d.f.	22,77			22,77			22,77		
p	<0·001			<0·001			<0·001		

Table 5.12. Discriminant analysis comparing all seventy-six subjects of study 1 with forty trade school students of study 2 on 'old', 'new', and all façades.

V	Old façades			New façades			All façades		
	mean 1	mean 2	difference	mean 1	mean 2	difference	mean 1	mean 2	difference
1	4·13	3·42	0·71	2·90	2·96	−0·06	3·51	2·97	0·54
2	4·11	4·11	0·00	2·64	3·60	−0·96	3·38	3·60	−0·22
3	4·11	3·80	0·31	2·72	3·00	−0·28	3·41	3·00	0·41
4	3·96	3·54	0·42	3·23	3·23	0·00	3·59	3·24	0·35
5	4·62	4·22	0·40	2·60	3·41	−0·81	3·61	3·42	0·19
6	3·96	3·55	0·41	3·31	3·26	0·05	3·64	3·26	0·38
7	4·23	3·54	0·69	2·67	2·75	−0·08	3·45	2·76	0·69
8	4·51	3·88	0·63	2·63	2·66	−0·03	3·57	2·68	0·89
9	4·60	4·02	0·58	2·28	3·03	−0·75	3·45	3·04	0·41
10	4·17	4·42	−0·25	2·09	2·40	−0·31	3·14	2·43	0·71
11	4·20	4·20	0·00	3·56	3·66	−0·10	3·88	3·65	0·23
12	4·27	4·06	0·21	2·82	3·17	−0·35	3·54	3·18	0·36
13	4·61	3·94	0·68	3·98	2·86	1·12	4·29	2·86	1·43
14	4·29	3·68	0·61	3·01	2·89	0·12	3·66	2·90	0·76
15	4·50	4·04	0·46	3·01	3·13	−0·12	3·76	3·14	0·62
16	4·62	4·00	0·62	2·55	2·59	−0·04	3·58	2·60	0·98
17	4·09	3·64	0·45	2·84	3·41	−0·57	3·46	3·42	0·04
18	4·49	4·15	0·34	2·25	3·19	−0·94	3·37	3·19	0·18
19	3·54	4·00	−0·46	2·35	3·42	−1·07	2·94	3·43	−0·49
20	4·25	3·54	0·71	2·76	3·08	−0·32	2·51	3·08	0·43
21	4·39	3·94	0·45	2·85	2·78	0·07	3·62	2·78	0·84
22	3·99	3·52	0·47	3·23	3·02	0·21	3·61	3·03	0·58
F	3·22			5·25			8·85		
d.f.	22,93			22,93			22,93		
p	<0·001			<0·001			<0·001		

5.4.5 Discussion

The results of this study clearly support the hypothesis that there is a difference in the perceptual reactions toward 'old' and 'new' façades of young subjects of different socioeconomic classes. This difference is in fact so pronounced that one could expect it to hold even if there was some 'noise' in the data. As was expected, the subjects of lower educational level (trade school students), although they were able to differentiate significantly between old and new façades, valued old façades lower and new façades higher than the subjects with high school degrees. They also used somewhat different semantic dimensions in making their distinction. This may indicate that they thought in more practical terms of the old façades as belonging to tenement houses which would be less functional from the point of view of comfort and installations than buildings constructed after 1945.

Could the difference between the two groups be due to age, rather than to educational level? It is true that the group of trade school students, though in their last year of schooling, was on the average younger (eighteen years old) than the group of army medical corps cadets of study 1 (not older than twenty-two). It could be argued that this difference in age accentuated the difference in reactions toward the façades. Although this may be true it is improbable that the trade school students after leaving their school would have the chance to acquire a different evaluation of 'old' and 'new' façades. The difference between the two groups in a few years time would probably become even greater as regards experience of objects of civilization, adding further support to the claim that it is not a 'natural' or genetic question of age, but a matter of cultural experience—of formal and informal education—which results in different ways of perceiving these objects. It is difficult to see how else the differences in perception of architectural objects, as they were presented in these studies, could have arisen, barring the sort of case where the mass media had run a campaign in favour of 'old' façades, to which the whole population had been exposed over a certain span of time. This was not the case in this instance. There is an unexplained phenomenon in the otherwise clear results of the two studies. This concerns the rôle of the two 'old' façades, 9 and 11, in the cluster of 'new' façades. Inspection of the pictures shows that they are located in the *Q* analyses of both studies near the new façades 3, 7, and 15. Façades 3 (1948) and 7 (1952) are the oldest of the 'new' buildings. By the presence of attics in one and different window sizes and other stylistic characteristics in both buildings some indication is given that might lead them to be associated with the old buildings. However, this is not a satisfactory explanation. Ultimately the positioning of these two façades in the wrong cluster may have to be attributed to random phenomena.

5.5 Methods of objective measurement of architectural objects

5.5.1 Visual objects and metalanguage

Up to this point only subjective reactions to architectural style have been considered. The SD has been introduced as a measuring instrument capable of tapping at least some (affective) aspects of the cognitive substance of the universe of the signified. The universe of the signifier must now be established to develop further the semiotic structure of architectural connotation. This may be done by 'objectively' analyzing visual architectural objects. A semiotic structure combining such an objective analysis with subjective reactions can only be postulated if the nature of objective visual stimuli can be established.

At this point, some reflection on the verbal processing of visual material might be appropriate, since a semiotic structure with a visual signifier is proposed. In linguistics the distinction between an object language and the metalanguage by which the former is scientifically processed is well established. There is no reason why the procedure applied to auditory material should not be applicable to visual material in order to establish meaningful nonlinguistic semiotic structures. The use of a verbal or symbolic metalanguage seems to be the only way to advance our knowledge of the semiological status and the degree of conventionality and organization of visual objects. Just as the behavioural psychologist, talking or writing about his observations, labels observed behavioural units by words, one can translate visual observations into a verbal metalanguage which in turn may be judged for its logical consistency in a meta-metalanguage (Greimas, 1966).

If buildings or cities, as hypothesized, are signifiers of semiotic structures these might be objectively tested by (1) taking a well-defined 'corpus' (sample) of buildings, (2) carrying out a repeatable analysis to extract the general features of their semiotic structures, (3) finding the appropriate metalinguistic (verbal or symbolic) terminology for these features, and (4) testing this terminology against other samples of buildings or cities. These analytical steps have been graphically expressed by Nattiez (1973) for codes in general. His graphic example has been applied to semiotic structures in architecture in figure 5.17.

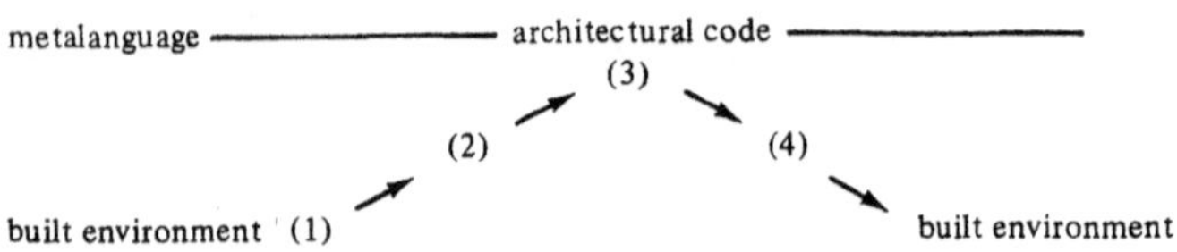

Figure 5.17. Sequence of steps in semiological analysis. This graphic representation has been adapted from Nattiez's (1973) general model for the special analysis of semiotic structures in architecture.

To carry out an investigation of the system of sets forming the universe of the signifier in architectural objects, the elements of these sets will have to be named (or expressed by symbols). One reasonable procedure could be to translate the visually distinguishable parts of an architectural object (say, a façade) into the natural language of architects; this would then constitute the metalanguage. Further analytic steps could then be applied to these language terms. Generally analytic operations are more easily carried out on the language terms than on the objects themselves. For instance, a hierarchical ordering of the components of an architectural object is definitely more difficult to accomplish on the component objects than on their symbolic substitutes. An interesting example of this substitution operation is Eco's (1971) "Componential analysis of the architectural element /column/" based on a pseudopoetic newspaper text discussing the 'beauty' of this architectural element. It is obvious that the analytical operations on the verbal terms must be logically compatible with the operations on their visual equivalents.

The proof of the logical consistency of a metalanguage and its pertinent operations must be given in terms of its (semiological) meta-metalanguage.

5.5.2 The type-token ratio applied to façades

In this book so far, a clear stand has been taken against linguistic analogy in architectural semiology. This position shall be maintained, even though the operations described in the following section could induce the reader to believe otherwise. What follows is the adaptation to architecture of a technique used previously in linguistics. By naming the architectural object parts in a natural language a metalanguage is formed that enables the linguistic technique to be applied. In this case the operations on the verbal metaterms are postulated to be logically compatible with potential operations on the visual architectural terms. The method to be discussed now has been called type–token ratio (TTR) in linguistics.

In psycholinguistic studies the 'flexibility' of linguistic sources has been measured by dividing the number of all different 'types' of words (nouns, verbs, adjectives) a source produced by the total number of 'incarnations' of these types, called 'tokens'. With equal-size samples of speech or text taken from two different sources, the TTR provides an index of variability of the language behaviour for comparison of these sources. This measure has been applied to the study of language since the late 1930s (cf Osgood, 1958, pages 719–721). Interestingly enough, it has found one of its major applications in the study of schizophrenic speech.

In a comparison of speech samples of schizophrenics and normal college freshmen, the TTR of the former was found to be significantly lower. This means that normal college freshmen are more capable of introducing variety into their speech production. Similar results were obtained in a comparison of the written language of schizophrenics and college freshmen.

Another comparison was made of schizophrenic speech, telephone conversation language, normal speech, and written material (Osgood, 1958).

This study showed that, in the case of schizophrenics, their 100 most frequently used words accounted for nearly 70% of total speech. It approached the low TTR of typically stereotyped telephone conversations. The highest TTR was found in written material. For college freshmen the 100 most frequent words accounted for 63% of total speech, indicating a considerably higher degree of flexibility than the percentage (70%) for schizophrenics.

To describe these results more tangibly, this means that in telephone conversations and in schizophrenic speech about thirty different words make up half of the entire speech production.

In further studies (Chotlos, 1944) it was established that the TTR increases with IQ and age, but is independent of sex and of such demographic data as the origin (city, town, or rural) of the subjects studied.

It is methodologically interesting to note that in these later studies, separate analyses of nouns, adjectives, verbs, and adverbs showed that the TTR for nouns (the number of different nouns divided by the total number of nouns) and for adjectives increases with IQ and age. On the other hand, low IQ and age are characterized by a high TTR in verbs. This indicates that the TTR of a subsample of nouns or adjectives only might be sufficient to generalize with respect to a total language sample—a considerable advantage in research economy if it is true.

In the present investigation of connotation in architecture a measure for the universe of the signifier has to be found. It is suggested that the TTR might be applicable to different compositions of particular architectural objects, for example, façades, as a measure of their intrinsic variability. Moreover, it is suggested that by analogy with the subsamples of nouns and adjectives in the Chotlos (1944) study, a subsample of architectural elements, perhaps windows and doors in the case of façades, should allow for the calculation of TTRs which clearly differentiate between the degrees of variability of different architectural compositions. It will be further suggested that significantly different TTRs might account for different architectural 'styles' in façade composition and might be hypothesized to correlate with different subjective reactions toward these façades in terms of a semiotic structure.

In order to measure façade TTRs the types of façade elements would have to be abstracted from real façades as classes listed verbally. Subsequently the different realizations of these types, namely the tokens, would be listed as frequencies. For example, a given façade might have four tokens of one type /window/, four tokens of another type /window/, one token of one type /door/, etc.

It should be remembered that one criticism of functionalist architecture which has been raised frequently during recent years has been the assertion that the 'expression' of functionalist architecture is 'rigid', not introducing enough visual variety into the urban environment, and has for this reason adverse psychological effects on citizens. Whether this lack of variability may be attributed to single architects or to a whole network of socioeconomic variables and whether the proposed analogy may be extended to suggest that being exposed to the monotony of a certain type of modern architecture might have a similar effect as being exposed to an impoverished, infantile or schizophrenic verbal environment is, for the moment, open to speculation only.

However, what is proposed here is to measure the difference in the TTRs of two different 'styles' of façade composition in a given sample of façades, the difference in style being caused by the fact that part of the sample was constructed before the turn of the century and part after the Second World War.

Theoretically, this application of TTR measurement to architectural façades could be generalized to treat entire streets and districts. Supposing that one could establish a certain finite number of façade types in a street or in a city (say, two types of old and two types of modern façade) one could measure the TTR of a street or district by assessing the occurrence of different types of façades within that street or district and dividing each type by its respective token occurrence. Conceivably the resulting TTR, as an index of the variability of that street or district, could be compared with TTRs of other streets or districts.

5.5.3 Information measurement of façades

Another possible way of obtaining objective indices of the universe of the signifier in architecture would be information measurement. Information measurement was first applied to architectural façades by Kiemle (1967). In analyzing modern buildings he utilized façade elements such as columns, windows, crossbars, panels, etc and calculated the amounts of information (in bits) from assumed subjective and objective (relative) probability distributions of these elements. He also calculated the amount of information on the basis of transitional probability distributions of the same façade elements, that is, on the probabilities of one type of façade element being followed by another, if the surface of the façade is scanned.

In advancing the concept of supersign formation by combination of façade elements into increasingly simpler grouping on the façade Kiemle showed that the amount of information decreases with supersign formation. Perception starts at the level of complexity of façade elements which can be digested by the human 'channel' (16 bits per second): this is the selective phase of perception. Subsequently, structural relations between the façade elements (supersigns) are perceived: this is the synthetic phase of perception. Finally, the supersigns are analyzed into their components: this is the analytic phase of perception.

By a similar method von Buttlar and Wetzig (1973) showed that façades of the late nineteenth century are richer in information than modern buildings. Modern façades do not contain enough detail for the analytic phase of perception and as a result become 'boring' after the synthetic phase of perception has come to a close.

A typical example of how von Buttlar and his collaborators Selig and Wetzig (von Buttlar et al, 1973) recorded a representative nineteenth century façade is given in table 5.13 (Wright, 1973). Their different stages of supersign formation are produced by a graphic manipulation of the façade very similar to that of Sieverts (1968) and his Berlin Seminar, who proposed some kind of a 'striptease' treatment of façades, defining 'primary signs' as all those structural elements of a building stripped of its finish (balcony, loggia, window, opening, gable, roof, etc) and 'secondary signs' as the ornaments applied as a finish, while the 'third layer of signs' consisted of verbal messages, for example window displays and their support (figure 5.18). A similar technique was used in the study of people's recognition of the social function of buildings (cf pages 97, 98). The work of von Buttlar and Wetzig has been critically reviewed by Wright (1973), that of Kiemle by Bortz (1972).

Bortz (1972) went about the objective assessment of architectural objects in different ways. At first he tried out a comparison of twelve architectural attributes rated by five independent subjects. The mean correlations between subjects' ratings of twenty-six façades ranged from 0·94 to 0·68 in the following ranked order of the architectural attributes

Table 5.13. Typical recording of façade elements according to von Buttlar and Wetzig (Wright, 1973).

Item	Occurrence
Round-headed windows	8
Tall rectangular windows	14
Square window	1
Cellar windows	2
T-shaped crossbars	22
Stepped gables	3
Wide triangular gables	2
Small triangular gables	2
Segmental gables	2
Sprockets in gables	20
Tympana above windows	12
Pilasters	33
Panels	10
Architraves	9
Entablatures over grouped windows	2
Moulded ledges	10
String courses	6
Brackets	27
Rusticated quoins	68
Door surrounds	6
Plinth	1

Number of different elements, $n = 21$
Total number of elements, $N = 260$

($r = 0{\cdot}49$ would be significant at the $0{\cdot}01$ level):

number of storeys,
number of flats,
number of balconies,
number of windows,
number of roof surfaces,
number of window forms,
estimated age of the building,
dominance of round features (measured on a six-point scale),
dominance of decoration (measured on a six-point scale),
number of silhouette corners,
dominance of regularity (measured on a six-point scale),
number of trees.

In a further attempt to reduce the data he had obtained from the façades, Bortz subjected the correlations between the façades in terms of their attributes to a factor analysis. Two resulting factors explained 75·7% of the variance. Bortz called his first factor 'architectural age'. On this factor old houses with many round features, decorations, different forms of windows, lack of regularity, high number of roof surfaces, high number of trees and high number of silhouette corners had high loadings, as did modern houses with the opposite attributes. The second factor had to do with 'architectural size'. On this factor houses with few windows, flats, storeys, and balconies (and houses with the opposite attributes) had high loadings. Bortz concludes that many architectural attributes seem to be ('objectively') covered by the factors of architectural age and size.

Figure 5.18. Five successive stages of supersign formation according to von Buttlar and Wetzig (1973).

He adds an interesting speculation: if modern façades would incorporate some of the stylistic attributes of the old façades, two architectural age factors should result. "The view is not convincing that modern façades should not use some of the stylistic means of the older ones. Certainly, architectural multiplicity, liveliness and variety in modern housing projects would be augmented."

But after his initial attempt with twelve architectural attributes Bortz proposed to increase the exactness of façade measurement by information measures. He projected slides of façades onto a standard grid of squares, covering each façade with approximately 250 grid units. Each unit was then inspected for the type of façade surface by which it was (mostly) covered. Bortz adopted six such types of façade surfaces:

wall surface,	window surface,	decorated surface,
roof surface,	balcony surface,	greenery.

By applying Shannon's formula (Shannon and Weaver, 1949) he calculated amounts of information for each façade on the basis of relative and conditional probabilities of occurrence. The façade measurements were then correlated with subjective SD ratings of the façades. One of the conclusions of the Bortz study was that information measurement seemed to enable good predictions to be made of subjective ratings of a façade's structural orderliness.

The information measurement for the façades studied in this book was carried out as follows. The façades were subdivided by grid units, then for each of the surface types a different typewriter sign was adopted. By this method it would be possible to 'write' down the information for each façade, grid unit by grid unit, in a code which could easily be transferred onto computer cards.

One of the practical difficulties of applying only one sign to a grid unit of the façade is to decide which sign should be chosen if one grid unit is covered by different types of façade units. In this study the rule was followed to apply that sign which covered over 50% of the grid unit. In case of doubt a coin was to be flipped and consequently one of the two possible surface types would be assigned randomly.

Once the entire façade is 'mapped' by the surface type signs, the calculation of the relative probability of a sign's occurrence is straightforward: the number of occurring signs of one surface type is summed up and divided by the total number of grid units covering the façade. The resulting fraction is substituted for p in the Shannon formula:

$$H = -p \log_2 p.$$

For practical calculation there are tables of values for $-p \log_2 p$ available (for example, Edwards, 1964, appendix).

To calculate the transitional probabilities, a list of all possible combinations of surface signs has to be made (including a sign's combination with itself). Then, by inspection of the 'mapping' of each façade, the number of occurrences of each of the possible pairs of adjacent surface types is counted. (Not all of these combinations are necessarily realized in a façade, as the rarity of the combination 'roof sign with door sign' indicates.) What makes the computation of transitional probabilities tedious is the fact that each grid unit has four transitions into neighbouring grid units: left, right, above, and below. All four transitional occurrences have to be counted; the resulting frequencies are then averaged (weighted mean) to obtain their mean frequency of occurrence. Each surface type sign is taken separately as a starting point of a transition (in this study there are nine such starting points), and the frequency of realized transitions to *other* signs of the repertoire is counted and then divided by the sum of *all* transitions realized from that particular starting-point sign. A starting-point sign's relative probability of transition to each other sign of the repertoire is thus obtained and can be substituted into Shannon's formula. All realized information values ($-\log_2 p$) of the transitions from one sign as a starting point to the other signs are then summed to give the transitional information of one particular sign. This operation is repeated, with each other sign being taken as a starting point. The entire calculation can best be understood by imagining a matrix in which the vertical column lists all starting-point signs (in our case nine) and the horizontal row lists the (nine) signs to which transitions may be realized. In the intersections of columns and any particular row the observed frequencies of transition of one particular sign to itself and to other signs of the repertoire are listed and then divided by the total number of realized transitions. Once the transitional information for each sign of the repertoire is computed, the average overall transitional information of a façade is computed by multiplying each sign's transitional information by its relative probability (a sign's weight) and summing the products. The sum of these products is the average amount of information in the façade, with the sequential dependencies of the signs in the façade repertoire taken into account.

The procedure is interesting also for its potential application to streets and districts. Suppose it was possible to identify four types of façades in one particular street. Each type would be identified by a sign: A, B, C, D. A particular observed façade constellation could be:

BACAABADACBADAABACAD (on one side of the street)
ABAACBABBADACBADAABA (on the other side of the street).

In this case it is like calculating the information of a string of forty signs from a four-letter alphabet. In this alphabet there are the following relative probabilities:

façade type A, 20/40 = 0·500,
façade type B, 10/40 = 0·250,
façade type C, 5/40 = 0·125,
façade type D, 5/40 = 0·125.

By the use of a table for $-\log_2 p$ values, these relative probabilities are easily converted into information values:

façade type A = 0·5000,
façade type B = 0·5000,
façade type C = 0·3749,
façade type D = 0·3749.

The average façade information H of the street is the sum of each particular façade type's information:

$H = 1{\cdot}750.$

To calculate the transitional probabilities one may consider, for the sake of simplicity, the whole street as one single string of letters. (One could also calculate the average transitional probabilities for façades on each street side, read from left to right, right to left, and across to each opposite façade.) In the simple case the matrix of transition frequencies is as in table 5.14. The average façade information of the street, with the transitions between façades taken into account, is $H_1 = 1{\cdot}235$ bits (table 5.15). Such an index may be interesting for comparing people's reactions to different streets or to a sample of streets from different districts, etc.

A further extension of this mapping technique to cover all façade types in entire town districts or towns might be considered. In this case the

Table 5.14. Matrix of transition frequencies between four elements occurring in a compositional string.

From	To				Row total
	A	B	C	D	
A	4	5	5	5	19
B	9	1	0	0	10
C	2	3	0	0	5
D	5	0	0	0	5

Table 5.15. Calculation of the transitional information measure H_1 of a compositional string of four elements.

Sign	Sum of transitional information (bits)	Relative probability	Product (bits)
A	1·994	0·500	0·997
B	0·469	0·250	0·117
C	0·971	0·125	0·121
D	0·000	0·125	0·000
Sum			1·235

different façade (or building) types may be mapped as different signs into a squared grid placed over district or town. The problem of transitional probabilities becomes then that of a surface of signs again, as in the case of façades. The results of such studies as the one on subjective ratings of eleven Berlin districts by Franke and Bortz (1972) might be better explained if each of the districts were indexed by an information value which could be correlated with the subjective ratings. The program SYMAP for mapping out planning data on geographic maps by Fisher (1964) from the Harvard Computing Center might be revised for this purpose.

5.5.4 Information aesthetics

The history and basic ideas of information aesthetics have been surveyed very clearly by Gunzenhäuser (1962). According to him, the information aesthetics of Max Bense (summarized in Bense, 1968a; 1969a) has three roots: information theory (Shannon and Weaver, 1949), semiotics (Peirce, 1960), analytic aesthetics (Birkhoff, 1950). In the context of this research particular attention should be paid to the analytic aesthetics of the American mathematician Birkhoff (1884-1944) which has become known through several publications since the beginning of the 1930s (Birkhoff, 1950). The work of Birkhoff in turn must be seen in connection with the earlier psychological theories of G T Fechner (1865; 1876) and H von Helmholtz (Gunzenhäuser, 1962, page 10).

Birkhoff speculates that perception and comprehension of an aesthetic object requires at first a 'sensory effort' on the part of the beholder. This effort—according to Birkhoff—is proportional to the complexity of the object: the more complex it is, the more effort must be needed to comprehend it. As a measure of complexity (*C*) Birkhoff designates the number of elements to be perceived (tones in music, phonemes in a poem, etc). Accordingly *C* varies with the number of artistically composed signs that make up an object.

How does one get, however, from an effort to that feeling of pleasure identical with aesthetic experience? This feeling obviously must stem from a process of communication going on between object and beholder. According to Birkhoff the intensity of this feeling is a measurable quantity—different aesthetic objects would be characterized by a different 'aesthetic measurement' (*M*) which varies directly with the feeling of pleasure in the beholder: the larger the 'aesthetic measurement' *M* of an object, the larger the 'feeling of pleasure' it should trigger.

The change from an effort to a feeling of pleasure can only come about through a third factor. This variable is, according to Birkhoff, the 'order' (*O*) of the signs of the object. The type of order which is perceived in terms of certain relationships by the beholder is of course different for different arts. The order that makes up a piece of music rests upon a relationship between tones and is different from the order of phonemes which makes up a linguistic object, and so on. The measure of order has thus to be defined empirically for the different arts. The relationship between the initial effort due to complexity, C, the aesthetic experience gained by the aesthetic measurement, M, and the compensatory function of the order of the sign elements, O, is summarized by Birkhoff in the formula:

$$M = \frac{O}{C}\,.$$

Since complexity is measurable as the number of sign elements, and order is a certain overall distribution of the sign elements, the product of the relation O/C—M— is also a numerical index which may therefore serve to characterize different objects and compare them for their aesthetic effect. The consequence of this formula is evident. At an equal level of complexity, M increases in value as order increases: the greater the order among the sign elements of an object the larger must be the 'feeling of pleasure'.

The transition from Birkhoff's formula to its interpretation by means of information theory is straightforward.

The information measurement of Shannon's formula,

$$H = -n\sum_i p_i \log_2 p_i,$$

results in index numbers called a 'bits', which increase with the number of elements in a message and with a decrease in the probability of their having an orderly distribution (that is, with their increasing 'disorder'). H has thus the same characteristics as Birkhoff's measurement of complexity C. The measurement $1-H$, which is called redundancy (R) in information theory, has the typical characteristics of Birkhoff's measurement of order O. One could thus define the measure of aesthetic information H_{ae}, by analogy with Birkhoff's formula, as

$$H_{ae,i} = \frac{R_i}{H_i}\,,$$

where

H_i is the mean information value of the individual information values of the different signs that make up the object message, and

R_i is the corresponding value of redundancy for a given object message.

In information aesthetics theory this formula gets somewhat complicated (Gunzenhäuser, 1962; Bense, 1968a; 1969a) if one takes into account the difference between the information H_i actually transmitted (that is, resulting from the composition of signs of a repertory in a concrete object message) and the information H_{max} which could potentially be transmitted (if the signs of the repertory were optimally composed). H_{max}, the highest information value could be reached, if the different signs were composed in such a way that all of them would occur as seldom as possible yet at the same time with equal frequency. In such a case the probability of their occurrence during the 'scanning' of the object message would be equal and there would be no redundancy. The relationship between H_i and H_{max}, H_i/H_{max}, is called 'relative information'.

Instead of the simple measurement of redundancy $(1-R)$ a redundancy term weighted by the difference between H_{max} and H_i is thus introduced:

$$R_w = \frac{H_{max} - H_i}{H_{max}} = 1 - \frac{H_i}{H_{max}} .$$

By substituting R_w into the original formula $H_{ae,\,i} = R_i/H_i$ the following formula is obtained:

$$H_{ae} = \left(1 - \frac{H_i}{H_{max}}\right) \Big/ H_i .$$

This formula can be simplified:

$$H_{ae} = \frac{H_{max} - H_i}{H_i H_{max}} = \frac{1}{H_i} - \frac{1}{H_{max}} .$$

If this formula is applied to any of the diverse messages transmitted by the environment, their aesthetic information measures should be calculable if their sign repertory is known. The problem would simply be to assess the statistical distribution of the signs in the object message and subsequently to calculate, by means of the Bense formula of information aesthetics, the 'objective' aesthetic numerical index of that object. A comparison between aesthetic indices of different objects (stemming from the same sign repertory) should permit us to predict which of the objects would trigger more and which less 'feelings of pleasure'.

5.6 Study 3: componential analysis of a façade

5.6.1 Introduction

Problem

The purpose of this preliminary study was to determine the feasibility of using common architectural terminology as a metalanguage for analyzing nonlinguistic building features. The success or failure of this inquiry would determine whether any 'objective' measurement could be applied to façades, since any such measurement must be based on the frequency of occurrence of visual façade components in linguistically determined categories.

Background of the problem: componential analysis

In the first two studies of this part of the book, differences in subjective impressions of pre-1900 and post-1945 façades were examined. The structure of the cognitive universe of the signified—reflected in the ratings of adjective pairs that gave the impressions of the subjects—was necessarily revealed. Now our knowledge of connotative semiotic structure in architecture must be advanced by investigating the structure of the universe of the signifier. That structure should be given by the presence or absence, and interrelations, of architectural components in façades. This step amounts then to a 'componential analysis' of façades, which will have to result in the demonstration of a correlation between the componential structure of the universe of the signifier and the oppositional adjectival structure of the universe of the signified. Is such a componential analysis of façades possible? To demonstrate this we would need to establish the 'vocabulary' and 'syntax' of façade design in verbal terms.

Eco's (1971) "Componential analysis of the architectural element /column/", carried out by extracting the necessary architectural terms and indications of their syntax from a newspaper article, was mentioned earlier (page 245). For façades von Buttlar and his collaborators have produced frequency counts of façade components (Wright, 1973; cf page 248 and table 5.13) without, however, demonstrating the objectivity of their 'vocabulary', or its 'syntax'.

The syntax structure of seventeen house and barn types occurring in North American folk architecture has been described by Glassie (1973). He proceeds along the lines of Chomsky's generative grammar by postulating successive levels of design organization to which designing rules are applied to derive particular types of buildings. The zero level of organization, comparable to that of kernel sentences, consists of 'base concepts'. The base concepts of the Anglo-American builder are, for instance, variations in the positioning of two axially ordered squares in the plan of the building. These two squares may be placed side by side or separated by an entrance space, etc. On the first level of organization the base concepts are conceived three-dimensionally (for example, two cubes placed side by side). Roofing is applied either perpendicular to or parallel with the entrance. The rules of 'piercing' and 'massing' are applied; in 'piercing', all the house parts are treated as alike, except for the central entrance part, which if it occurs may be treated differently but must remain symmetrical. In 'massing', the chimneys, for example, may be placed in the centre or at both ends of the roof. The second level of organization is optional, and results from the application of rules for vertically or horizontally doubling the parts of the first level (storeys or lateral extension), and subsequent rules for piercing and massing. On a third level, further diversification may be obtained by subtraction or addition of whole components of the base concept followed in turn by rules of optional extension, roofing, piercing and massing, etc to provide final subtypes. This generative approach of Glassie's is interesting, especially for the builder's way of conceptualizing buildings (performance), but seems to be less adequate for the analysis of the building user's point of view (competence), which must be based more on perceptual than on constructional rules.

5.6.2 Method

In this study the members of the 'objective' façade vocabulary were obtained by taking the verbal labels printed on a façade drawing of a typical nineteenth century house. This drawing was found in an old edition of the French illustrated dictionary *Petit Larousse Illustré* under the heading 'maison' (house). It is given in figure 5.19(a).

If one erases the drawing of this figure conserving only the verbal labels a 'cloud' of fifty-eight façade terms in French is left. The result of this procedure is shown in figure 5.19(b). One can now start to manipulate these verbal labels which stand for the erased visual parts of the façade. The labels could be arranged for instance in alphabetical order to become a dictionary of French façade terms. This dictionary could be translated into English. Also, a rough syntactical hierarchy could be established among the terms by observing that some of them are printed in upper-case letters to indicate their major importance, while others are not. In the drawing at hand (figures 5.19a and 5.19b) the terms CAVE (cellar), REZ-DE-CHAUSSEE (ground floor), COMBLE (attic), TOITURE (roofing), etc are printed in upper-case letters, whereas such terms as ancre (brace), chaperon (coping), croisée (casement windows), are not. In addition, the façade vocabulary could serve as categories by which the frequency of occurrence of façade components observed in a sample of real façades could be grouped.

It seems that different operations can be carried out on objects or parts of objects with such a collection of verbal equivalents for these objects or parts which they describe.

The following componential analysis of façade terms standing for façade components was carried out in the order of the levels by which a building must be erected (from the cellar to the roof). De Saussure (1976) had used exactly this vertical constraint in building when he tried to illustrate the difference between the linear structure of syntagms and the structure of associative relationship: "a linguistic unit is comparable with a given part of a building, for example a column. For one thing the physical relationship of the column to the architrave which it carries calls to mind the syntagmatic relationship of words. For another, the associative relationship of, say, the Doric style with Corinthian or Ionic styles recalls associations of words."

With modern construction techniques there may be less clearly constrained sequences. Sometimes a skeleton of pillars is erected first, then the roof added and curtain walls filled in later. Yet there are still constraints, albeit different ones, and all techniques, traditional or modern, must start from the ground. For the present analysis of the façade, the traditional sequence may be followed without any problem.

(a)
(b)

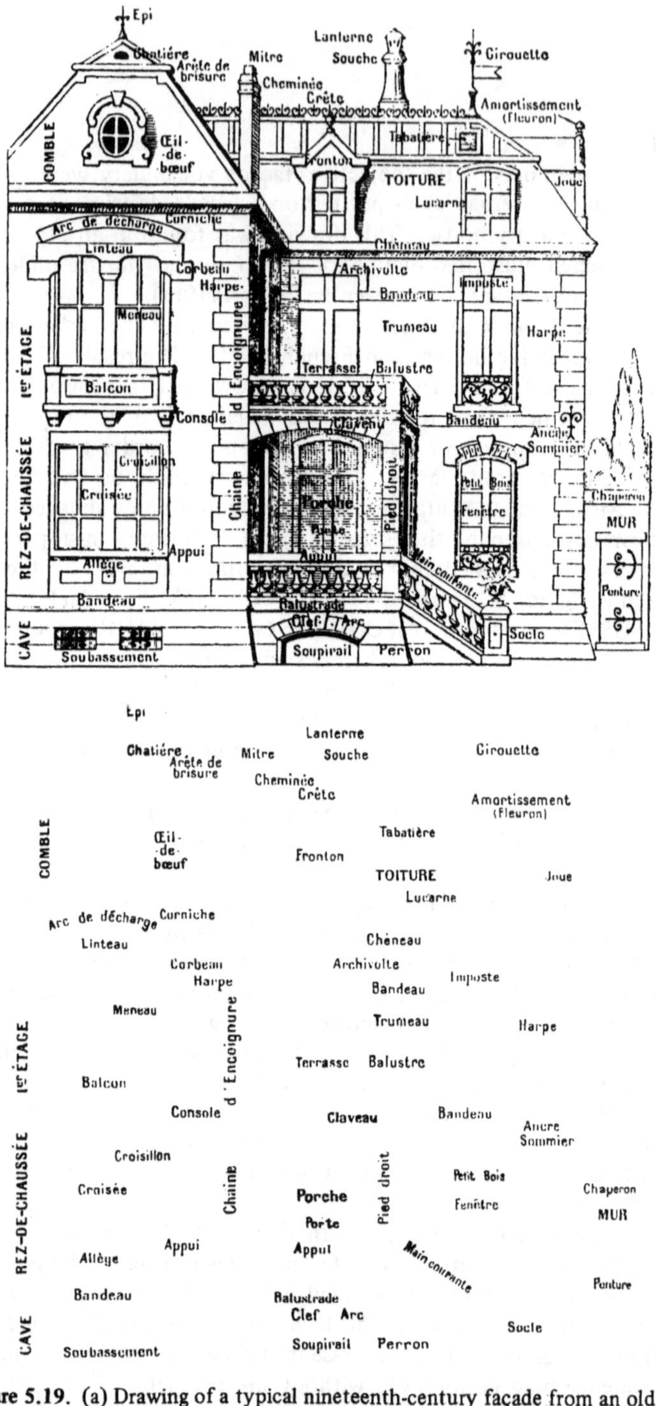

Figure 5.19. (a) Drawing of a typical nineteenth-century façade from an old edition of the French illustrated encylcopaedia *Petit Larousse Illustré*. This illustration is found under the heading 'maison' (house). (b) The fifty-eight French façade terms remaining after the drawing of the façade has been erased.

5.6.3 Results of the componential analysis

The componential analysis operated on the façade vocabulary resulted in three hierarchical trees of components. These trees are given in figures 5.20–5.22. The first two trees, concerning the lateral and forward extensions of the building's volume, must be considered on one level, the third concerning the roof stands syntagmatically on a second level. The first tree subdivides into a syntagm of enclosure (walls and mouldings) and of piercing (windows). The second tree introduces the vertical extension of a building above the level of the cellar, not unlike Glassie's (1973) level of vertical and horizontal 'doubling' of the 'base concept'.

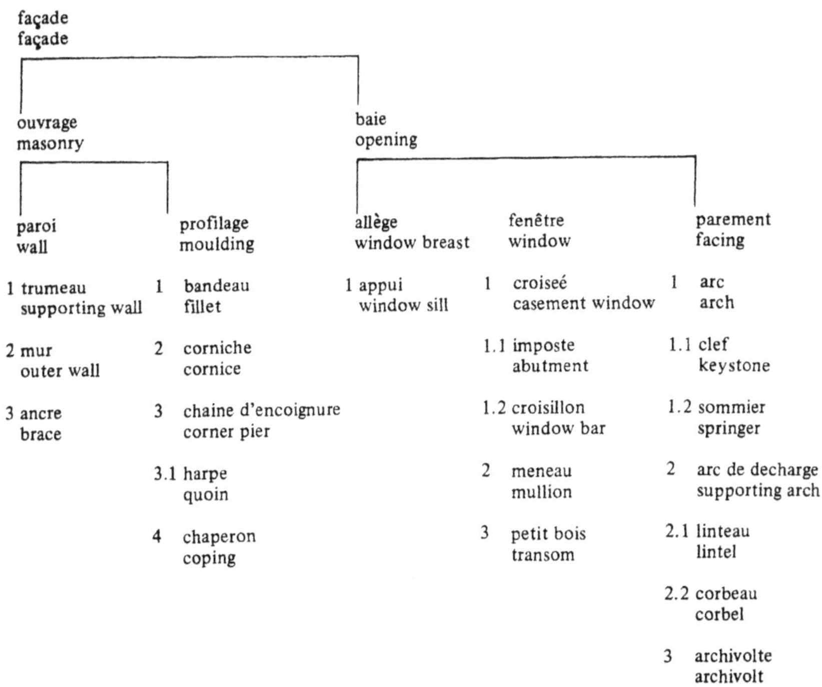

Figure 5.20. The componential analysis of a façade: walls, and piercings, forming a 'syntagm of enclosure'.

- échellonement / relief
  - podium / podium
    - cave / cellar
      - 1 soubassement / foundation wall
      - 2 soupirail / vent
  - avant corps / front structure
    - rez-de-chaussée / ground floor
      - 1 porche / porch
        - 1.1 perron / front steps
          - 1.1.1 socle / pedestal
          - 1.1.2 main courante / hand rail
        - 1.2 balustrade / balustrade
          - 1.2.1 balustre / baluster
          - 1.2.2 appui / window sill
        - 1.3 porte / (front) door
          - 1.3.1 penture / hinge
    - premier étage / first floor
      - 1 terrasse / terrace
        - 1.1 pied-droit / post
        - 1.2 claveau / voussoir
        - 1.3 balustrade / balustrade
          - 1.3.1 balustre / baluster
          - 1.3.2 appui / window sill
      - 1 balcon / balcony
        - 1.1 console / console
        - 1.2 balustrade / balustrade
          - 1.2.1 balustre / baluster
          - 1.2.2 appui / window sill

Figure 5.21. The componential analysis of a façade: vertical and horizontal extension of the 'base concept' (Glassie, 1973).

- comble / attic
  - toit / roof
    - 1 toiture / roofing
    - 2 cheneau / eaves
    - 3 arrête de brisure / arris
  - baies/toit / openings
    - 1 lucarne / dormer
      - 1.1 fronton / fronton
      - 1.2 joue / cheek
    - 2 oeil-de-boeuf / bull's eye
    - 3 autres: / others:
      - 3.1 tabatière / skylight
      - 3.2 chatière / ventilation hole
  - cheminée / chimney
    - 1 souche / chimney stack
    - 2 lanterne / lantern
    - 3 mitre / chimney top
  - decorations / decorations
    - 1 crête / crest
    - 2 amortissement (fleuron) / pinnacle
    - 3 epi / spike
    - 4 girouette / weather vane

Figure 5.22. The componential analysis of a façade: the roofing.

5.6.4 Discussion

These results (based on the labels of a drawing) have a certain similarity to Glassie's results based on a generative structure—at least as far as the 'deeper' (hierarchically higher) levels of the trees are concerned.

It could be concluded most generally that an architectural object can be viewed in terms of a syntagm of enclosure (the wall or screen of elements preventing physical or visual penetration), another of piercings (doors, windows, etc allowing physical or visual penetration), a third of roofing, and a fourth of interconnections between these syntagms. But, at the same time, this 'deep' structure is covered by a more 'superficial' structure which probably allows for the differentiation of styles. One need only compare the fifty-eight words of the façade vocabulary of the nineteenth-century house used in this study with the vocabulary needed for an average contemporary house to see that the latter has a reduced vocabulary by comparison and therefore a relatively impoverished 'surface' structure.

A special problem is encountered with the use of architectural vocabularies. This has to do with the influence of language on perception. It becomes clear if one tries to translate a façade vocabulary from one 'natural' language into another (although both 'natural' languages will serve in this case as metalanguages). The lower the words rank on the hierarchical trees (the more 'superficial' their architectural equivalents), the more difficult they are to translate. This increase in difficulty may result from either of two facts:

the lower-ranking 'technical' terms may be less familiar (even in one's own language);

there may be no verbal equivalents in the two languages.

Given this alternative one could be tempted to adapt the 'Whorfian hypothesis' to state that: observers can perceive façades only to that level of detail for which they possess verbal labels. Whorf (1953) has presented similar arguments regarding Hopi architecture. On the other hand one could advance a familiarization-by-work hypothesis similar to that of Rubinstein (1972): the fact of working with façades either practically (as a mason or an architect) or theoretically (as a translator or semiologist) increases one's vocabulary of façade terms (in different languages) very rapidly and, at the same time, sharpens perception of façade detail. This would be like the manner in which a widened taste vocabulary acquired from a wine taster sharpens one's attention to the distinctive features of different wines.

5.7 Study 4: type-token ratios (TTRs) and information measurements of two styles of façades

5.7.1 Introduction

Problem

The purpose here was to further clarify the visual design properties of façades. This was done to reveal the oppositional structure of the universe of the signifier that is, by hypothesis, present in the eighteen tenement house façades examined previously. In this study two types of objective measurements, TTRs and information measures (bits), will be compared.

Background of the problem: TTRs in façade measurement—a pilot study
While making the componential analysis of a façade, it was noticed that quite apart from whether the fifty-eight terms in the French façade vocabulary might be considered more or less rich, most of them would simply find no application with modern buildings. With the decrease in the architectural detail to be labelled on a 'modern' façade, the corresponding vocabulary shrinks to very few useful terms. It is at this point that analogies with the study of language behaviour come to mind. It will be remembered that low TTRs were indicative of telephone conversation and schizophrenic speech (pages 245, 246).

In a pilot study (Krampen, 1974a) it was hypothesized that there would be a significant difference in the TTRs of façades of buildings constructed before 1900 and after 1945. The TTRs of the façades of the latter period would be significantly lower than those of the former, indicating their distinctly less varied composition. Documentary photographs of buildings (from which the eighteen façades of studies 1 and 2 were drawn) were used to study the TTRs of façade openings (windows and doors). The photographs are given in figure 5.23. The buildings were all situated in Karlsruhe (FRG) in a zone designated for urban renewal. Of total sample of seventy-five documented façades, four were eliminated because they were built between 1900 and 1945 and two because they consisted only of a makeshift ground-floor shopping level, one even without indication of the year of construction. The remaining total of sixty-nine façades was examined for the TTR of each individual building's front windows and doors.

The façades lay in two particular streets. In street 1, out of the thirty-three buildings used in the analysis, fourteen were pure apartment houses and nineteen contained (small) shops at the ground floor level. This subsample had only three buildings erected after 1945, while the rest were built before 1900. Street 2 was a shopping street in which the ground floor level of each house was occupied by a shop or some other business. Seven of the thirty-six buildings chosen in this street, were erected after 1945, the rest before 1900.

All the openings in a façade (for example, windows and doors) were defined to be elements whose TTR was representative of the overall TTR of that façade. This assumption was based on an analogy with linguistic studies. As will be remembered, Chotlos (1944) found that the TTR of nouns and adjectives exhibited the same characteristics (variation with IQ and age) as the entire verbal material. By analogy it may be claimed that windows and doors, the main visual elements of a façade could define the overall degree of flexibility or variability of a façade as well as could the complete number of all elements.

One could object that balconies are important visual elements of the façade, which should have gone into the calculation of the TTRs. In this study, however, balconies were not added to the elements making up the TTR of a façade, for the following reason: balconies protrude from the basically flat façades and thus constitute a three-dimensional addition to the otherwise two-dimensional surface of a façade. This confers on them a special visual attractiveness which renders a comparison between buildings with and without balconies practically impossible. Part of the special visual role balconies play in a façade has, however, been captured in this study since balconies, even if they are not counted as such, add to the TTR of the façade because of doors and windows usually associated with them.

With the elements of the TTR defined, some indication of what would constitute a difference between any pair of elements (windows or doors) had to be given. This difference results from the superimposing of two elements of the same façade (a sort of visual equivalent of the 'commutation test' in linguistics). If two superimposed elements, say windows, exactly coincide in size, decoration of the outer frame, and other details, they are defined as equal; if not, they are said to be different. With this operational definition, the procedure of assessing the TTR of a single façade becomes quite simple.

The analysis was carried out separately for the ground-floor level and for the rest of the façade. For reasons of convenience the roof level was added to the rest of the façade, although it could have been analyzed separately.

At first all different types of openings were counted for one level of the façade. The difference between a door and a window lies, of course, in their functions and this was counted in addition to the differences of size and detail already explained.

The different types of openings in the façade level are summed and this sum is divided by the overall number of openings at that façade level. Finally, the TTRs for the different levels of the façade are added together to give the TTR of the overall façade.

The distribution of TTRs for façades within one 'style' is not known. It is not likely to be normal, since within one style period there may exist a 'rich' and a more economical version of that style. By the same token, the 'rich' version of one style may exhibit the same TTR as the economical version of another. Therefore, ordinal measurement and the corresponding nonparametric statistical tests (Siegel, 1956) seem to be indicated.

In order to test for the difference between the TTRs of two 'styles', all TTRs are ranked regardless of their 'style' in one common rank order. The hypothesis was that the bulk of the TTRs of one of the two 'styles' which were being compared would tend to be at the top of the rank order. Whether this is the case can be decided by the median test, or by the Mann–Whitney *U* test, the latter being more 'powerful' and using more information from the data. In the case of this study it was generally assumed that the bulk of TTRs from buildings constructed before 1900

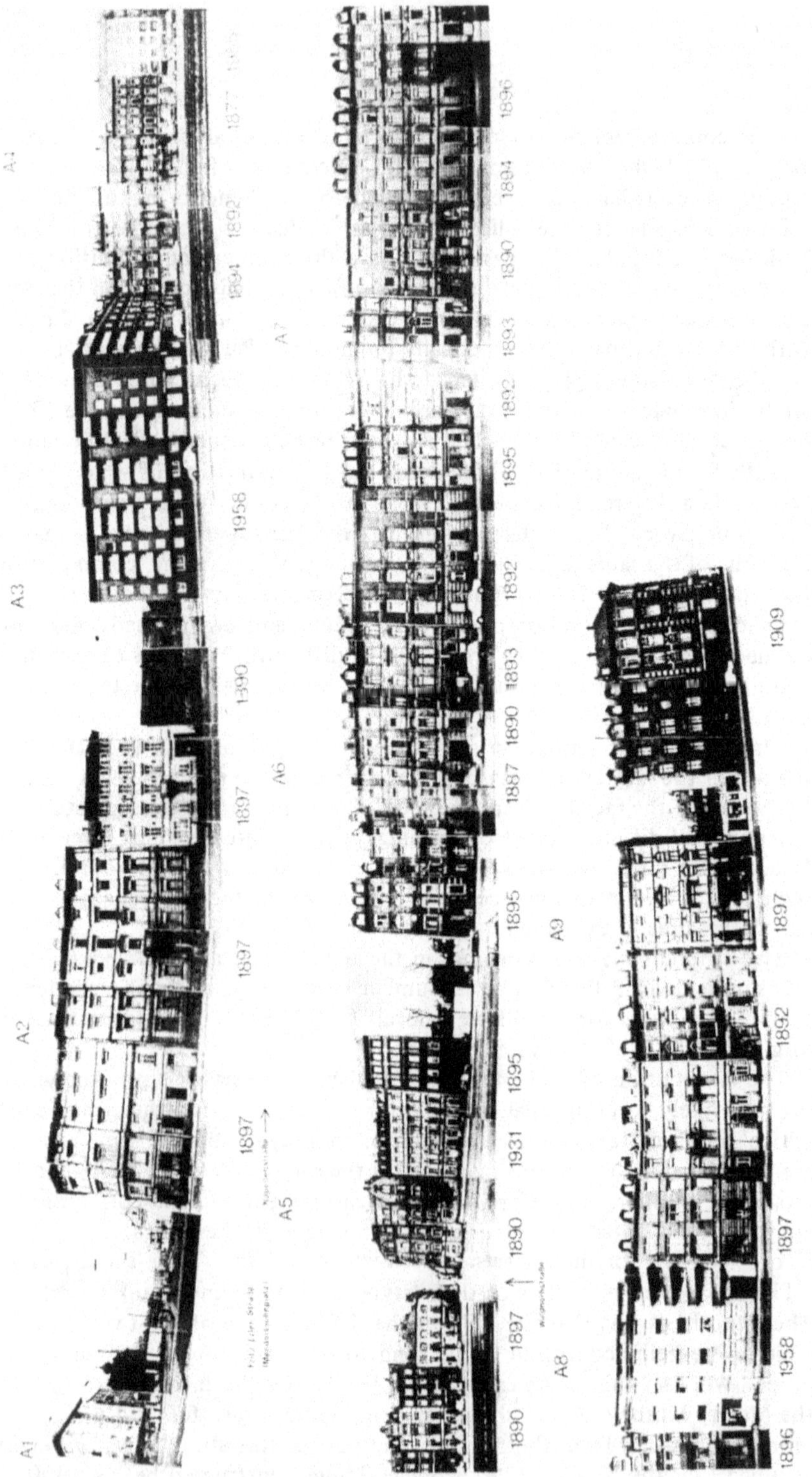
A1
1897
A2
1897
1897
1890
A3
1958
1894
1892
A4
1877
A5
1890
1897
1890
1931
1895
1895
A6
1887
1890
1893
1892
1895
1892
A7
1893
1890
1894
1896
A8
1896
1958
1897
1892
A9
1897
1909

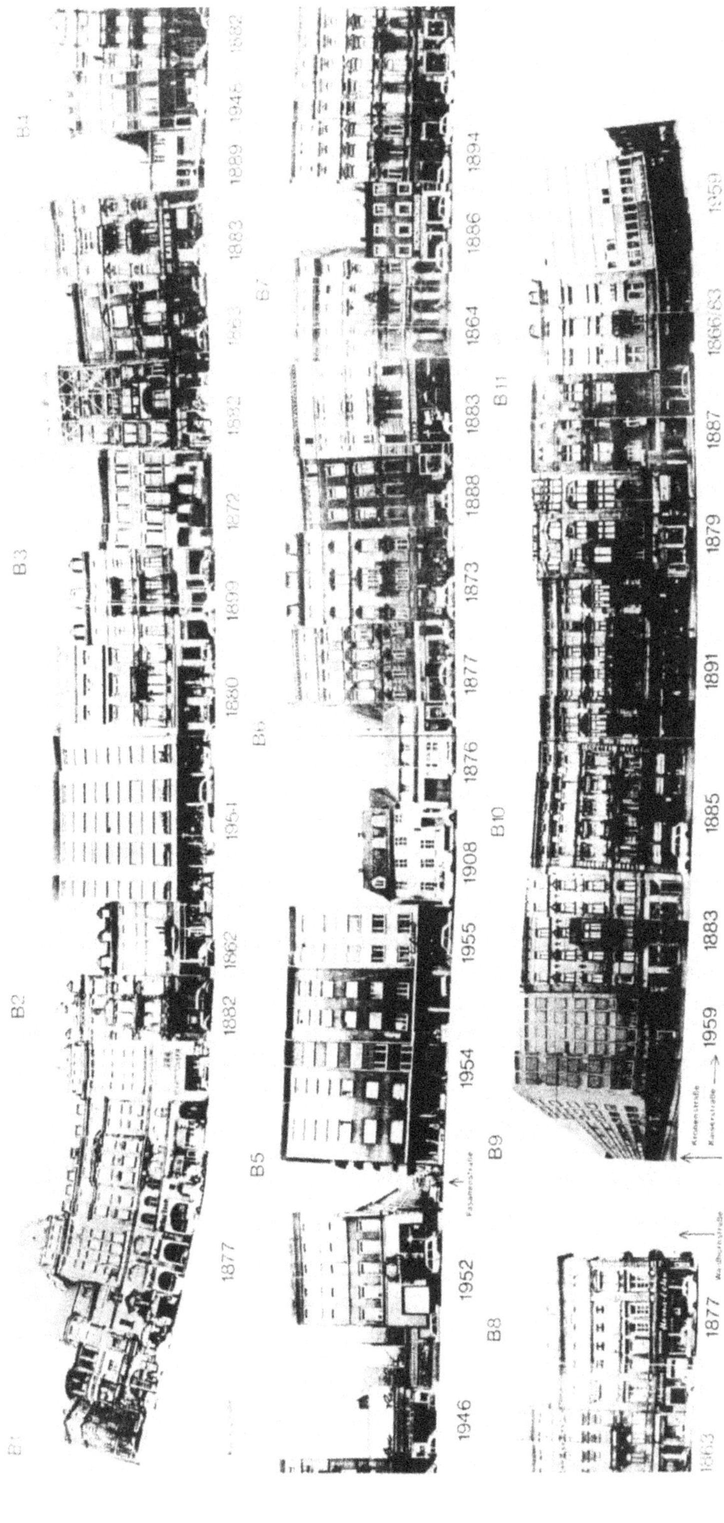

Figure 5.23. Documentary photographs of buildings situated in Kaiserstraße and Fritz-Erler-Straße, Karlsruhe. A total of sixty-nine of these buildings were used in a pilot study for the calculation of the façades' type-token ratios. These buildings were situated in a marginal zone of the city which was scheduled for urban renewal. The eighteen photographs used in studies 1, 2, 4, 5, and 6 were selected from this sample.

would tend to gravitate toward the top when the buildings are put in rank order with TTRs from buildings constructed after 1945.

The results generally confirmed the hypothesis. Although no overall difference in the TTRs was found for the ground-floor levels of old and modern buildings, a highly significant difference was found for the higher parts of the façades. There was also no difference found between the TTRs of the fronts of old and modern ground-floor shops.

These results are presented together in table 5.16. The results show that there is a highly significant difference between the TTRs of old and modern buildings, although the façades at the ground-floor levels are, overall, not significantly different. In fact, when the ground-floor levels of apartment houses and shops were compared, a highly significant difference in the TTRs appears, but no such difference was found between old and modern shop façades. Shops with their particular necessity to display merchandise and to attract customers have high TTRs, regardless of whether they were built before 1900 or after 1945.

These results have established that the old-façade composition in the sample was more varied than its modern counterpart.

For all the apparent 'concreteness' of the TTR there remains still a high degree of abstraction in this measurement of the variety of façades. The effect of the overall state of the building materials, the changing of light and shadows during the day, the different curtains of the individual inhabitant, colourful awnings against the direct rays of the sunlight, flower pots on the window sills, posted flags, faces appearing in windows and, at night, the randomness of lit versus dark windows—all these dynamic variables play an important role in the objective variability of façades (even if this applies only to certain seasons, days, or hours of the day). The study of the recognition of social functions of buildings, for example, showed that subjects tended to use—apart from balconies—the fact that there are individual curtains at the windows of a façade to identify a modern-style building as an apartment house rather than an office building which characteristically has standard curtains. The TTR is then a measure 'ceteris paribus'—if one succeeds in taking the other variables into account or controlling them in the experimental design.

Table 5.16. TTRs of 'old' and 'new' buildings compared by means of a Mann-Whitney *U* test.

Rank order of TTRs: 'old' versus 'new' buildings	Mann-Whitney *U* test
Entire façades	$p = 0{\cdot}0013$
Ground-floor levels only	n.s.
Façades without ground-floor level	$p < 0{\cdot}00003$
Tenements with shops at the ground-floor level	$p < 0{\cdot}0005$
Ground-floor levels with shops	n.s.
Façades without shops at the ground-floor level	$p < 0{\cdot}00003$

5.7.2 Method

Selection of the building photographs studied

A comparison of the results obtained by TTR and information measurement was made by applying both methods to the eighteen façades of studies 1 and 2. The nine old façades had originally been chosen by their high TTR (for windows and doors) revealed in the pilot study. The nine new façades were chosen from the ten postwar buildings in the streets examined.

Computing TTRs

The method of computing TTRs by using façade openings only (TTRo) has already been described in the pilot study (pages 262–266). For each of the buildings three TTRs were obtained: one for the entire façade, another for the façade without the ground floor, and a third for the façade without the ground-floor level or roof.

But TTRs were also computed for each of the eighteen façades with all visible façade elements considered, including window size, balcony, window mouldings (including window sill, post, and lintel as one unit), panels, ledges, decorative iron braces, etc (TTRe). Again three counts were made for each building, the first for the whole façade, the second for the same without the ground floor, and the last without the ground floor or roof. Thus each façade was indexed by three TTRs for openings only and a further three for all façade elements.

Computing information measurements

Following the method of Bortz, reported above (pages 250, 251), slides of all eighteen selected façades were projected on a standard squared grid of 13 x 17 = 221 units. An attempt was made to use as many units of the grid as possible for each façade. (Obviously, this was difficult for very tall and narrow façades.) In this study nine kinds of façade surface were used:

sky surface (in the case of gabled roofs),
roof surface,
wall surface,
window surface,
balcony surface,
decoration surface,
display-window surface,
door surface,
advertising surface.

The latter three kinds served almost exclusively to quantify the ground-floor level.

The nine kinds of surface were coded by different typewriter signs. In this way it was possible to 'write down' each façade and transfer the code onto punch cards. Different types of codes (letters, numbers, symbols) were tried out. A symbol code, containing the typewriter symbols /–/ for sky, /U/ for roof, /+/ for wall, / / for window, /I/ for balcony, /&/ for decoration, /./ for display window, /:/ for door, /*/ for advertising, gave a more realistic impression of the façades than coding by letters or numerals. Figure 5.24 shows different versions of coding for all façades, figure 5.25 gives an example of additional numerical coding for two façades.

Façade 1

Façade 2

Figure 5.24. The eighteen façades rated by the subjects in studies 1 and 2, coded 'objectively' into nine kinds of façade surface: sky, roof, wall, balcony, decoration, display window, door, advertising. Two different types of code are shown: the more 'iconic' typewriter symbols, and uppercase letters.

Façade 3

```
– U U – U – U U –
U     U   U   U U
+ + + + + + + + +
+   +         + +
+   +         + +
+ + + + + + + + +
& & + & & & + & &
&   +         + &
&   +         + &
& & & & & & & & &
&   +         + &
&   +         + &
&   +         + &
+ + + + + + + + +
+ + + + + + + + +
+ : + · · · + : :
+ : + · · · + : :
+ : + · · · + : :
```

```
– B B – B – B B –
B C C B C B C B B
A A A A A A A A A
A C A C C C A C A
A C A C C C A C A
A A A A A A A A A
E E A E E E A E E
E C A C C C A C E
E C A C C C A C E
E E E E E E E E E
E C A C C C A C E
E C A C C C A C E
E C A C C C A C E
A A A A A A A A A
A A A A A A A A A
A G A H H H A G G
A G A H H H A G G
A G A H H H A G G
```

Façade 4

```
– – – – & & – – – –
& & & + & + & & U U
+ & + &   & & & & +
    &       &     +
& & & & & & & + & +
    +   &   & +   +
    & I I I & +   +
& & I I I I I & & +
  & &   &   & &   +
  + +   +   + +   +
& + I I I I & & & &
  + & I I I + & & +
  + +   +   +     +
  + +   +   +     +
+ x x x x x x & & +
&     & &     & : &
      : :     : : :
      : :     : : :
+ + + : : + + : : :
```

```
– – – – E E – – – –
E E E A E A E E B B
A E A E C E E E E A
C C E C C C E C C A
E E E E E E E A E A
C C A C E C E A C A
C C E D D D E A C A
E E D D D D D E E A
C E E C E C E E C A
C A A C A C A A C A
E A D D D D E E E E
C A E D D D A E E A
C A A C A C A C C A
C A A C A C A C C A
A J J J J J J E E A
E C C E E C C E G A
C C C G G C C G G C
C C C G G C C G G G
A A A G G A A G G G
```

Figure 5.24. (continued).

Façade 5

```
U U U U U U U U U U U U U U U
+ +   + +     & & + +   +   +
+ + + + + + +     & & & + & +
+ +   + +     &   +     +   +
+ + + + + + + & & + + + + + +
+ +   + +   +     +     +   +
+ + + + + + + & & + + + + + +
+ +   + +   +     +     +   +
+ + + + + + + & & x + + + + +
+ +   + +   +     x     +   +
. . . . + + + & & & & & & & &
. . . . . : & : & . . . . . +
. . . . . : & : : . . . . . +
```

```
B B B B B B B B B B B B B B B
A A C A A C C E E A A C A C A
A A A A A A A C C E E E A E A
A A C A A C C E C A C C A C A
A A A A A A A E E A A A A A A
A A C A A C A C C A C C A C A
A A A A A A A E E A A A A A A
A A C A A C A C C A C C A C A
A A A A A A A E E J A A A A A
A A C A A C A C C J C C A C A
H H H H A A A E E E E E E E E
H H H H H G E G E H H H H H A
H H H H H G E G G H H H H H A
```

Façade 6

```
U U U U U U U U U U U U U U
+   +     +   +     +   + +
+   +     +   +   + + + + +
+ + + + + +   +   + +   + +
    +     +   +     +   + +
+ + + + + + + + + + + + + +
    +         +     +   + +
+ + + + + + + + + + + + + +
    +         +     +   + +
& + + & + + & + & + + & + +
. . . : : . . . :   . . . .
. . . : : . . . : : . . . .
+ + + : : + + + : : + + + +
```

```
B B B B B B B B B B B B B B
A C A C C A C A C C A C A A
A C A C C A C A C A A A A A
A A A A A A C A C A A C A A
C C A C C A C A C C A C A A
A A A A A A A A A A A A A A
C C A C C C C A C C A C A A
A A A A A A A A A A A A A A
C C A C C C C A C C A C A A
E A A E A A E A E A A E A A
H H H G G H H H G G H H H H
H H H G G H H H G G H H H H
A A A G G A A A G G A A A A
```

Figure 5.24. (continued).

Façade 7

```
U U U U U U U U U U U U U    B B B B B B B B B B B B B
& & & & & & & & & & & & &    E E E E E E E E E E E E E
+   +     +   +     +   +    A C A C C A C A C C A C A
+   +     +   +     +   +    A C A C C A C A C C A C A
+ + + + + + + + + + + + +    A A A A A A A A A A A A A
+   +     +   +     +   +    A C A C C A C A C C A C A
+   &     &   &     &   +    A C E C C E C E C C E C A
+ + + + + + + + + + + + +    A A A A A A A A A A A A A
+ & + & & + & + & & + & +    A E A E E A E A E E A E A
+   +     +   +     +   +    A C A C C A C A C C A C A
+   +     +   +     +   +    A C A C C A C A C C A C A
+ + + + + & x & + + + + +    A A A A A E J E A A A A A
+ + + + + & x & + + + + +    A A A A A E J E A A A A A
+ · · · + : : + · · · · +    A H H H A G G A H H H H A
+ · · · + : : + · · · · +    A H H H A G G A H H H H A
+ · · · + : : + · · · · +    A H H H A G G A H H H H A
& & & & & : : & & & & & &    E E E E E G G E E E E E E
```

Façade 8

```
U   U U   U U U   U    B C B B C B B B C B
U   U U   U U & & &    B C B B C B B E E E
& & + & & & + & & +    E E A E E E A E E A
+   + &     + &   +    A C A E C C A E C A
& & & & & & & & & &    E E E E E E E E E E
+   + &     + +   +    A C A E C C A A C A
+   + I I I + +   +    A C A D D D A A C A
& & & I I I & & & &    E E E D D D E E E E
+   + &     & +   &    A C A E C C E A C A
+   + &     & +   +    A C A E C C E A C A
& & + I I I I + & &    E E A D D D D A E E
+ & + &     & +   +    A E A E C C E A C E
+   + &     & +   +    A C A E C C E A C A
+   + +     & +   &    A C A A C C E A C A
+ x x x x x x x x +    A J J J J J J J J A
· · · · : · · · · +    H H H H G H H H H A
· · · · : · · · · +    H H H H G H H H H A
· · · x : · · · · +    H H H J G H H H H A
+ + + + : + + + + +    A A A A G A A A A A
```

Figure 5.24. (continued).

Façade 9

```
U U   -   - - - - - - - -      B B C - C - - - - - - - -
& & & & & & & & & & & & &      E E E E E E E E E E E E E
+   + +   +   +   + +   +      A C A A C A C A C A A C A
& & & & & & & & & & & & &      E E E E E E E E E E E E E
+   + +   + & + & + + & +      A C A A C A E A E A A E A
+   + +   +   +   + +   +      A C A A C A C A C A A C A
+ & + + & + & + & + + & +      A E A A E A E A E A A E A
+   + +   +   +   + +   +      A C A A C A C A C A A C A
+   + +   +   +   + +   +      A C A A C A C A C A A C A
& & & & & & & & & & & & &      E E E E E E E E E E E E E
+   + +   & & & & +   & +      A C A A C E E E E A C E A
+   + +   & : &   + +   +      A C A A C E G E C A A C A
& & & & & & : & & & & & &      E E E E E E G E E E E E E
```

Façade 10

```
U U I I I U U I I I U I U I U I U      B B D D D B B D D D B D B D B D B
+ + I I I + + I I I + + + + + + +      A A D D D A A D D D A A A A A A A
  + +   + + + +   + +   +   + + +      C A A C A A A A C A A C A C A A A
+ + I I I + + I I I + + + + + + +      A A D D D A A D D D A A A A A A A
  + +   + + + +   + +   +   + + +      C A A C A A A A C A A C A C A A A
+ + I I I + + I I I + + + + + + +      A A D D D A A D D D A A A A A A A
  + +   + + + +   + +   +   + + +      C A A C A A A A C A A C A C A A A
+ + I I I + + I I I + + + - + + +      A A D D D A A D D D A A A A A A A
  + +   + + + +   + +   +   + + +      C A A C A A A A C A A C A C A A A
+ + I I I + + I I I + + + - + + +      A A D D D A A D D D A A A A A A A
  + +   +   + : : : + · + · + · +      C A A C A C A G G G A H A H A H A
+ + +   + + + : : : + · + · + · +      A A A C A A A G G G A H A H A H A
+ + + + + + + : : : + + + + + + +      A A A A A A A G G G A A A A A A A
```

Figure 5.24. (continued).

Façade 11

```
U  –     U  –     U  –  –     U  –        B  –  C  B  –  C  B  –  –  C  B  –
&  &  &  &  &  &  &  &  &  &  &  &        E  E  E  E  E  E  E  E  E  E  E  E
+  +           +        +  +        +     A  A  C  C  A  C  C  A  A  C  C  A
+  &  &  &  &  &     &  &        &        A  E  E  E  E  E  C  E  E  C  C  E
+  &     &  +  &  &  &  +  &  &  +        A  E  C  E  A  E  E  E  A  E  E  A
+  &           +        +  +        +     A  E  C  C  A  C  C  A  A  C  C  A
+  &  &  &  +  &     +  +        +        A  E  E  E  A  E  C  A  A  C  C  A
+  &  &  &  +  &  &  +  &  &  &  +        A  E  E  E  A  E  E  A  E  E  E  A
+        &  +        &  &        +        A  C  C  E  A  C  C  E  E  C  C  A
+  &  &  &  +  I  I  +     &     +        A  E  E  E  A  D  D  A  C  E  C  A
+  &  &  +  &  I  I  +  +  &  +  +        A  E  E  A  E  D  D  A  A  E  A  A
+        +  &        +  &  &  &  +        A  C  C  A  E  C  C  A  E  E  E  A
+        &  +        +        &  +        A  C  C  E  A  C  C  A  C  C  E  A
&  x  x  x  x  &  &  &  &     &  &        E  J  J  J  J  E  E  E  E  C  E  E
&  &  &  &  &  &  &  &  &  &  &  &        E  E  E  E  E  E  E  E  E  E  E  E
+  +  +  +  +  +  x  x  ·  ·  ·  +        A  A  A  A  A  A  J  J  H  H  H  A
·  ·  ·  ·  ·  ·  :  :  x  x  ·  +        H  H  H  H  H  H  G  G  J  J  H  A
·  ·  ·  ·  ·  ·  :  :  ·  ·  ·  +        H  H  H  H  H  H  G  G  H  H  H  A
+  +  +  +  +  +  :  :  ·  ·  ·  +        A  A  A  A  A  A  G  G  H  H  H  A
```

Façade 12

```
–  –  –  &  &  &  –  –  –        –  –  –  E  E  E  –  –  –
U  &  –  &     &  –  &  U        B  E  –  E  C  E  –  E  B
&     U  &     &  &     &        E  C  B  E  C  E  E  C  E
&  &  &  &  +  &  &  &  &        E  E  E  E  A  E  E  E  E
&     +     &     +     &        E  C  A  C  E  C  A  C  E
&  &  +  &  &  &  +  &  &        E  E  A  E  E  E  A  E  E
&     +     &     +     &        E  C  A  C  E  C  A  C  E
&     +     &     +     &        E  C  A  C  E  C  A  C  E
&  &  &  I  I  I  I  &  &        E  E  E  D  D  D  D  E  E
&  &  &  I  I  I  &  &  &        E  E  E  D  D  D  E  E  E
&     +     &     &     +        E  C  A  C  E  C  E  C  A
&     +     &     &     +        E  C  A  C  E  C  E  C  A
&  &  &  I  I  I  I  &  &        E  E  E  D  D  D  D  E  E
&  &  &  &  &  &  &  &  &        E  E  E  E  E  E  E  E  E
      +     &     +  :  :        C  C  A  C  E  C  A  G  G
      +     &     +  :  :        C  C  A  C  E  C  A  G  G
&  &  &  &  &  &  +  :  :        E  E  E  E  E  E  A  G  G
&  &  &  &  &  &  &  :  :        E  E  E  E  E  E  E  G  G
```

Figure 5.24.
(continued).

Façade 13

–	–	–	&	&	–	–	–	–
U	&	–	&		–	U	&	–
&		&	&	&	&	&		&
&	&	&	&	&	&	&	&	&
		&		&				&
		&		&				&
&	&	&	[illegible]	&	&	&	&	&
		&			&	&		+
		&			&	&		&
&	&	I	I	I	I	&	&	+
&	&	&	I	I	&	&		+
		&			&			+
		&			&			+
&	I	I	I	I	I	&	&	+
+	&	&	&	&	&	&	+	+
:	x	+	+	:	:	&	:	+
:	+	+	x	:	:	+	:	+
:	+	+	x	:	:	+	:	+
:	&	+	+	:	:	&	:	+

–	–	–	E	E	–	–	–	–
B	E	–	E	C	–	B	E	–
E	C	E	E	E	E	E	C	E
E	E	E	E	E	E	E	E	E
C	C	E	C	E	C	C	C	E
C	C	E	C	E	C	C	C	E
E	E	E	E	E	E	E	E	E
C	C	E	C	C	E	E	C	A
C	C	E	C	C	E	E	C	E
E	E	D	D	D	D	E	E	A
E	E	E	D	D	E	E	C	A
C	C	E	C	C	E	C	C	A
C	C	E	C	C	E	C	C	A
E	D	D	D	D	D	E	E	A
A	E	E	E	E	E	E	A	A
G	J	A	A	G	G	E	G	A
G	A	A	J	G	G	A	G	A
G	A	A	J	G	G	A	G	A
G	E	A	A	G	G	E	G	A

Façade 14

U	&	&	&	U	&	U	&	U	U	U	U	U
+		+	&		&		&	&	&	&	&	&
&	&	&	&		&		+			+		+
+	&	&	&	&	&	&	&	&	&	&	&	&
+		+		&	&		+	&	&	&	&	&
+		+		&	&		+	&		+		+
&	&	&	&	&	&	&	&	&		&		&
&		&	&	&	&	&	+	&	&	&	&	&
&		&		&	&		+		&	+		+
&	&	&		&	&		+		&	&		+
&	&	&	&	&	&	&	&	&	&	&	&	&
&		+		+	&		+	&	&	&	&	&
&		+		+			+		&	&		+
&	&	&		&			&		&	&		&
&	&	&	&	&	&	&	&	&	&	&	&	&
&	&	&	&	&	+	+	x	x	x	+	+	+
&	:	&	:	&		·	·	:	·	·	+	+
&	:	&	:	x	·	·	·	:	·	·	x	+
&	:	&	:	+	·	·	·	:	+	+	+	+

B	E	E	E	B	E	B	E	B	B	B	B	B
A	C	A	E	C	E	C	E	E	E	E	E	E
E	E	E	E	C	E	C	A	C	C	A	C	A
A	E	E	E	E	E	E	E	E	E	E	E	E
A	C	A	C	E	E	C	A	E	E	E	E	E
A	C	A	C	E	E	C	A	E	C	A	C	A
E	E	E	E	E	E	E	E	E	C	E	C	E
E	C	E	E	E	E	E	A	E	E	E	E	E
E	C	E	C	E	E	C	A	C	E	A	C	A
E	E	E	C	E	E	C	A	C	E	E	C	A
E	E	E	E	E	E	E	E	E	E	E	E	E
E	C	A	C	A	E	C	A	E	E	E	E	E
E	C	A	C	A	C	C	A	C	E	E	C	A
E	E	E	C	E	C	C	E	C	E	E	C	E
E	E	E	E	E	E	E	E	E	E	E	E	E
E	E	E	E	E	A	A	J	J	J	A	A	A
E	G	E	G	E	H	H	H	G	H	H	A	A
E	G	E	G	J	H	H	H	G	H	H	J	A
E	G	E	G	A	H	H	H	G	A	A	A	A

Figure 5.24. (continued).

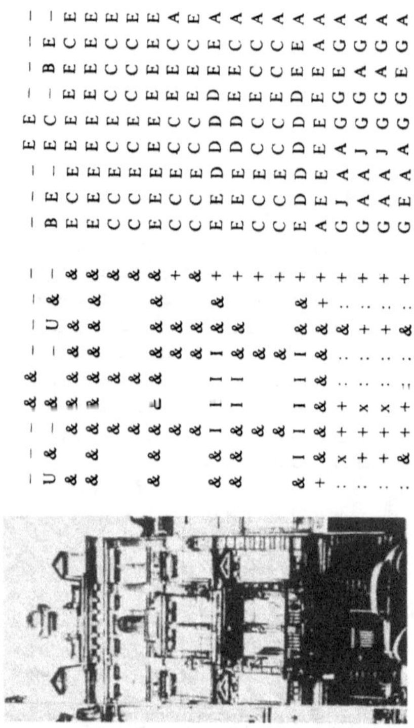

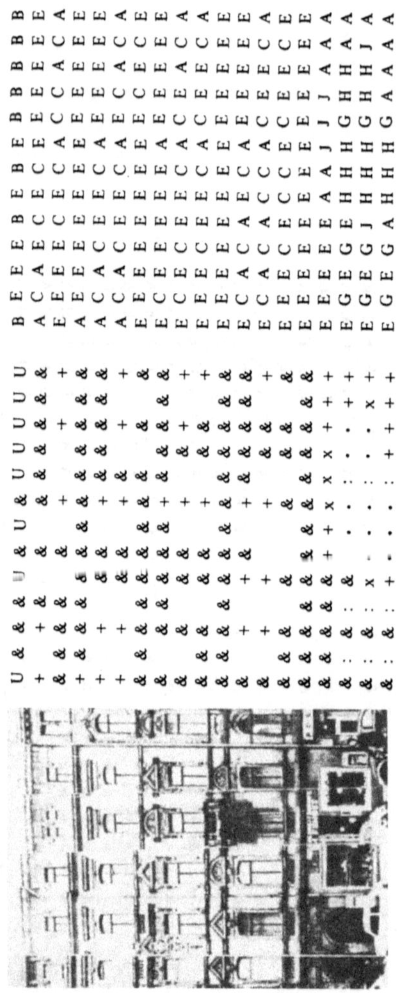

Façade 15

```
U U U U U U U    B B B B B B B
+ + + + + + +    A A A A A A A
+     +     +    A C C A C C A
+ + + + + + +    A A A A A A A
+     +     +    A C C A C C A
+     +     +    A C C A C C A
+ + + + + + +    A A A A A A A
+     +     +    A C C A C C A
+ + + + + + +    A A A A A A A
+ + + + + + +    A A A A A A A
+     +     +    A C C A C C A
+ + + + + + +    A A A A A A A
+     +     +    A C C A C C A
+     +     +    A C C A C C A
+ x x x x + +    A J J J J A A
+ + + + + + +    A A A A A A A
: : . . . . .    G G H H H H H
: : . . . . .    G G H H H H H
: : . . . . .    G G H H H H H
```

Façade 16

```
U & & U & & & & U &    B E E B E E E E B E
+   & +     & &   &    A C E A C C E E C E
& & & & & & & & & &    E E E E E E E E E E
& & + & & & & & & &    E E A E E E E E E E
+   & +     + +   +    A C E A C C A A C A
&   & &     & &   &    E C E E C C E E C E
& & & + & & & & & &    E E E A E E E E E E
&   & &     & &   &    E C E E C C E E C E
&   & &     & &   &    E C E E C C E E C E
& & & I I I I & & &    E E E D D D D E E E
& & & & I I & & & &    E E E E D D E E E E
+   & &     & &   &    A C E E C C E E C E
&   & &     + &   &    E C E E C C A E C E
& & & & & & & & & &    E E E E E E E E E E
& & & & & & & + + +    E E E E E E E A A A
. . + : . . . & : &    H H A G H H H E G E
. . + : . . . & : &    H H A G H H H E G E
. . + : . . . & : &    H H A G H H H E G E
+ + + : : + + & : &    A A A G G A A E G E
```

Figure 5.24. (continued).

Façade 17

```
+  +  +  +  +  +  +  +  +  +  +  +      A A A A A A A A A A A A
+                                +      A C C C C C C C C C C A
+  +  +  +  +  +  +  +  +  +  +  +      A A A A A A A A A A A A
+                                +      A C C C C C C C C C C A
+  +  +  +  +  +  +  +  +  +  +  +      A A A A A A A A A A A A
+  +  +  +  +  +  +  +  +  +  +  +      A A A A A A A A A A A A
+                                +      A C C C C C C C C C C A
+  +  +  +  +  +  +  +  +  +  +  +      A A A A A A A A A A A A
+                                +      A C C C C C C C C C C A
+  +  +  +  +  +  +  +  +  +  +  +      A A A A A A A A A A A A
+                                +      A C C C C C C C C C C A
+                                +      A C C C C C C C C C C A
+  +  +  +  +  +  +  +  +  +  -  +      A A A A A A A A A A A A
+                                +      A C C C C C C C C C C A
+  +  +  +  +  +  +  +  +  +  -  +      A A A A A A A A A A A A
+  +  +  +  +  +  +  +  +  +  -  +      A A A A A A A A A A A A
·  ·  :  ·  ·  ·  :  :  ·  ·  -  +      H H G H H H G G H H H A
·  ·  :  ·  ·  ·  :  :  ·  ·  ·  +      H H G H H H G G H H H A
+  +  :  +  +  +  :  :  +  +  +  +      A A G A A A G G A A A A
```

Façade 18

```
U     U     U           U     U        U  U      B C B C B C C C B C B C C B B
I  I  I  I  I  I  I  I  I  I  I  I     I  I      D D D D D D D D D D D D C D D
+        +     +        +     +  I  I  I  I      A C C A C A C C A C A D D D D
+  +  +  +  +  +  +  +  +  +  +  +               A A A A A A A A A A A A C C C
+        +     +        +     +  I  I  I  I      A C C A C A C C A C A D D D D
+  +  +  +  +  +  +  +  +  +  +  +               A A A A A A A A A A A A C C C
+        +     +        +     -  I  I  I  I      A C C A C A C C A C A D D D D
+  +  +  +  +  +  +  +  +  +  -  +               A A A A A A A A A A A A C C C
+        +     +        +     -  +     I  I      A C C A C A C C A C A A C D D
+  +  +  +  +  +  +  +  +  +  +  +  I  I  I      A A A A A A A A A A A A D D D
+  :  :  +  ·  +  ·  ·  ·  ·  +  ·  :  :  +      A G G A H A H H H H A H G G A
+  :  :  +  ·  +  ·  ·  ·  ·  +  ·  :  :  +      A G G A H A H H H H A H G G A
+  :  :  +  +  +  +  +  +  +  +  +  :  :  +      A G G A A A A A A A A A G G A
```

Figure 5.24. (concluded).

Façade 1883

```
B B B B B B - - - - - - -    2 2 2 2 2 2 9 9 9 9 9 9 9    U U U U U U - - - - - - -
E E E E E E E E E E E E E    5 5 5 5 5 5 5 5 5 5 5 5 5    & & & & & & & & & & & & &
A C A C A A C A C A A C A    1 3 1 3 1 1 3 1 3 1 1 3 1    +   +   + +   +   + +   +
A E E E E E E E E E E E A    1 5 5 5 5 5 5 5 5 5 5 5 1    + & & & & & & & & & & & +
A C A C A A C A C A A C A    1 3 1 3 1 1 3 1 3 1 1 3 1    +   +   + +   +   + +   +
A C A C A D D D C A A C A    1 3 1 3 1 4 4 4 3 1 1 3 1    +   +   + I I I   + +   +
A E E E A D D D E A E E A    1 5 5 5 1 4 4 4 5 1 5 5 1    + & & & + I I I & + & & +
A C A C A C C E C A A C A    1 3 1 3 1 3 3 5 3 1 1 3 1    +   +   +     &   + +   +
A C A C A C C E C A A C A    1 3 1 3 1 3 3 5 3 1 1 3 1    +   +   +     &   + +   +
E E E E E E E E E E E E E    5 5 5 5 5 5 5 5 5 5 5 5 5    & & & & & & & & & & & & &
E C E C E E C E C E E C E    5 3 5 3 5 5 3 5 3 5 5 3 5    &   &   & &   &   & &   &
A C A C A A C A C A A C A    1 3 1 3 1 1 3 1 3 1 1 3 1    +   +   + +   +   + +   +
E E E E E D D D E E E E E    5 5 5 5 5 4 4 4 5 5 5 5 5    & & & & & I I I & & & & &
A A A A A A A A A A A A A    1 1 1 1 1 1 1 1 1 1 1 1 1    + + + + + + + + + + + + +
A G A H H H G H H A G G A    1 6 1 7 7 7 6 7 7 1 6 6 1    + : + . . . : . . + : : +
A G A H H H G H H A G G A    1 6 1 7 7 7 6 7 7 1 6 6 1    + : + . . . : . . + : : +
A G A H H H G H H A G G A    1 6 1 7 7 7 6 7 7 1 6 6 1    + : + . . . : . . + : : +
```

Façade 1955

```
B B B C B B B B B B C B B    2 2 2 3 2 2 2 2 2 2 3 2 2    U U U   U U U U U U   U U
B B B B B B B B B B B B B    2 2 2 2 2 2 2 2 2 2 2 2 2    U U U U U U U U U U U U U
D D C D D C D D C D C D D    4 4 3 4 4 3 4 4 3 4 3 4 4    I I   I I   I I   I   I I
A A A A A A A A A A A A A    1 1 1 1 1 1 1 1 1 1 1 1 1    + + + + + + + + + + + + +
A C C A C C A C C A C C A    1 3 3 1 3 3 1 3 3 1 3 3 1    +     +     +     +     +
A C C A A A A A A A C C A    1 3 3 1 1 1 1 1 1 1 3 3 1    +     + + + + + + +     +
A A A A A A A A A A A A A    1 1 1 1 1 1 1 1 1 1 1 1 1    + + + + + + + + + + + + +
A C C A C C A C C A C C A    1 3 3 1 3 3 1 3 3 1 3 3 1    +     +     +     +     +
A C C A A A A A A A C C A    1 3 3 1 1 1 1 1 1 1 3 3 1    +     + + + + + + + + + +
A C C A C C A C C A C C A    1 3 3 1 3 3 1 3 3 1 3 3 1    +     +     +     +     +
A C C A C C A C C A C C A    1 3 3 1 3 3 1 3 3 1 3 3 1    +     +     +     +     +
A A A A A A A A A A A A A    1 1 1 1 1 1 1 1 1 1 1 1 1    + + + + + + + + + + + + +
A C C A C C A C C A C C A    1 3 3 1 3 3 1 3 3 1 3 3 1    +     +     +     +     +
A C C A A A A A A A C C A    1 3 3 1 1 1 1 1 1 1 3 3 1    +     + + + + + + +     +
A A A A A A A A A A A A A    1 1 1 1 1 1 1 1 1 1 1 1 1    + + + + + + + + + + + + +
G G G A C C C C C C C C C    6 6 6 1 3 3 3 3 3 3 .3 3 3   : : : + +
G G G A A A A A A A A A A    6 6 6 1 1 1 1 1 1 1 1 1 1    : : : + + + + + + + + + +
```

Figure 5.25. Three different ways of 'coding' various kinds of façade surface: uppercase letters, numerals, and 'iconic' typewriter symbols. The 'iconic' coding brings out the difference between a typical 'new' and a typical 'old' façade: the symbol '&' for 'decorative surface' is lacking in the 'new' façade.

From figures 5.24 and 5.25 it can be seen that 'new' façades seem to be characterized by a complete absence of the sign for 'decorative surface' (a belated, though doubtful, success of Ruskin, Morris and Loos).

Two separate information values (H and H_1, the latter taking into account the transitions between signs) were obtained with a special computer program for each of the three different aspects of the eighteen façades for which TTRs were also obtained, that is, whole façade, the same without the ground floor, and the same without the ground floor or roof. (The author is indebted to M Sund, formerly at the Computing Centre, University of Ulm, for writing the program.)

Comparison of TTRs and information measurements
To compare the two types of measurement, twelve pairs of façade rankings were obtained for comparison by Spearman's correlation coefficient (ρ). Thus for the whole building face the 'façade opening' and 'all feature' TTR measures were each compared with both H and H_1. Similarly, four pairs were obtained from the figures for the façade without the ground floor, and four from the figures without either ground floor or roof (Krampen, 1974b).

5.7.3 Results

TTRs for eighteen façades
The TTRs for the sample of eighteen façades indexed by their year of construction, with separate listing according to whether they were calculated for all façade elements or façade openings only, are given in their rank order in table 5.17. It can be seen from the table that both kinds of TTRs are generally higher for 'old' buildings.

Information measurements for eighteen façades
The information measurements H and H_1 for the sample of eighteen façades indexed by their years of construction are given in their rank orders in table 5.18. Again, the table shows that 'old' buildings generally have higher H and H_1 values than 'new' buildings.

Correlations between TTRs and bits
The correlation coefficients between the two types of façade measurement are given in table 5.19. All correlations are significant. However, among the twelve correlations between TTRs and bits, the façade-opening TTRs correlate better with H and H_1 values for the façade and the façade plus roof than do the TTRs based on all façade elements. Also correlations of H_1 with TTRs for openings are higher than the corresponding H correlations in the pairings of façade and façade plus roof. Where the entire building front is concerned, TTRs, computed on all façade elements correlate better with both H and H_1 than do façade-opening TTRs, the best results being obtained with H_1.

Table 5.17. Rank orders of TTRs for façade elements (e) and façade openings only (o) of eighteen façades.

Rank	Façade only						Façade plus roof						Façade plus roof plus ground floor					
	façade[a]		TTRe	façade		TTRo	façade		TTRe	façade		TTRo	façade		TTRe	façade		TTRo
1	4	1882	5·27	8	1883	0·67	4[b]	1882	7·27	8	1883	0·67	8	1883	9·93	8	1883	0·65
2	8	1883	5·25	4	1882	0·53	13	1882	5·97	4[b]	1882	0·53	4	1882	9·78	16	1880	0·60
3	13	1882	5·24	11	1866/ 1883	0·50	8	1883	5·93	12	1896	0·53	13	1882	7·97	12	1896	0·58
4	11	1866/ 1883	4·98	12	1896	0·50	11	1866/ 1883	5·42	13	1882	0·50	11	1866/ 1883	7·42	11	1866/ 1883	0·56
5	2	1883	4·78	16	1880	0·50	2[b]	1883	4·78	16[b]	1880	0·50	16	1880	7·07	4	1882	0·55
6	16	1880	4·07	13	1882	0·47	16[b]	1880	4·07	11	1866/ 1883	0·47	2	1883	6·11	13	1882	0·55
7	12	1896	3·18	3	1948	0·44	12	1896	3·96	14[b]	1894	0·43	9	1892	5·20	3	1948	0·47
8	15	1955	2·10	14	1894	0·43	9	1892	2·21	3	1948	0·42	3	1948	5·14	14	1894	0·46
9	14	1894	1·97	2	1883	0·29	15[b]	1955	2·10	2[b]	1883	0·29	7	1952	4·81	2	1883	0·35
10	9	1892	1·67	9	1892	0·27	14[b]	1894	1·97	9[b]	1892	0·29	14	1894	4·64	9	1892	0·32
11	6	1954	1·52	7	1952	0·20	18[b]	1958	1·69	7[b]	1952	0·24	12	1896	4·36	7	1952	0·30
12	10	1958	1·39	6	1954	0·36	3	1948	1·64	18[b]	1958	0·17	15	1955	4·10	6	1954	0·27
13	5	1954	1·32	18	1958	0·15	6[b]	1954	1·52	6[b]	1954	0·16	5	1952	3·32	1	1955	0·25
14	3	1948	1·28	1	1955	0·13	10[b]	1958	1·39	1[b]	1955	0·15	6	1954	3·19	15	1955	0·25
15	7	1952	1·27	5	1954	0·12	5[b]	1952	1·32	10[b]	1958	0·13	10	1958	2·84	10	1958	0·24
16	1	1955	1·15	15	1955	0·10	7[b]	1954	1·31	5[b]	1952	0·12	1	1955	2·24	5	1952	0·23
17	18	1958	0·69	17	1959	0·10	1[b]	1955	1·18	15[b]	1955	0·10	17	1959	2·17	18	1958	0·22
18	17	1959	0·50	10	1958	0·09	17[b]	1959	0·50	17[b]	1959	0·10	18	1958	1·89	17	1959	0·20

[a]Number and date of construction.
[b]Photograph shows no roof.

Table 5.18. Rank orders of the information measures H and H_1 for eighteen façades.

Rank	Façade only						Façade plus roof						Façade plus roof plus ground floor					
	façade[a]		H	façade		H_1	façade		H	façade		H_1	façade		H	façade		H_1
1	4	1882	1·9290	4	1882	1·7825	4[b]	1882	2·1951	8	1883	1·8731	8	1883	2·6217	4	1882	1·9807
2	8	1883	1·8480	8	1883	1·7543	2[b]	1883	2·1082	2[b]	1882	1·8660	4	1882	2·5060	8	1883	1·9429
3	2	1883	1·7771	2	1883	1·6666	8	1883	2·1077	11	1866/ 1883	1·7735	2	1883	2·4492	11	1866/ 1883	1·8514
4	11	1866/ 1883	1·7712	11	1866/ 1883	1·6415	12	1896	2·0437	2[b]	1883	1·7343	11	1866/ 1883	2·4430	2	1883	1·7979
5	12	1896	1·7509	12	1896	1·6093	11	1866/ 1883	2·0241	12	1896	1·7130	13	1882	2·3919	13	1882	1·7179
6	13	1882	1·6379	9	1892	1·4872	3	1948	1·9723	9	1892	1·6317	3	1948	2·2900	12	1896	1·6672
7	3	1948	1·5438	7	1952	1·4368	13	1882	1·9354	13	1882	1·5423	7	1952	2·2173	9	1892	1·6368
8	9	1892	1·5092	13	1882	1·4224	9	1892	1·8707	3	1948	1·5357	5	1954	2·1915	14	1894	1·6314
9	18	1958	1·4872	3	1948	1·3964	7	1952	1·7849	7	1952	1·4520	12	1896	2·1103	7	1952	1·6119
10	7	1952	1·4799	16	1880	1·3052	18	1958	1·7085	18	1958	1·4318	18	1958	2·0849	16	1880	1·6047
11	16	1880	1·3842	14	1894	1·2895	5[b]	1954	1·6753	14	1894	1·4020	6	1954	2·0226	18	1958	1·5773
12	5	1954	1·3403	18	1958	1·2742	1	1955	1·5808	5[b]	1954	1·3993	14	1894	2·0192	3	1948	1·5447
13	14	1894	1·3235	5	1954	1·2442	14[b]	1894	1·4966	16[b]	1880	1·3752	16	1880	1·9833	5	1954	1·5394
14	10	1958	1·2289	10	1958	1·1720	10[b]	1958	1·4960	10[b]	1958	1·3366	9	1892	1·8845	6	1954	1·4621
15	1	1955	1·1844	1	1955	1·1086	6[b]	1954	1·4841	6[b]	1954	1·2543	15	1955	1·8806	10	1958	1·4280
16	6	1954	1·1279	6	1954	1·0970	16[b]	1880	1·4643	1	1955	1·1791	10	1958	1·7265	1	1955	1·2047
17	15	1955	1·0888	15	1955	1·0298	15[b]	1955	1·3696	15	1955	1·1286	1	1955	1·6895	15	1955	1·1956
18	17	1959	0·9464	17	1959	0·9318	17[b]	1959	0·9464	17[b]	1959	0·9318	17	1959	1·4297	17	1959	1·1441

[a]Number and date of construction
[b]Photograph shows no roof.

Table 5.19. Correlations (ρ) between TTRs and information measures over eighteen façades; $p < 0{\cdot}05$ for all correlations.

	H-TTRe	H-TTRo	H_1-TTRe	H_1-TTRo
Façade	0·678	0·829	0·630	0·84
Façade plus roof	0·641	0·719	0·759	0·81
Façade plus roof plus ground floor	0·740	0·604	0·839	0·77

Comparison of TTRs and bits

Comparing the rank orders of TTRs and bits the former measurement seems to differentiate a little better between 'old' and 'new' façades. Of the six bit-rankings in table 5.18 there are in four cases two, in one case three, and, in one case one new façade which ranks among the upper nine of the eighteen façades.

Of the six TTR rankings in table 5.17 there are in only one case two new façades (in all other cases only one) which rank among the upper nine of the eighteen façades.

In conclusion, then, the TTR in both its forms appears to be a somewhat better measure for indexing flexibility or heterogeneity of façades as is information measurement. If the ground-floor level is excluded, the simple form of the TTR based on façade openings only should yield façade measures very similar to those of information measurement, the best correlation being with H_1. If the ground-floor level is included, TTRs based on all façade elements yield closer results to those of information measurement, especially if compared with H_1.

5.7.4 Discussion

Similarities and differences between information measurement and TTR

The comparison of information measures and TTRs obtained from façades has shown that the two measures are significantly correlated. In what respects are they similar and in what respects are they different?

They are similar in that they are both based on some proportional realization of façade elements in relation to the total surface of a façade. However, the TTR must apparently be considered more sensitive to façade detail. If TTR calculation were applied to the 'mapping' of a façade, different ornaments on the façade within the various grid units would be assessed as different 'types' of decoration to be divided by the total number of grid units covering decorated façade surface. In information measurement, at least in the form proposed here, all decorated surfaces are treated as if they were homogeneous. Therefore it will not discriminate so sensitively between the various types of objects which may be found on the surface of a façade.

The problem of transitional probabilities

The TTR (at least in its classical form) does not include transitional probabilities in the assessment of the flexibility of the façade design. This can be done, on the other hand, by information measurement, as has been shown by the H_1 measures. An example of a 'rigid' and a 'flexible' composition, based on the same TTR, which underlines the importance of transitional probabilities of occurrence, is given in figure 5.26. It shows two types of windows, one square, the other rectangular, distributed over façades. In one case nine square windows and six rectangular windows are located together in a regular fashion, the square windows in one half, the rectangular windows in the other half of the façade of three storeys. This façade gives an extremely regular impression. In the other case, the same number of square and rectangular windows are displayed in a random order over the three storeys, rectangular windows being followed by square and vice versa in an irregular fashion. This façade makes a quite irregular impression, although the TTRs of the two façades are the same.

At first glance this example seems to defeat the TTR as a potential predictor of the subjective judgement of 'interest' or 'monotony'. However, this example, although it demonstrates the importance of transitional probabilities, does not capture the essence of the TTR, which is to measure the variability of a vocabulary rather than of a given phrase. Clearly a more irregular distribution of only two types of element on the surface of a façade will produce subjective judgements describing it as 'more varied' than a more regular distribution of the same elements. The phrase 'a rose is a rose is a rose' will have a low TTR as compared with the phrase 'a rose is one flower among others', although the former line from a poem of Gertrude Stein will produce stronger subjective responses. The TTR is correctly applied to cases with rather more types such as the following. One façade contains six rectangular and nine square windows on three storeys, distributed in an irregular fashion. The other façade also contains fifteen windows, but varying considerably in shape: two rectangles, three squares, three circles, one triangle, two diamonds, two parallelograms,

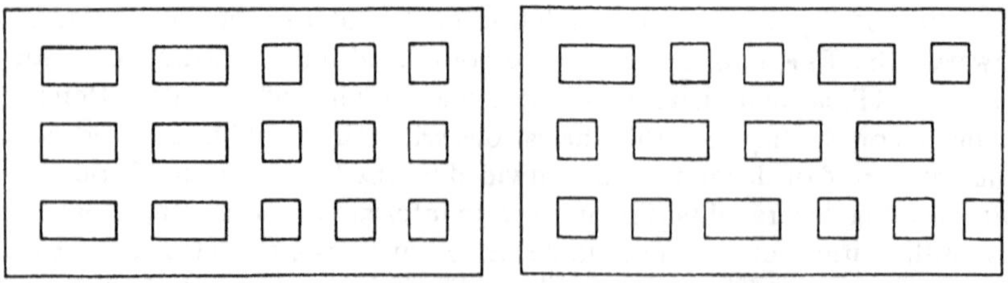

(a) (b)

Figure 5.26. (a) A 'rigid' composition of a façade with two types of windows; type-token ratio = 2/15 = 0·14. (b) A more flexible composition of a façade with the same two types and the same number of window tokens as in (a), and thus the same type-token ratio.

and two rectangles with rounded corners. This situation is depicted in figure 5.27. In this case the two façades would have very different TTRs.

Nevertheless, its inability to respond to distribution is an important shortcoming of the TTR: it is probably not the only source of a hypothesized difference in subjective judgements of the 'interestingness' of a façade.

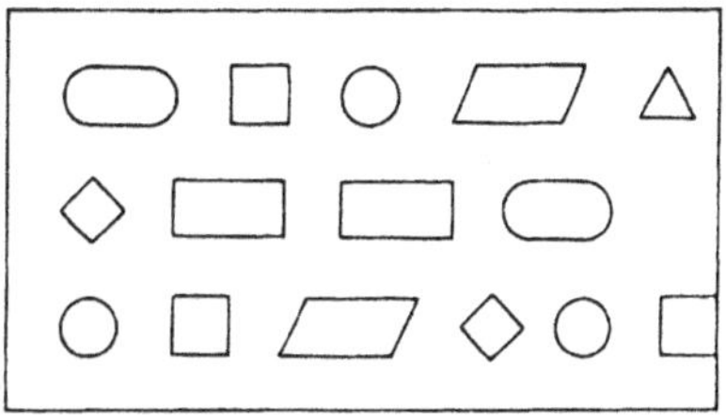

Figure 5.27. A flexible composition of a façade with seven types of windows. The type-token ratio in this example ($7/15 = 0{\cdot}47$) is superior to those given in figure 5.26.

5.8 Study 5: an application of information aesthetics to two styles of façades

5.8.1 Introduction

Problem

The theory of connotation of architectural objects must include the case in which the manner of conceiving a particular object was refunctionalized to deliver an artistic message. In this case the theory of information aesthetics would become relevant. The aim of the following study was to compare information measurements derived from the Shannon formula with measures of aesthetic information derived from formulae developed by Bense (1968a; 1969a).

Background of the problem

As has been mentioned above (page 247), Kiemle (1967) dealt with information measurement of façades. Although he discussed Birkhoff's (1928; 1932) formula of aesthetic measure in this context and showed that it can be expressed in information-theory terms, he applied a differently based theory of information aesthetics to the town hall of Kurashiki (Japan) designed by Kenzo Tange. The version of information aesthetics which he applied was developed by Frank (1968). It postulates the three successive stages of aesthetic perception mentioned before (page 247): the selective phase adapting a segment of reality to the capacity of the human channel; the synthetic phase reducing the complexity of an object by supersign formation; and the analytic phase scrutinizing the structure of the supersigns. Kiemle is interested in demonstrating that aesthetically satisfying buildings allow for several stages of supersign formation. Supersign formation is accomplished by reducing the number of objectively present façade elements into unifying classes and by the grouping of unified element classes into 'complexes', that is, into unified 'zones' on the façade surface. This process of simplification may be carried out graphically, by the same technique used later by von Buttlar et al (1973). In four steps of complexity reduction by class and complex formation the town hall of Kurashiki was simplified to emanate 160 bits of information, an amount just fitting the capacity of the human channel, which cannot keep more than 160 bits at a time present in consciousness. Then the aesthetic information is calculated as the difference between the information measure passing the human channel and the information measure of the next higher level of simplification—in this case the fifth level. The introduction of the human channel capacity into the calculation serves to introduce a 'subjective' element into the otherwise 'objective' aesthetic calculus. The stages of supersign formation result in an increase in redundancy or order. Kiemle goes on to show by this method that an architecture that does not allow successive steps of complexity reduction is too simple to be of interest to the human observer since the aesthetic event would be accomplished after one or two steps of reduction. The human mind would have to turn to the appreciation of another object immediately after the simple object has been exhausted. This would not be the case with the town hall of Kurashiki—hence it is aesthetically valid. The problem with this procedure lies in the arbitrariness of the graphic simulation of supersign formation. Also, with Kurashiki town hall, Kiemle had only estimated the transitional probabilities between the elements of its façade. Although Kiemle's results are interesting and illuminate the aesthetic behaviour of subjects, the information theory of the Birkhoff formula will be put to the test in this study.

Hypothesis
If the Birkhoff formula correctly mirrors the aesthetic behaviour of subjects, it should distinguish by high-information aesthetic values significantly the same façades from the sample used in studies 1 and 2, that were subjectively valued most highly for pleasingness, friendliness, etc in those studies. As has been pointed out before, this null hypothesis will in fact be rejected, since the Birkhoff formula is biased over the importance of order in aesthetic perception.

5.8.2 Method
The information-aesthetics measurements for the eighteen façades used in the previous studies were calculated by the formula

$$H_{\text{ae}} = \frac{1}{H_i} - \frac{1}{H_{\max}} ,$$

where

$$\frac{1}{H_{\max}} = 0 \cdot 315 \text{ bits,}$$

on the basis of the information measurement obtained in study 4. $1/H_{max}$ in the formula is a constant since H_{max} is a constant of 3·17 bits for nine kinds of façade surfaces occurring with equal probability and without redundancy (the nine kinds having constituted the repertoire given for information measurement of the eighteen façades).

5.8.3 Results

The resulting measures of the aesthetic information in the eighteen façades, permitting a comparison of classical information measurements in table 5.18 and measurements of aesthetic information are given in table 5.20.

Table 5.20. Rank orders of the information aesthetics measures H_{ae} for nine 'old' and nine 'new' façades.

Rank	Façade only			Façade plus roof			Façade plus roof plus ground floor		
	number	year	H_{ae}	number	year	H_{ae}	number	year	H_{ae}
1	17	1959	0·742	17[a]	1959	0·742	17	1959	0·384
2	15	1955	0·603	15[a]	1955	0·415	1	1955	0·277
3	6	1954	0·572	16[a]	1880	0·368	10	1958	0·264
4	1	1955	0·529	6[a]	1954	0·359	15	1955	0·217
5	10	1958	0·499	10[a]	1958	0·353	9	1892	0·215
6	14	1894	0·441	14[a]	1894	0·352	16	1880	0·189
7	5	1954	0·431	1	1955	0·318	14	1894	0·180
8	16	1880	0·407	5[a]	1954	0·282	6	1954	0·179
9	7	1952	0·361	18	1958	0·270	18	1958	0·165
10	18	1958	0·357	7	1952	0·245	12	1896	0·157
11	9	1892	0·348	9	1892	0·220	5	1954	0·141
12	3	1948	0·333	13	1882	0·202	7	1952	0·136
13	13	1882	0·296	3	1948	0·192	3	1948	0·122
14	12	1896	0·256	11	1866 1883	0·179	13	1882	0·103
15	11	1866/ 1883	0·250	12	1896	0·174	11	1866/ 1883	0·094
16	2	1883	0·248	8	1883	0·159	2	1883	0·093
17	8	1883	0·226	2[a]	1883	0·159	4	1882	0·084
18	4	1882	0·203	4[a]	1882	0·141	8	1883	0·066

[a] Photograph shows no roof.

Comparison of information and aesthetic information

If one analyzes the rank order of information measures presented in study 4, it becomes evident that the older buildings hold, on the whole, the upper ranks while buildings constructed after 1945 show lower information values. This trend is highly significant for H_1 values ($p \leqslant 0\cdot01$ for the whole façade; $p \leqslant 0\cdot001$ for the façade minus the ground floor; $p \leqslant 0\cdot001$ for the façade minus roof minus ground floor). In effect the façades used in these studies which were built before 1900 have a higher information value than those built after 1945.

If one compares the information values of study 4 with the information-aesthetics measures of this study, one notices that the façades of buildings constructed after 1945 hold the highest ranks. Since the measures of aesthetic information result simply from subtracting the constant $1/H_{max} = 0{\cdot}315$ bits from the H values, this trend must be as significant as the reciprocal and opposite trend for information measurements: the façades of the buildings constructed after 1945 possess, on the whole, higher aesthetic information than those of buildings constructed before 1900.

5.8.4 Discussion

This result is surprising and the opposite of what one would expect if one trusts a formula specially devised to measure aesthetic properties in objects. If one reflects, however, on the origin of the information-aesthetics concepts in Birkhoff's (1928; 1932) work, the ideological starting point from which this result necessarily follows becomes evident: the larger the relationship of order between the sign elements of an object, the larger becomes the aesthetic measure M_O. In the case of this study this leads to the absurd statement that the façades used in these studies have a higher aesthetic measure, the more 'orderly', that is, the more boring and monotonous, they are. Since, according to Birkhoff, the aesthetic measures derived directly from the structure of the object would relate directly to a 'feeling of pleasure', the paradox results that an object triggers more pleasure the more boring it is.

The aesthetic ideology appearing here is that of 'purity of form' (Gestaltreinheit). An opposite aesthetic idea, presented by the Gestalt theorist Christian von Ehrenfels (1967), is that of 'height of form' (Gestalthöhe). This ideology would have better fitted the data of study 1. The same position is taken also by Arnheim (1971), who differentiates between 'order' and 'orderliness'. Arnheim criticizes Birkhoff by defining the latter's concept of order O as one of 'orderliness', basically identical with simplicity. The term 'order' is reserved by Arnheim for a clear organization of complex structures.

Other aesthetic theories take a middle road between 'Gestaltreinheit' and 'Gestalthöhe' (see Moles, 1958; 1971): an object exhibiting too little complexity is 'banal', an object with too high a complexity is 'threatening'. Day (1965) has pointed out that the degree of aesthetically pleasing complexity depends on the personality structure. Individuals scoring high on the authoritarian dimension of a personality test (for example, on the F scale) would be able to accept less complexity than nonauthoritarian personalities. In this study, however, only a negative correlation between information measurement and information-aesthetics measurement has been established.

The problem with information aesthetics, already apparent in the comparison of Kiemle's (1967) approach and Bense's (1968a; 1969a) adaptation of the Birkhoff formula, is that there are two opposite uses of the term 'order' and 'information'. The first use of the terms is based on a given combination of object elements or signs as a repertoire. There may be regularities in a given sequence of these object elements or signs, that is, there may be connection rules governing the composition of elements or supersign formation, so that the sequence contains transitional probabilities. In this case, higher-order repertoires of object parts or supersigns would be formed and new information values would have to be calculated on the basis of these superordinate repertoires. Evidently, with this concept the information is lower, the higher the order is (that is, order is not equivalent to information). This use of the terms goes back to Moles (1958) and Frank (1968) and is the one adopted by Kiemle. The second use of the terms is based on all possible combinations of a given repertoire of object elements or signs: the larger the repertoire the higher the number of possible combinations according to the factorial ($n!$) calculation. Highly ordered (internally constrained) sequences of elements would occur in only a few cases in this universe of possible combinations. Hence a highly ordered sequence has a low probability of occurrence and therefore possesses a high degree of information (novelty, surprise value, etc). This second use of the terms (order is information) goes back to Wiener (1954) and is adopted by Bense (1956).

According to the first concept, the aesthetic object must blend order and variety. The second concept sees aesthetic objects as islands of information in a sea of chaos.

The direct transformation of the Birkhoff formula into information measure might do its author an injustice. Birkhoff (1928; 1932) stipulates five fundamental postulates which restrict the application of his aesthetic measure.

1 If the object contains symbol value (semantic information), the aesthetic measure is falsified, since it is determined syntactically only.
2 If a given element of order (for example, symmetry) occurs repeatedly it should not be considered simply according to its frequency of occurrence, since its information value diminishes with increasing repetition.
3 If the relationships of order in an object are not clearly recognizable there will be a discrepancy between the aesthetic measure and subjective pleasure responses.
4 If the aesthetic object has no focus, its measure will deviate from subjective feelings.
5 If two objects have the same relationships of order, but different degrees of complexity, the aesthetic measure of the more complex object is relatively lower than would be expected from the formula.

It is difficult to say whether the sample of façades used in the present study conforms to the first postulate. There is no obvious application of symbols to the old façades, unless one were to claim symbolic value for the mere presence of decoration. In that case the method would provide no possibility for studying the difference between the two styles. The second postulate is fulfilled if transitional probabilities are used, since they reduce the average amount of information of an object due to repetitive transitions between the elements, that is, due to elements or order. The applicability of the third and fourth postulates to the present study is again difficult to judge. Generally one would be justified in assuming that façades are sufficiently clear in their bilateral and translational symmetries, and thus both recognizable order and a (central) focus would seem to obtain. The fifth postulate, like the first, cannot be considered satisfied in this study if the mere presence of decoration is considered an index of complexity.

There has been a recent attempt to correct the simplistic application of the Birkhoff formula to design objects. Dörner (1976), basing his work on that of Garnich (1968), has suggested that the Birkhoff formula O/C, being a purely syntactic measure, must be complemented by semantic and pragmatic measures to obtain a viable definition of the 'objective' power of appeal built into an object. The syntactic aspect is represented in Dörner by the original Birkhoff quotient, where O is the number of ordering determinants (symmetries, etc) and C is the number of points (corners, turning points, intersections, etc) constituting the structure of the object.

The semantic aspect is given by the quotient B/G. B is the sum of meanings, where one principal meaning is considered to be attached to each segment of the object silhouette (vertical is 'pointed', horizontal is 'resting', straight is 'hard', etc). G is the sum of meanings, where one principal meaning is considered to be attached to each of the forms constituting the surface parts of an object (square is 'static', horizontal rectangle is 'resting', parallelogram is 'dynamic', etc). This quotient pits the number of meanings due to linear segments against the number of meanings due to shapes.

The pragmatic aspect of objects is calculated by S/K. S is the number of partial functions (including psychological ones) an object performs (for example, a ventilator cools by circulating air, its motor is silent, it stands securely on the table, etc). K is the number of object parts necessary to fulfill the partial functions listed in S (four rotor blades, one electrical motor, one base, one connection between base and motor part, etc).

Since there are three quotients, the syntactic (Birkhoff), the semantic, and the pragmatic, there are several possible relationships between these quotients.

If the syntactic aspect O/C is combined in a quotient with the pragmatic aspect S/K, a scale of formal combinations between 'stingy' and 'rich' forms results. A 'baroque' form would be realized if the syntactic quotient were composed of relatively few ordering determinants (for example, symmetries) over a form with many turning points, edges, etc, and at the same time the pragmatic quotient were composed of many functions fulfilled by few object parts. A 'purist' form would be realized if the syntactic quotient contained many ordering determinants (symmetries) over a basically simple form (few edges, turning points, etc), while the pragmatic quotient had few functions fulfilled by many distinct object parts (ideally perhaps each function by a part).

If the syntactic aspect O/C is combined into a quotient with the semantic aspect B/G, a scale of formal combinations between 'iconic' and 'symbolic' (in the Peircean sense of the terms) results. A very 'iconic' form will have a low number of ordering determinants (symmetries) over a large number of turning points, edges, etc and at the same time a large number of meanings attached to the silhouette segments, with relatively few meanings attached to few form surfaces. A very 'symbolic' form will have a high number of ordering determinants over a low number of edges, turning points, etc, together with few meaningful silhouette segments and a relatively large number of meaningful surface shapes.

The combination of the extremes of the scales resulting from the syntactic–pragmatic and syntactic–semantic quotients should provide four basic object 'styles': the iconic–baroque, the iconic–purist, the symbolic–baroque, and the symbolic–purist. The sample of façades investigated here should be distinguished mainly along the baroque–purist axis, old façades exhibiting fewer ordering determinants over more turning points, edges, etc than new façades. But conceivably the old façades could be even more iconic than the new façades if with few ordering determinants and many turning points and edges they had more meanings attached to the many silhouette segments in comparison to fewer meaningful shapes.

All three aspects of object form, the syntactic, the semantic, and the pragmatic, combine according to Dörner (1976) in the aspect of form innovation. A form may be situated on a scale between 'original' and 'banal'. An 'original' form tends to deviate from the norm and presents an irregular order; a 'banal' form, on the other hand, presents a regular order, which is predictable from the norm. The originality or banality of an object can thus be calculated by first calculating the combined value of the syntactic, semantic, and pragmatic quotients of the normative object in a category of objects to which the combined three values of the object in question can be related. The possibility of an original object becoming known and accepted (as a new norm) depends on the number of its kind, the duration of its exposure, and the population size of the space in which it is distributed. The latter reasoning of Dörner brings the economic and historical conditions of style diffusion, which were briefly discussed in connection with the historical relativity of subjective appreciation of façade styles (page 226), into the picture.

5.9 Correlation between subjective reactions and objective façade measures

Semiotic structure has been defined in this book (pages 51-60) as a coordination or correlation between two universes which are each structures of opposed sets of characteristics. In studies 1 and 2 the oppositional structure of subjective impressions of façades was measured by the SD technique. It was found that subjective impressions tended to be more positive for 'old' façades than for 'new' façades. This contrast was, however, less pronounced (although still significant) with subjects of lower educational level.

In studies 3-5 an attempt was made to apply two objective types of measurements—the TTR and information measurement—to façades, and to compare their efficiency. It was shown that the two types of measurements, although they gave comparable results, had their drawbacks. The universe of the signifier of these façades could, with these measures, be constructed as an oppositional structure of high and low objective measurements. Thus it has become possible to consider the question of whether the two universes—the signified of subjective impressions and the signifier of objective indices—are coordinated, that is, constitute a semiotic structure. Since in both groups of studies the results are series of measurements taken on the same objects, the obvious way to test the significance of any coordination between the two universes would be to calculate the covariance of the two sets of measurements. A test of the null hypothesis (that there is no semiotic structure) may thus be obtained by the calculation of correlation coefficients for the paired subjective and objective sets of façade measurements. However, before the relevance and the results of these calculations are shown, an interesting study from the literature will be discussed which used a similar approach.

In that study Bortz (1972) used—as has been reported earlier in this book (pages 205-207)—subjective measurements of house façades taken with the SD. He also calculated a whole series of objective façade measures (including, among others, information measures), which were subsequently tested for correlation with the subjective measures. The results of his study will be summarized in the following pages.

5.9.1 Twelve architectural attributes and three subjective dimensions

First Bortz had expert raters catalogue twelve architectural attributes on twenty-six façades. These attributes were (see study 4, page 249): number of storeys, number of flats, number of balconies, number of windows, number of roof surfaces, number of window forms, estimated building age, predominance of round features (measured on a six-point scale), predominance of decoration (measured on a six-point scale), number of silhouette corners, predominance of regularity (measured on a six-point scale), and number of trees. Since Bortz found three SD dimensions of subjective façade experience (evaluation, structural order, stimulation) in his study, he used the factor scores on these three dimensions for tests of correlation with the twelve architectural attributes. He found that the evaluation dimension only correlated with the architectural attributes of estimated age and number of window forms: the subjective evaluation of a building was less positive, the older a building was estimated to be, and the greater the number of its window forms. In contrast there were many more significant correlations between the subjective experience of structural order and the architectural attributes (for example, estimated age; predominance of round features, decoration, regularity; number of roof surfaces, window forms, flats, and silhouette corners).

This implies that dominance of round features, decoration, irregularity, and many roof surfaces and window forms contributes to an impression of low structural order. The factor of subjective stimulation was only correlated with the number of silhouette corners.

To further reduce his data Bortz tested the twelve architectural attributes for correlation over the twenty-six buildings to which they were applied in this study. When he subjected the resulting matrix of correlation coefficients to factor analysis he found that two ('objective') factors explained 75% of the variance: building age and building size.

Estimated age, round features, decoration and regularity, number of window forms, roof surfaces, silhouette corners and trees were the attributes that loaded highest on the age factor whereas number of windows, flats, storeys, and balconies had high loadings on the size factor.

Further comparison of the factor scores for the three subjective dimensions with the two dimensions of architectural attributes (age and size) revealed only one significant correlation, that between architectural age and subjective experience of structural order. Thus the greater the estimated building age the less the impression of structure and order. Size of building has apparently no connection with subjective evaluation or feelings of order or stimulation.

5.9.2 Relative frequency of occurrence of types of façade surface and three subjective dimensions of façade ratings

By projecting façades onto a standard grid of squares, as described previously in study 4 (page 250), Bortz counted the frequency with which each of six kinds of façade surface occurred on a building. The surface types were: wall surface, roof surface, window surface, balcony surface, decorated surface, and greenery. The relative frequency (Bortz called it 'probability') of each kind of surface was calculated, by dividing the frequency of its occurrence on a given façade by the total number of square units covered by that façade.

The relative frequencies of the surface types in façades were then tested against the factor scores of the three subjective dimensions of building experience. Bortz found significant correlations between:

a high relative frequency of wall surface and negative evaluation,

a presence of balconies and positive evaluation, an absence of balconies and negative subjective evaluation,
an increasing presence of decoration and feelings of low structural order,
a presence of greenery and feelings of low structural order,
a high relative frequency of balconies and feelings of structural order,
a larger roof surface and feelings of decreasing structural order.

5.9.3 Differences between kinds of façade surface

The relationship of the relative frequency of one kind of surface to another can be computed by subtracting the number of grid units of one kind of surface from that of the other and dividing the difference by the total number of grid units covered by a given façade. Bortz tested some of these differences for correlation with his three subjective dimensions. He found that subjective negative evaluation is correlated with:
a high wall-surface : roof-surface ratio,
a high wall-surface : window-surface ratio,
a high wall-surface : balcony-surface ratio.

5.9.4 Relative frequency of transitions between façade surface types and subjective façade dimensions

Given a facade projected on a standard grid, the count of combinations of adjacent kinds of surface (including the combination of each kind with itself) divided by the total number of all realized transitions in a facade, may be used to index the extent of the continuity of distribution of a surface type over a given façade. A façade being a two-dimensional object, Bortz counted, of the four possible transitions from a given grid unit to its neighbour (left–right, right–left, up–down, down–up), only the left–right and up–down transitions. Significant correlations of types of surface transitions with subjective evaluations of façades were found between:
a relatively frequent wall-to-wall transition and negative evaluation of the façade,
a relatively frequent wall-to-window transition and negative evaluation of the façade,
a relatively frequent window-to-balcony transition and positive evaluation of the façade.

Significant correlations of types of surface transitions with the subjective experience of structural order of façades were observed between:
a high relative frequency of wall-to-decoration transition and decreasing structural order,
a high relative frequency of decoration-to-decoration transition and decreasing structural order,
a high relative frequency of balcony-to-balcony transition and increasing structural order,
a high relative frequency of window-to-balcony transition and increasing structural order.

High frequencies of decoration-to-decoration and wall-to-decoration transition are related, in the Bortz study, to change in the feeling of stimulation. Bortz accepted, however, that the generation of student subjects he used might not be influenced by façade decoration in the same way as the generation of the time in which decorated façades were built.

5.9.5 Information measures and subjective façade dimensions

If it may be assumed that the façades in a sample are (relatively) representative, each relative frequency of façade surface type (found by the grid unit count described above) may be treated as a probability of the occurrence of that surface type. The resulting probability values may be used to calculate the average information of a façade by the well-known Shannon formula (cf page 250).

On the assumption of a representative façade sample, each relative frequency of façade type-transition coupling (found by the grid count) may be treated as the probability of its occurrence. The transitional probability values that result may be used to calculate the average transitional information of a façade by the appropriate formula:

$$H_1 = -\sum_{i=1}^{k}\sum_{j=1}^{k} p(y_i)p(y_iy_j)\log p(y_iy_j),$$

where

$p(y_iy_j)$ is the probability that an element y_i is followed by an element y_j.

The more couplings of or transitions between *different* kinds of surface are realized on a façade the larger H_1 will be. The difference between H and H_1 is called H_{digram}. The larger H_1 with a given H, the smaller H_{digram} will be. Conversely, a large H_{digram} indicates that a façade offers little variety (in transitional couplings).

Bortz tested seven information measures for correlation with the factor scores of the three dimensions of façade experience (evaluation, structural order, stimulation). The information measures were: H, H_1 for the horizontal left–right count, H_1 for the vertical up–down count, H_1 for an average of the horizontal and vertical counts, H_{digram} derived from H and H_1 horizontal, H_{digram} derived from H and H_1 vertical, and H_{digram} derived from H and the average horizontal–vertical H_1.

None of the seven information measures correlated significantly with the factor scores on the evaluation dimension. Bortz explains this lack of linear correlation by reference to the inverse U-function which has been variously found by Berlyne and his coworkers (for example, Day, 1965) and postulated by theorists of objective aesthetics (for example, Moles, 1958; 1971).

However, the subjective experience of structural order is highly correlated with all H_1 and H_{digram} values, especially those taking into account an average of horizontal and vertical count. Façades are apparently felt to be less structurally ordered, the more transitional couplings composed of different surface types are realized. Conversely, increasing H_{digram}, that is, increasing repetition of transitions on a façade, correlates strongly with the experience of structural order. The subjective dimension of stimulation does not—at least in this study—significantly correlate with information measurement.

5.9.6 Subjective dimensions and the total of all frequency and information measurements

In the calculations of his study Bortz used twenty-four objective façade measures: six surface types (wall, window, roof, etc), four differences between kinds of surfaces, eleven relative transitional frequencies, and three information measurements (H, H_1, H_{digram}). To reduce this complex of variables he factored the twenty-four measures to find their common dimensions. The best solution (on the basis of those factors which optimally predict subjective façade experience) consisted of six 'objective' factors. Bortz called these factors (the sign '-' is used for surface-type difference, '/' for surface-type transition):

F1: wall versus window surface—the variables wall, wall-roof, wall-window, wall-balcony, wall/wall, wall/window had high positive loadings, the variables window and window/window had high negative loadings on this factor.

F2: Balcony surface—the variables balcony, balcony/balcony, window-balcony were loaded on this factor.

F3: Roof surface—the variables roof and roof/roof were loaded on this factor.

F4: H_1—this factor differentiated between façades with few and façades with many different surface-type transitions, that is, between monotonous and diverse façades.

F5: H—this factor differentiated between façades with approximately equal amounts of different surface types and those in which certain surface types dominate; decoration and decoration/decoration are also loaded on this factor, which means that increasing decoration increases H.

F6: greenery surface—the variables greenery and greenery/greenery were loaded on this factor.

The canonic correlation between these six façade measure dimensions and the three subjective rating dimensions shows that the correlation is almost exclusively produced by the dimension of subjectively felt structural order.

A second canonic correlation carried by subjective evaluation is mainly weighted on F6 (greenery) and F2 (balcony). This means that façades with increasing greenery in their surrounds and large balcony surfaces are positively evaluated. The contribution of F4(H_1) and F5(H) to evaluation is not significant.

5.10 Study 6: the semiotic structure of two façade styles: correlations between SD measures and objective measures

5.10.1 Introduction

Problem

The aim of this last study was to complete the hypothesized semiotic structure consisting of the universe of the signified, containing opposed subjective impressions on one side, and the universe of the signifier, containing opposed distributions of façade parts, on the other. This may be achieved by calculating correlation coefficients between the two kinds of measurements involved—SD measures of façade impressions, and TTRs or information measures of variety and complexity of façade composition.

Background of the problem: monotony versus complexity as a stimulus variable of psychological experience

The reason for this last step is more than the academic demonstration that there is a semiotic structure between subjective impressions and façade composition. This demonstration is practically relevant since it can be shown that façade composition covaries with subjective feelings in a special way.

The two sets of data which will be correlated in this study are a rank order from high to low of subjects' evaluative reactions toward facades (reflected by the SD profiles of these façades) and a ranking from high to low of the same façades according to the relational complexity of their elements.

Informal evidence that there is an association between negative evaluation and low visual complexity of architecture has accumulated in recent years and contributed to the 'crisis of functionalism' in architecture. Functionalist architecture, according to these critics, has contributed to the experience of monotony and dullness in the urban environment of recent construction. The separation of different human activities sanctioned by the Charter of Athens, rationalization, standardization and industrialization in the construction industry, and all this combined with the ruthless profiteering of real-estate speculators are mentioned as reason for the 'crisis of the urban environment'. These accusations seem quite plausible on inspection—in any event as far as the visual aspects of tenement-house façades are concerned. However, there have been few studies which could empirically confirm these critical claims.

Monotony versus variety

In experimental psychology the role of visual complexity has been described as a concomitant variable of the orientation reflex by Sokolov (quoted in Berlyne, 1960) and as a concomitant variable of manipulation behaviour and curiosity in numerous investigations by Berlyne (1960), his collaborators, and many other researchers (for example Fowler, 1965). The concept of excitation and also that of aesthetic pleasingness has repeatedly been connected with simple versus complex distribution of environmental stimuli. Parr (1966), Rapoport and Kantor (1967), and other authors have stressed the importance of complexity as a variable of environmental perception and behaviour. Rapoport and Hawkes (1970), after summarizing these ideas, proposed the concept of 'useful amount of information' as a possible psychological measurement for an optimal distribution of stimulation (between the extremes of 'too simple' and 'too complex').

It is in this sense that 'meaning in the urban environment' should be understood from the psychosemiological point of view—the meaningful urban environment must be viewed as a behavioural setting providing useful stimulation and information (Barker, 1968; Prochansky et al, 1970).

Hypothesis
A first, though general, hypothesis then is that the null hypothesis of no significant correlation between these two sets of measurements, be it between information measurement or TTRs on one hand and subjective ratings on the other, may be rejected. In fact, there should be significant correlations between the two sets of measurements. It is predicted that the subjective cluster of older façades with higher ratings will prove to be significantly the same façades that are distinguished by high objective measurements (both bits and TTRs). From the results of study 4 it may be predicted that such correlation would be somewhat better with TTRs than with information measurement. Significant correlations between the two sets of measurements would demonstrate the existence of connotative semiotic structure between two styles, that is, between two manners of conceptualizing façades and two kinds of impressions coordinated with them.

5.10.2 Method
Correlation between 'objective' façade measurements and subjective façade clusterings
In studies 1 and 2 the two-factor solutions obtained by factoring the balanced sums of cross products of each pair combination of eighteen façade profiles, clustered the old façades into one cluster and the new façades (including two exceptional 'old' ones) into another. All new façades were highly loaded on factor 1 in both studies. In effect, factor 1 loadings of the unrotated factor solutions, when ranked, distinguished new façades from old façades. This rank order is shown in table 5.21.

A rank-order correlation coefficient (Spearman's ρ) can be computed between the objective façade measurements obtained in study 3 and the façade profile loadings on F_1 obtained in studies 1 and 2. If the hypothesis is correct the resulting correlations must all be negative since the rank order of the factor 1 loading in both studies goes from high loadings on new façades to low loadings on old façades.

Table 5.21. Ranked factor 1 loadings (unrotated two-factor solution, no iterations) of eighteen façades for study 1 and study 2.

Study 1				Study 2			
Rank	façade	loading	year	rank	façade	loading	year
1	5	0·96913	1954	1	7	0·95863	1952
2	15	0·96463	1955	2	15	0·95858	1955
3	1	0·96232	1955	3	5	0·95602	1954
4	7	0·94347	1952	4	1	0·92956	1955
5	10	0·90270	1958	5	3	0·88905	1948
6	17	0·88003	1959	6	18	0·88282	1958
7	3	0·87093	1948	7	17	0·84291	1959
8	18	0·78585	1958	8	9	0·76565	1892
9	11	0·77854	1866/ 1883	9	11	0·75154	1866/ 1883
10	6	0·77846	1954	10	10	0·74751	1958
11	9	0·76113	1892	11	6	0·53337	1954
12	2	−0·92133	1883	12	14	−0·55043	1894
13	8	−0·93232	1883	13	12	−0·64081	1896
14	16	−0·97038	1880	14	16	−0·85383	1880
15	14	−0·98488	1894	15	4	−0·92014	1882
16	12	−0·98752	1896	16	2	−0·92367	1883
17	4	−0·98936	1882	17	13	−0·95735	1882
18	13	−0·99314	1882	18	8	−0·97737	1883

5.10.3 Results

Correlations of objective façade measurements with the factor loadings of façades on factor 1 (cluster of 'old' façade profiles)

The correlation coefficients calculated between the ranked orders of different façade measurements for different portions of façade and the ranked order of the loading of the façades on factor 1, are given in tables 5.22 and 5.23. All correlations are statistically significant beyond the 0·05 level ($\rho > 0\cdot399$) and most even beyond the 0·01 level ($\rho > 0\cdot564$), thus confirming the hypothesis of the relationship between façade measurements and subjective façade clustering. The correlation coefficients for the

Table 5.22. Correlations (ρ) between the façades rank-ordered according to the factor 1 loadings of the subjects' SD ratings and the same façades rank-ordered according to the 'objective' façade measures: TTRs.

	TTRe		TTRo	
	study 1	study 2	study 1	study 2
Ground floor plus roof	−0·746	−0·765	−0·769	−0·684
Ground floor	−0·773	−0·752	−0·845	−0·730
Whole façade	−0·628	−0·655	−0·730	−0·658

subjective measure of study 2 tend to be lower than the coefficients for study 1, confirming again the greater difficulty subjects with lower educational background had grasping connotative semiotic structures of style. When the correlations for the different façade measurements are compared, the predicted superiority of TTR correlation coefficients over information measures seems to obtain in most instances for TTRos. It will be remembered that TTRos are TTRs computed by taking into account façade openings only, whereas TTRes are computed on the basis of all façade elements (cf page 267). Of the information measures, H_1 yields higher correlations than H, of the TTRs the better results are for TTRos. It must follow from study 5 on information aesthetics that the correlations of H_{ae} measures with the subjective façade clustering should be positive, though identical in numerical value to the correlation coefficients for H.

In relation to the other façade studies, especially studies 1 and 2, this result seems to confirm the assumption that there is a semiotic structure in which two re-cognized manners of conceptualizing tenement house façades are correlated with two kinds of façade design. To the universe of the signified (the two re-cognized 'styles) and to the universe of the signifier (two kinds of design) two oppositional systems of properties appear to correspond. The system structuring the universe of the signifier

Table 5.23. Correlations between the façades rank-ordered according to the factor 1 loadings of the subjects' SD ratings and the same façades rank-ordered according to the 'objective' façade measures: H and H_1.

	H		H_1	
	study 1	study 2	study 1	study 2
Ground floor plus roof	−0·608	−0·573	−0·614	−0·556
Ground floor	−0·486	−0·469	−0·598	−0·548
Whole façade	−0·451	−0·442	−0·738	−0·631

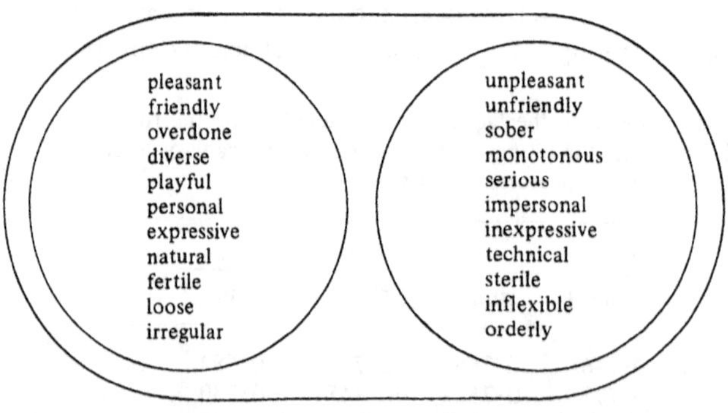

Figure 5.28. Structure of the universe of the signified, opposing adjectives descriptive of the impressions subjects had of 'old' and 'new' façades.

must consist of such oppositions as 'presence of decoration' versus 'absence of decoration' and 'heterogeneous distribution' versus 'homogeneous distribution' of façade elements. The system structuring the universe of the signified consists of the oppositions between adjectives reflecting the impressions subjects had of the façades. How the system might look is shown in figure 5.28. The semiotic structure would thus be of the variety in which one system of sets is given its relevance by a single second system of sets (cf page 44):

$$u\bar{a}b \longleftrightarrow y\bar{n}m, \qquad u\bar{a}\bar{b} \longleftrightarrow y\bar{n}m, \qquad u\bar{a}b \longleftrightarrow y\bar{n}m, \qquad uab \longleftrightarrow ynm,$$

The complete semiotic structure is given in figure 5.29.

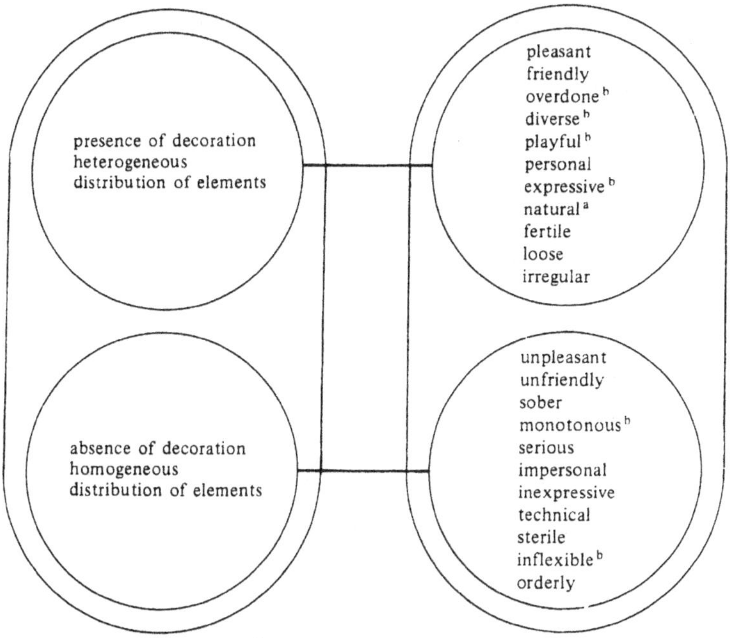

[a]Very high loadings, ~90. [b]High loadings, ~70+.

Figure 5.29. Semiotic structure of architectural connotation. In this structure the universe of the signified shown in figure 5.28 is coordinated with the universe of the signifier in which two kinds of façade design are opposed.

5.10.4 Discussion

Why are TTRs better predictors of subjective façade ratings than information measurements?

In most instances TTRs, especially TTRos correlated better with subjective façade ratings than did information measurements. By comparing what goes into the computation of TTRs and information measurement an attempt to find a reason for this important sensitivity of TTRs will be made. Information measurement, at least as it is proposed by Bortz (1972) and von Buttlar and Wetzig (1973), takes into account a repertoire of different signs. Bortz's signs were based on six different surface types (wall, roof, window, balcony, decoration, and greenery), which were increased in study 4 to nine surface types (sky, roof, wall, window, balcony, decoration, display window, door, advertising). Von Buttlar and Wetzig sorted façade items into many different classes (such as round-headed windows, tall rectangular windows, square windows, cellar windows, T-shaped crossbars, stepped gables, wide triangular gables, small triangular gables, segmental gables, sprockets in gables, tympana above windows, pilasters, panels, architraves, entablatures over grouped windows, moulded ledges, string courses, brackets, rusticated quoins, door surrounds, plinths).

In any particular study the repertoires may be amplified or restricted according to the desired degree of precision. However, once an observed architectural detail of a given façade is attributed to any one of the categories proposed by Bortz, by study 4, or by von Buttlar and Wetzig (see also Wright, 1973) it is not distinguished any longer from the other observed elements in that category. Thus placing a first observed decoration surface in the same category as a second different decoration surface neglects the differences which may exist between the details of the two decorative surfaces. Von Buttlar and Wetzig, although working with many more categories, neglect nevertheless the distinctions between items within their categories. If they list, for example, eight occurrences of round-headed windows, they take no account of whether these eight windows have the same kind of round head.

TTRs, either dealing with a complete list of façade detail or with façade openings only, differentiate between items in a given category by expressing the proportion of each different item type in relation to all items in that category. Thus, whereas von Buttlar and Wetzig lump together two different types of round-headed windows, each type occurring four times, into one category containing a total of eight, the TTR would distinguish 4/8 + 4/8.

If—as one might deduce from Bortz's (1972) study—objective façade structure (and the subjective feeling of structural order) is made up mainly of a window-balcony-wall relationship (including the presence of decorative window frames and different window size), the count of the different types of façade openings divided by their total number will contain more 'information' than the absolute number of windows, doors and balconies divided by the sum total of these elements, this proportion being entered as a 'probability of occurrence' into Shannon's (Shannon and Weaver, 1949) information formula.

It is not asserted here that information measurement cannot be sensitive to difference in detail, but that researchers using information measurement have overlooked this difference so far. It is clear, however, that the larger a list of architectural detail is, the better will it approximate the results of recordings taking those differences of detail into account.

Different stimulus sources of reactions toward façades
In this study only one possible stimulus source of people's reactions toward façades has been studied. This concerned the number of architectural elements in a façade's repertoire and the relationship between the occurrence of elements on a façade and the entire repertoire of façade elements. What was studied then, was the *'homogeneous versus heterogeneous' distribution* of those elements as a possible stimulus for different subjective impressions.

Furthermore, by also considering the transitional probabilities between façade elements the intrinsic structure of façade elements of a given façade was taken into account. As indicators of façade structures the measures H_1 and H_{digram} are especially effective.

Kiemle (1967) mentions another aspect of the perception of architecture which he calls '*transinformation*' (cf pages 114, 115). Looking at a building, one may guess at its internal and constructional properties and some of them may even be apparent on the surface of a façade. For instance, static properties of posts and beams may not be hidden but shown in a façade. Also, when Bortz (1972) had his raters estimate how many flats were in a building these raters then used 'transinformation' of the façade to make their estimates (cf page 249). Incidentally, the correlation between these raters on the architectural attributes 'number of flats' was second highest in the rank order of twelve attributes (page 249). This result indicated that 'transinformation' could be a very reliable source of differential appreciation of the outside of a building. 'Transinformation', however, would be largely 'functional' in nature, telling the observer by the configuration of posts and beams something about the static structure of the building, by the size and number of windows something about its inside illumination, and by the number of balconies something about the number of flats in the building.

Since at present it is still very difficult to conceive of a repertoire of transinformational elements which could be quantified, this aspect has not been taken into consideration (nor has it been controlled) in the present study.

Proportion should be another stimulus source of subjective reaction toward façades. In fact, since the Pythagoreans, proportion in the different art fields (including music) has been cited as the basis of subjective pleasure. In music, chords produced by proportionally subdivided strings of instruments were said to have therapeutic effects. The 'golden section' is a particular proportion (related to proportions of the human body) which allegedly elicits aesthetic pleasure almost like a reflex. However, the studies carried out since Fechner (1865; 1876), on preference for rectangles with different proportions, indicate that there is by no means an exact 'golden section' eliciting aesthetic preference but a clustering of proportions around the 'golden section' and in addition, around the square. Nevertheless, proportion and especially proportions based on measurements of the human body may very well influence aesthetic evaluation by cultural habit.

Concerning a further possible source of subjective reactions toward façades, the following suggestion was made by the Geneva architect Vicari in a personal communication. Architects reacted very late to the introduction of *conventional measurement* based on the metre, which since the French Revolution had become the abstract common equivalent for spatial measurements (as money had become much earlier in history the same for merchandise). Until the 1930s, many architects on the European continent measured in feet and other such measures based on the human body. The disappearance of this type of organic measurement and the adoption of conventionally based measurement may have resulted in an unpleasant break in the proportional continuity of our cities.

5.11 Summary and discussion of all façade studies

5.11.1 Summary

The empirical results of these studies on architectural connotation tend to support the hypothesis that, by means of the polar adjective scales of an SD, subjects are capable of perceiving the difference between building styles within a given category of social function. This should, however, be qualified by the empirical results of the second hypothesis that lack of stylistic knowledge due to insufficient education tends to diminish the differences between building styles perceived by the subjects.

It was also possible to measure differences in façade style 'objectively' by expressing in terms of TTRs and information measures, the variety and complexity of visible architectural design detail and the occurrence of particular details within the overall distribution of design features. When the rank order of these measures was tested against the subjects' impressions, as recorded by the SD, significant correlations were found between the subjective impressions of façade complexity and variety and high TTRs and information measures, and again between subjective impressions of order, amorphousness, and impersonality and low information measures. These results contradict the theory of information aesthetics based on Birkhoff's (1928; 1932) quotient of aesthetic measure. The quotient is obtained by dividing a coefficient of order by a coefficient of complexity, and yields a higher aesthetic index as order increases.

5.11.2 Discussion of the studies—two different styles of façades

Despite the relatively unequivocal results of the subjective, objective and correlational studies, their limitations should also be stressed. First of all, there are probably further design variables not investigated in this study which could conceivably be correlated with subjective impressions of façades. As has been mentioned, one such variable might be proportion. Another variable could be the decimal system of the metre coming into use on the European continent since the French Revolution and playing the role of a 'general and abstract equivalent of space'. The buildings constructed before 1900 probably were not dimensioned in metres since the use of that measure penetrated relatively late into the practice of architecture and the building trade. They are therefore still dimensioned after measures derived from the human body (foot, thumb, etc) and could therefore conceivably express a different dimensional quality.

Another shortcoming is that TTRs and information measurement, used in these studies to describe differences in the structure of the 'universe of the signifier', are based only on the frequency and the relations of architectural elements. This type of information might be called 'syntactic' since it is concerned with the distribution and composition of elements only. Not taken into account in these studies was what might be called the 'semantic' information of architectural elements. This kind of information would be based on the frequency and the relations of the 'meanings' of the architectural elements, such as their functional but also their symbolic meanings, which would concern the 'universe of the signified'.

By 'functional meaning' the technical function of an architectural element is meant (for example, the functional meaning of a beam is to carry a load). The symbolic meaning of an architectural element is its potential associations for the viewer (a Grecian column at the entrance of a theatre or bank building is not only carrying a load but reminding the onlooker of its use in Greek or Roman temples). The present incapacity of information theory (and especially of information aesthetics) to deal with 'semantic information' has been criticized by many authors (for example, Rauen, 1971; Becker and Keim, 1972). It is, however, perfectly feasible to calculate semantic information if a repertory of semantic units can be established in advance.

As the difference between subjects with high school diploma and trade school students in the above studies appears to show, the re-cognition of style might not be the only manner of perceiving façades. It is quite possible that the subjects not educated to deal with stylistic differences should look at façades in an entirely different way. As a matter of fact, their subjective reaction to buildings might have little or nothing to do with the façades. Bortz's (1972) finding that high subjective evaluation of buildings may be predicted by the presence of balconies or greenery indicates that the imagining of such advantages as trees and greenery and private access to open air and sun might be appreciated more than façade design. In addition, old buildings, recognizable when their façades differ from those of new buildings, might be regarded as simply lacking in the conveniences of new buildings. In this respect, the criticisms of Winkel (1972) about housing satisfaction studies carried out by having small groups of college students rate coloured slides of façades on an SD, is worth considering. As to possible sources of housing satisfaction, Winkel speculates:

> "housing satisfaction is obviously much more clearly related to the types of behaviour which internal space arrangements make possible,

attitudes toward and use of space external to the dwelling unit, the ease with which needed goods and services can be obtained, the judged desirability of the social and physical characteristics of the neighbourhood, whether the house is owned or rented, the factors surrounding the move into the house (was the resident forced to move or did he move voluntarily) and finally the person's judgement concerning overall life satisfaction. The latter question must be asked because variations in the housing-related behaviour listed above might be better accounted for by overall satisfaction than by any characteristics of the immediate housing environment."

If housing satisfaction is related to such factors as Winkel suggests it is conceivable that the subjects project their desires and attitudes with respect to housing onto the SD scales, reacting to the façade stimuli in a similar way as they would to Rorschach inkblots in a projective test.

There is still another potential shortcoming of studies like the above to be discussed: although 'synchronic' investigations of architectural codes are necessary and legitimate, the consideration of semiotic structures in architecture and urban design remains incomplete without a 'diachronic' (historical) perspective.

Regarded from a historical point of view, the sample of façades used in the studies stem from two different phases of the development of a particular capitalistic system. Both kinds of façades must thus be considered as a result and expression of the forces of production (technical advancement) and the social circumstances of production under capitalism. How then could the recognizable differences in façade design be explained, if both kinds are brought about by the same historical economic system?

A possible explanation of this difference could be based on the concept of "inequality in the developmental relationship between material and artistic production" (Marx: Marx and Engels, 1970, Volum 13, pages 640–642; Volum 23, pages 535–537). This hypothesis maintains that artistic production may develop more slowly than material production since it is dependent on cultural and symbolic processes which maintain a certain inertia in comparison with the more speedy and less complex development of technique.

According to this concept the architecture of the original phase of industrial capitalism based on competition, namely architecture before 1900, would result from a manner of designing and constructing buildings according to materially new ways, but 'semantically' (formally) still adhere to models of the feudalistic era which had historically already been superseded by capitalism.

Some support is lent to this hypothesis by the fact that the decorative designs featuring Greek, Roman, or Renaissance motives, which were used on façades in the second half of the nineteenth century, were increasingly prefabricated industrially. Kürth and Kutschmar (1969, page 168) write about façade design during this period:

> "For the external design of the buildings the different parts of the façade, columns, mouldings, architraves and other decorations were mass produced by industry and could be commissioned from catalogues. They were frequently made out of inferior building materials" (cf Lessing, 1882; 1884).

In this early phase of industrial capitalism architecture had thus already found some ways of industrially prefabricating elements but had not yet developed an 'artistic' form to correspond to this development of material production (figure 5.30; Monke, 1968). How would such a 'capitalist' artistic form have to look? In the ultimate stage of development it should reflect the mentality, the basic motivations of the capitalist building proprietor, which were not yet visible in the sumptuous villas (and in their modest copies, the workers' homes) of the nineteenth century. This mentality has been aptly described by Marx in the following way: "The scorn of theory, of art, of history, of man as an end in himself, that is

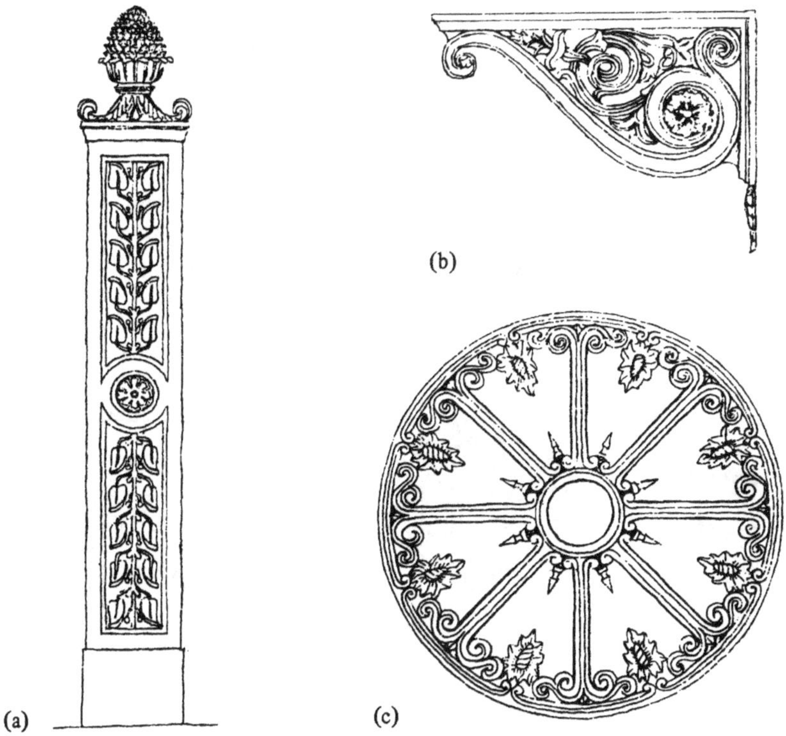

Figure 5.30. Zinc castings for architects by M Geiß, Berlin: (a) pillar of a balcony railing after Knoblauch; (b) console after Stüler; (c) door decoration after Schadow; (figure continued overleaf).

the real conscious standpoint, the virtue of the money-man" (Marx and Engels, 1970, Volum 3, page 212) "... To the capitalist method of production corresponds a manner of spiritual production which is different from that of the middle ages... for example, capitalist production is hostile to certain branches of spiritual production, e.g. to art and poetry" (Marx and Engels, Volum 26, page 255).

A comparison of the architecture of the precapitalist and the capitalist eras would tend to confirm this description. This comparison in the West would show that architecture on the whole has become even less 'artistic' than it used to be in the earlier phase of capitalism. (Such a comparison would have to be made on large random samples of buildings and treated statistically.) It seems that, after all the visual 'noise' resulting from the escalation of copying all architectural styles of the past which architects could put their hands on in the early phase of monopoly capitalism, the turn of the century saw an uneasy visual 'silence' strike in: the decorations literally drop off the façades. Where architecture holds on to 'meaning' it now copies the machine, for instance the ocean steamer (figure 5.31). It is in this phase that "the old local and national moderation and isolation" is succeeded by "universal communication, universal dependency of the nations upon each other. This applied in material, as well as in spiritual, production. The spiritual products of the single nations become public property. National one-sidedness and narrowness is made increasingly impossible, and out of many national and local literatures comes one world literature" (Marx: Marx and Engels, 1970, Volum 4, page 466), "and the 'international style' in architecture"—one is tempted to add, to complete the description of this process.

It is true that the development of capitalist functionalistic architecture out of the narrow-mindedness of copying the styles of the past represents a certain progress. On the other hand, the highly developed and all-pervasive functionalism of modern architecture shows its true face in the

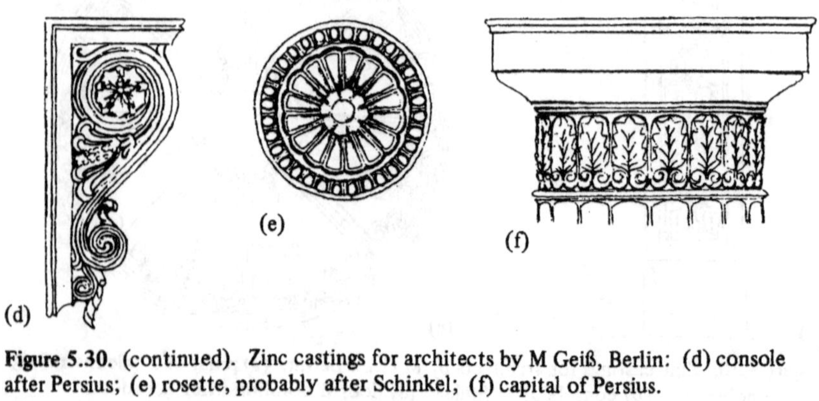

Figure 5.30. (continued). Zinc castings for architects by M Geiß, Berlin: (d) console after Persius; (e) rosette, probably after Schinkel; (f) capital of Persius.

Figure 5.31. Pavillion Le Corbusier, Zürich (photo' by courtesy of Urs Janger, Zürich).

anti-artistic emptiness or in the abstract decorativism of the façades. It took the architecture of industrial capitalism a century to reveal its basic motives in its external form: stinginess. One common reply at this point is that architecture in the Socialist countries is no better than architecture in capitalist countries. Apparently this is true—at least at the moment. But could it not be that the real socialist architecture has not yet developed just as nineteenth century capitalist architecture had not yet developed its own models and lived copying the style of feudalism? It is only fair to concede the same amount of time for the development of an architecture of socialism, as capitalist architecture has had in order to come up with functionalism. When the architecture of developed socialism has found its own expression it should certainly not be anti-artistic and empty of content if it correctly expresses the mentality of the people who live according to its principles.

6

Synopsis of the research as a whole

6.1 The fallacy of linguistic analogy in object semiology

The survey at the beginning of this book, of the literature on object and architectural semiology, showed that the notion of the universal semantization of objects implies the reduction of material objects to signs and the undervaluation of the notion of material practice. It can therefore be criticized on the grounds of philosophical idealism. A particular variant of this tendency is the reductionist analogy between 'sign systems' (particularly language) and 'object systems'. The result of this has been a search for linguistic equivalents in certain object systems with a view to constructing semiologies of these systems, for example, semiologies of architecture or the urban environment. The explanatory and practical value of regarding a building as consisting of phoneme-like or moneme-like units (bricks or walls and floors) and a city as a text is doubtful to say the least.

Rossi-Landi (1975) proposed instead that rather than treating linguistics and object systems as analogous, one recognizes that between material artefacts and linguistic artefacts there exists a constitutive homology based on their common genetic root in production. This homology he called the homology of production. Whereas analogy traces a posteriori similarities between objects of any sort, not united in any necessary way, homology reveals a common anthropogenic root existing between objects which have since come to be seen as distinct. Linguistic analogy is based simply on assumed structural equivalences between language as a system and any other system. The logical errors which this gives rise to have already been discussed (pages 34–36). Linguistic homology on the other hand recognizes the common origin of material and linguistic artefacts in the ability of men to engage in cooperative labour for their common survival.

Prieto's (1975a; 1975b) 'semiology of instruments' presented another alternative. Prieto proposed that, instead of reducing objects to signs, one reduces signs to objects. Signs would then become a particular set in the universe of instruments, serving the function of communication.

The method finally adopted in this book to construct a semiology of architectural and urban activity took as its methodological basis a line of research which had been first adopted by linguistics (phonology). This was to study how people differentiate between different speech sounds, an area of research which leads to the more general question of how people distinguish and conceptualize any type of material reality, whether perceived acoustically, visually, or otherwise.

6.2 The utility of the adopted theoretical framework

In following Prieto's (1975b) general framework, it was first established that the perpeptual conceptualization of material reality proceeds by 'constructing' sets of objects distinct from other complementary sets of objects within a universe of discourse. This is done on the basis of an oppositional structure of the material characteristics of the objects involved. The utility of the notion that perception involves classification is in its general applicability. Different spatial visual patterns must be as subject to analysis into oppositional structures as are linear sequences of different acoustic utterances.

In accordance with Prieto, the correlation of two mutually relevant oppositional structures was then considered a semiotic structure. One of these oppositional structures assumes the function of 'universe of the signifier', the other that of 'universe of the signified'. Again the value of the concept of semiotic structure is its generality. A relationship of signification is not confined to semiotic structures of communication (where the universe of the signifier is a system of sets of signal variants correlated by means of a 'code' to a system of sets of 'meanings'), but can be guaranteed whenever two mutually relevant oppositional structures are correlated. A universe of the signifier might thus be composed of the variants of any kind of tool (for example, an architectural or urban support), while the universe of the signified be composed of the set of all functions which can be implemented by that kind of tool (for example, production or distribution of products carried out by cooperative groups).

It was seen further that a semiotic structure is, as a principle, viable in two directions. If one regards it from the standpoint of the producer, one proceeds from a classificatory activity conceptualizing an oppositional structure between abstract functions and their characteristics to the realization of concrete objects. The consumer proceeds instead from a classification of concrete objects to the re-cognition of their correlated system of abstract functional properties. This twofold viability of the semiotic structure was seen to be particularly useful in the construction of a theory of architectural and urban production and consumption. The case of the sender discovering the appropriate concrete signal for what he wants to say, or that of the receiver inferring from the signal what the sender wants to say, is only a special application of the 'double access' to semiotic structures.

Also of much practical value was the concept of double classification. Here the universe of the signified is classified twice, first 'notatively' in order to detach the set of functions to be realized from its complement (the set of alternative functions which could have been realized instead within a given material culture), and then 'connotatively' in order to establish one method of realizing that set of functions in contrast to other possible ways of realizing it. In architectural and urban activity this double classification forms a necessary sequence in the production and consumption of design. It has to be established first what has, or could have been, designed in order to then establish according to what method it was, or could have been, designed.

The method of realizing a function (subsequent to conceptualizing that function) has been regarded as the problem of 'connotation' or 'style'. The concept of 'style' has a fundamentally heuristic value. For every function conceptualized there are theoretically different methods of realizing it. This provides the free space necessary for creativity. With regard to architectural and urban semiology it is important to understand that this logic regards functionalism as a style as much as any other.

Another important aspect of style is that it may be functionalized in turn by being deliberately contrasted with other styles of the same function, becoming thus notatively the signifier of a message. The author of the choice of a particular style may wish to communicate that this style was deliberately chosen in opposition to 'normal' styles and should therefore be considered an 'innovation' or 'work of art'.

Deliberate choice of a style could also be used to indicate class distinctions, etc. This notion of style-as-a-message has been useful in determining whether an architectural or urban object can be regarded as a 'message' and under which circumstances this is the case. Generally an architectural or urban object would be regarded as pertaining to a universe of the signifier, which is correlated in a semiotic structure with a universe of the signified of social functions (human group activities). Such a structure is significative, but not in the sense of intentional communication. The manner of realizing the object—that is, its style—may, however, be functionalized, becoming the signifier of a message, for example about its particular innovative or artistic quality. In this case the style of the object enters a semiotic structure of communication.

It has been pointed out that a particular style operates as a message only on the basis of the receiver's ability to conceptualize a system of sets in which its special characteristics are classified in opposition to the characteristics of other styles. Since this ability is a function of socialization and formal education, it serves frequently to distinguish the culturally and economically privileged elite from the dominated class.

Whenever communication occurs, several semiotic structures are involved. The mere production of signals becomes a signifier in a semiotic structure, the signified being the intention to communicate. Whether this communication is actually understood or not makes no difference. Besides the code that indicates which messages are permitted, there is the semiotic structure that indicates which of the permitted messages are favoured by the circumstances and which are not (Prieto, 1966). It is possible that the role of context has so far been neglected in the semiology of signification. It has been suggested in this research that the context of the neighbourhood or city should play an important role in the conceptualization or recognition of architectural functions, in the same way that the regional context would be important in the conceptualization or recognition of urban functions.

Finally, an important consideration of style appears to be that sort of semiotic structure where the universe of the signifier consists of a system of sets made up of opposed formal design characteristics and a universe of the signified consisting of a system of sets of opposed 'impressions' as they could be verbally expressed by means of adjectives. It can be seen that the semantic differential (SD) (Osgood et al, 1957) is particularly appropriate for measuring such hypothetical variations of impression correlated with variations produced by the manipulation of form. The SD should thus be particularly useful as a measuring instrument if variations of form are the independent variables, and variations of impression the predicted dependent variables in architectural or urban design.

A result which should have interesting implications for the application of the semiological framework proposed in this book is that architectural practice and urban design practice have been established to be logically interdependent as a result of the interdependence of the two semiotic structures that underlie them. It was argued that an optimal urban design is an 'integrated whole'. A whole is integrated when each of its parts has its own meaning by virtue of its own semiotic structure, much as a word has its own meaning in a sentence. The integrated whole should at the same time be an independent unit by virtue of its own semiotic structure, much as a sentence has an independent status in relation to other sentences. For this reason the semiotic structure of the whole (the urban design) should ideally be established prior to that of its parts (the architectural objects). Although these (architectural) parts can have meanings of their own, they become 'integrated' only in terms of the whole, remaining otherwise merely 'juxtaposed'. Designing an architectural object should thus amount in practice to a sequence of two double classifications. First it would have to be established what kind of urban design is warranted, which is a notative process, then how the urban design is to be composed, which is a connotative process. Second it has to be established what kind of architectural object is to be designed, which is notative, then finally how this object is to be composed, which is again connotative. This interdependence should have interesting implications for architectural education and practice.

6.3 Methodological considerations

Although the results of the empirical studies are valuable to those interested in the particular research questions asked, the methods applied in the studies are probably of more general importance.

One of the possible aims of architectural and urban research has now been clearly formulated. This was to establish the semiotic structures underlying architectural and urban activity. This search for semiotic structures entails a precise procedure.

1 The universe of the signified in architectural and urban activity is analyzed twice, first in terms of a system of notative functions, then in terms of a system of connotative functions.

2 The universe of the signifier, containing the system of design variations by which the above functions are implemented, is then analyzed.

A number of empirical methods were proposed in conjunction with this general procedure. These were:

1 The indexing of functional identity by similarity measures applied to subjects' sorting of visual stimuli, then the clustering of these measures into categories of functions.
2 The indexing of stylistic differences by recording subjects' impressions of these differences on appropriate SD scales.
3 The description of architectural objects in terms of the morphology of buildings: an index of perceived size derived from the number of storeys and the size and number of windows per storey; an index of perceived homogeneity derived from the degree of unity of building volume and the degree of variation of window form.
4 The description of architectural objects in terms of TTR and information measures applied to the classes and distribution of architectural design detail visible on a building façade.

The first method has been applied previously to the study of word meaning. Each subject sorts words with similar meanings into groups. The degree of similarity is calculated in terms of either the distance between the most similar members or the distances between the most dissimilar members, similarity or difference being defined as the maximum or minimum frequency of association of a word with other words in the pooled word-by-word matrices. Similarity can thus be expressed as a hierarchical system of clusters, with increasingly dissimilar items towards the apex of the hierarchy or as a plot in which the distances between points in an *n*-dimensional space are analogous to the degree of similarity between the items represented by the points (pages 118ff). There is no reason why this method should not be equally successfully applied to measure similarity between picture meanings, the functions of pictured objects or the objects themselves.

The second method involved the SD, which was described as a high 'objectivity' instrument in persistently yielding the same (E, P, and A) dimensions applicable cross-culturally to a wide variety of concepts and objects. It was found to have similarly high 'reliability' as far as correlation between subjects within the same group is concerned. Specific attention has not been paid to the 'validity' of the instrument, although the face validity of Orlik's (1965) imaginary experiment can hardly be doubted. The question of validity is less satisfactory in the case of polarity scales which are only metaphorically applicable. According to Osgood (1971) it is probably these scales which are responsible for the inflation of the EPA dimensions. But it has been suggested that this problem can be avoided by selecting only scales which are appropriate to the objects or concepts concerned.

Furthermore, Lindekens has shown that metaphorical scales do not produce any significant effects when applied to visual objects. It has to be admitted, however, that Lindekens did not use factor analysis in his studies, and the effect of metaphorical scales has generally been accounted for by inflated proportions of variance in the EPA dimensions, resulting from factor analysis. This does not exclude investigation of the EPA dimensions of objects or concepts as long as these dimensions are measured on nonmetaphorical scales. In general a mixed procedure is advisable: in preliminary investigations adjectives should be chosen which are appropriate to the subjects doing the experiment; if possible, these adjectives should be classified in advance as to their E, P, or A pertinence and should be represented in equal proportion with other adjectives appropriate to the objects concerned. For problems involving architectural objects, lexicons which are related to the research topic should be chosen from Hershberger's (1972) collection. In any event, a classical item analysis, after a pilot study, should reveal irrelevant items and should lead to a sharpening of the discriminatory powers of the instrument.

The objectivity, reliability, and—with appropriate cautions—the validity of the SD are overshadowed by the impressive 'utility' of the instrument. The low cost of its production and administration and the wide scope of its applicability would make the SD an important psychometric device even if its precision were relatively limited. Since the production, administration and simple techniques of data analysis (for example, mean profiles) of the SD are well within the range of those concerned with architecture and urban planning, it can be seen to be of considerable value in these fields.

The third method, the morphological description of architectural objects, has not yet been systematically developed. Some possibilities for the construction of a documentary language have been mentioned (pages 115-118). It is not clear, however, which terms in this language could be regarded as equivalent to distinctive features in the recognition of architectural function. The best method of obtaining information about these features (which could differ from individual to individual) should be a careful ontogenetic study of architectural perception in children. A more practical approach would be to carefully interview subjects during an experiment on the classification of buildings according to function.

The fourth method, the information-theory approach to measuring objective features of architectural design, was approached in this research by a specific mapping technique. The architectural objects to be compared, whether façades of single buildings or of a street front or of an entire district, are projected onto a standard grid of squares. By labelling the visual content of each square the repertory of architectural design elements as well as the frequency and distribution (contiguity, dispersion) of these elements can be obtained. The results can be fed into the Shannon (Shannon and Weaver, 1949) information formula so as to derive index measures for the architectural objects in question. A count of the visible elements of a building which can be labelled verbally can be used in a similar fashion to the mapping method. Here façade elements may be counted, the different labels forming the repertory (that is, the façade 'vocabulary'), and the frequency of occurrence of each label weighted by the total of labels and entered into the Shannon formula. It should be difficult, however, to establish transitional probabilities between labelled parts with this method. Aside from the information-theory approach, TTRs appear to give quick and useful estimates on the objective side of façade design.

6.4 Unsolved problems

One of the questions remaining open for further research is the difficulty of assessing the importance of individual differences in classifying the semiotic structures underlying the urban environment. It has been deduced that the subject is the final member of the semiotic chain (pages 58, 59); hence each classificatory perspective depends on the subject. But as each subject is socialized in a given society according to the educational goals and values of that society, there should be some common principles underlying subjective variations. Since they constitute learned classification behaviours, individual differences in perceiving the urban environment should thus be related to differences in social class and socialization. This relationship would constitute a broad and useful field of research.

A further problem is integrating the different branches of semiotics which have developed. The two main currents, the Saussurean and the Peircean, have yet to be systematically analyzed for similar content structures. It could be proposed that Peirce's diad of the material and object relations of the sign corresponds to de Saussure's diad of signifier and signified. The interpretive dimension in which the diad of material and object relations is nested in Peirce's system would correspond to the classificatory procedures of the producer/consumer in semiology—although in the Peircean school the interpretant is identified with meaning itself.

Probably the most important problem remaining in urban and architectural semiology is that of the historical changes which take place in the semiotic structures underlying the urban environment. One of the most important distinctions introduced by de Saussure into linguistics was the distinction between the diachronic and the synchronic study of language. By diachronic analysis of a language is meant the description of its historical evolution through time. By synchronic analysis is meant the description of the language as it exists at a certain time in history. This distinction is equally applicable to the subject matter dealt with in this book. Here the synchronic approach has been adopted, describing and explaining the semiotic structures concerned as they are at present internalized by members of a specific urban culture. The diachronic approach would analyze the structures in terms of the changes which they permit on the basis of the structural calculus by which they are defined. It is important to realize that the latter approach is not simply an historical account of the development of the structures but a search for the changes which are potentially permissible in them if the correlation between the two oppositional structures, each based on its own calculus, should persist. The historical development of a semiotic structure is only part of the search and can at times only be reconstructed by consideration of the potential changes built into it. Nevertheless, potential changes must be explained in terms of the functions these changes fulfill. And since semiotic structures imbue human activity with meaning, their structures and potential for change have to be considered within the realm of these activities and their potential for change.

Ultimately, the potential for change of human activity is rooted in the concrete history of men who perform meaningful activities. In each 'act' on the 'stage' on which this history unfolds, certain social classes are posed in opposition of interest to other classes. Instruments and signs are the props of these protagonists, differentiated by their different roles in the class conflict which is evolving with a dialectical rhythm.

The correlation between 'protagonist opposition' and 'opposition of protagonist props' could be considered a diachronically unstable semiotic structure. Krina Costa-Gomes-Ber et al (1973a; 1973b) investigated a limited episode of such an opposition within an area of the Swiss city of Lausanne. They developed a formula for describing such diachronically unstable semiotic structures in which the opposition of protagonist props is divided by the opposition of the protagonists. For example, in a particular section in a medieval town a social opposition can be seen between merchants and master craftsmen and between domestic servants and apprentices, while there may be no opposition between their props, that is, the section of the town consists of building of 'mixed functions'. This could be expressed in the formula:

$$\frac{\text{dwelling} + (\text{workshop or store})}{(\text{merchant vs master craftsman}) \text{ vs } (\text{domestic servant vs apprentice})}\cdot$$

A transformation of that state of the semiotic structure (the above formula being a representation of a semiotic structure) to a state prevalent in contemporary city centres can be represented in the following formula:

$$\frac{\text{dwelling} + (\text{workshop or store})}{(\text{merchant vs master craftsman}) \text{ vs } (\text{domestic servant vs apprentice})}$$

$$\rightarrow \frac{\text{office} + \text{department store}}{\text{professional vs (lower manager vs employee)}}\cdot$$

The development of separation of classes and functions can likewise be expressed in a general formula:

$$\frac{(I_{C_1} + S_{C_1}) + (I_{C_2} + S_{C_2})}{C_1 \text{ vs } C_2} \rightarrow \frac{(I_{C_1} + S_{C_1})}{C_1} \text{ vs } \frac{(I_{C_2} + S_{C_2})}{C_2}$$

where C is social class, I is instrument, S is sign.

The history of urban configurations could thus be described as a result of changing oppositions between social classes as manifested materially in the urban environment. It is important to realize that change in class oppositions, for example by the descent of a dominating to a dominated class, is not necessarily accompanied by a sudden change in the 'props' of the victorious class. Instead, the instruments and signs of the subdued class exist on as 'models', until the development of the forces of production in combination with the new social conditions allow the victorious class the elaboration of new instruments and signs.

It is an interesting fact that the above formulae developed to describe such changes merely present an alternative method to the method set out in this book to represent semiotic structures. The relation between the sections above and below the fraction lines would correspond to the relations between the universes of the signifier and the universe of the signified in the schematic representations of semiotic structures in this book.

6.5 Conclusions: implications for professional practice and education

If it is true that the integrated urban whole is logically prior to the architectural object, then architectural practice should be carried out in much closer connection to and cooperation with urban practice. Since urban practice concerns the geographical and topological distribution and density of architectural (that is, social) functions within a given urban project, integrated architectural practice should not be developed independently of urban practice. This dependence of architecture on urban design should extend even into the choice of style. The consideration of a 'local' stylistic context in the design of a building may occasionally influence the choice of architectural design solutions. But occasional consciousness of the immediate urban context is not enough to produce 'meaning in the urban environment'. The entire 'image of a city' has to be planned and realized on the level of architectural practice as well.

A further point emerging from the theory of this research, which could have significant implications for professional practice and education, is the notion of 'double classification' of the universe of the signified. This results in a double working cycle in design. This double cycle is characterized by the question 'what' is to be designed and 'how' it should be designed. First it has to be established what 'type' of urban environment is to be planned before one can establish how to plan it. The information from this cycle is then fed as an input into the next cycle. This involves establishing first the 'type' of function that has to be fulfilled by the building, then the realization of that function. This more fundamental paradigm of the 'design process' should be superimposed onto its current models, which generally contain an analytic information-gathering phase followed by a synthetic 'realization' phase. Each of these phases should be applied to each of the two double working cycles and there should be feedback loops connecting all the phases in all parts of the working cycles. This indicates that the main task of urban planners and architects is the maintenance of the flow of information between the various phases of the various cycles; that is, that their task is less oriented toward physical implementation—as has been believed in the past—than toward communication and negotiations concerning alternatives.

Yet another implication which the theory set out in this book has for professional practice is the notion of 'connotative' classification of the system of stylistic alternatives in fulfilling a particular function. This notion could lead to a definition of professionalism derived from the task of producing stylistic variants. In fact, historically, the profession of the 'free architect' detached itself from the narrow organization of craftsmen in guilds as a result of the boom in the demand for architectural alternatives caused by the concentrated appearance of the 'nouveau riche' Renaissance bourgeoisie on the building market (Hauser, 1951). This boom in the 'business of stylistic alternatives' gave way to a corresponding slump. Today the style of functionalism—the choice of no variation in style—is practiced for reasons of economy and industrialization. The question may actually be asked, in fact, whether stylistic variants in architecture are necessary, which leads to the next question: is the architectural profession necessary. A tentative positive answer can be derived from the research: the polysemic uniformity of the urban environment requires the careful and deliberate introduction of meaningful variation.

Closely related to this problem is consideration of context in urban and architectural design. The context of urban design is regional, that of architecture urban. As has been shown (page 66), failures in recognizing urban and architectural functions can be caused in part by neglect or incorrect prediction of context. The notion of context has a double significance for architectural and urban practice: first, it stimulates the conceptualization of regional and urban adaptation of design projects for better recognition; second, it permits 'semantic economy' by allowing the omission of certain significants in a design where the context already carries sufficient information about the 'correct' re-cognition of its functions. This is similar to the situation in communication where, for example, it would be sufficient to say /it is coming/ rather than /the bus is coming/ in the context of a bus stop.

These implications for professional practice imply consequent implications for professional education. If the goal of professional education in the field of urban and architectural planning and design is to prepare for the production and maintenance of a 'meaningful urban environment', the appropriate knowledge, attitudes (professional and political), and skills could be efficiently acquired by engaging in the type of research outlined theoretically in this investigation. The didactic method proposed here is 'learning by research'. The training of future professionals would thus not confine itself solely to the acquisition of the skills necessary to produce urban and architectural variants (that is, 'how' to solve problems), but would encompass the knowledge and attitudes necessary for critically selecting 'what' problems in the urban environment deserve attention. This would require a study of the historical developments which the present economic, political, and social factors determining every concrete situation have undergone, in order to plan their stabilization or change. Learning to produce design variants should thus proceed by investigating concrete, situational forms of the hypothesis that the manipulation of design variables is not without consequences for the citizen-consumer. The questions to be answered would then be what the precise consequences of these manipulations are and who benefits from them.

Appendix

Table A1. Subjects' comments on the six building types at the four levels of detail.

Subject	Comment
Office buildings, minimum level	
1	'piled-up' form, indicates decisionmaking centre
7	form, context, city centre, car
8	uniformity, scale
9	severe lines, grandeur
14	height
15	shape (Manhattan)
16	height, notion of scale
18	articulation of volumes
19	heights like 'New York'
21	articulation of volumes, scale
27	New York City, proportions, car
31	Manhattan, big volumes
32	suspension, city centre, skyline like Manhattan
33	prestigious volume
34	indentations, scale, the wall
Office buildings, second level	
3	shape
8	articulation of volumes
12	height, form
17	nothing special
21	form
28	height, continuous façade
31	cut of the volumes (Manhattan), verticals
33	'showy' volume, pretentious
34	general silhouette
Office buildings, third level	
3	height, volume
7	framework of walls: a lot of windows, no balconies
9	coldness, monotonous, monumentality (actively)
10	form like a skyscraper
15	shape of the buildings, windows
16	height, continuous glazing, articulation of volumes
17	nothing special
18	relation of volumes
28	height, glass surface, old buildings rehabilitated
30	prestige
31	vertical volumes (Manhattan)
32	form of skyscraper
33	prestige, standing
34	form of the building
35	archetype of skyscraper
36	the rhythm of the façades
Office buildings, photographic level	
1	entry (pillars), windows, white curtains
2	sequence of like elements, no personal elements, entry
3	module, openings

Table A1. (continued)

Subject	Comment
Office buildings, photographic level	
13	traditional façade of modern office building
15	windows and ground floor
19	uniformity, little decoration
21	windows, façades
22	resemblance, functionality, penetrating light
23	oblong building, form of windows
29	form of windows, system of window opening, great repetitiveness, subdivision of interior space unlikely to be for dwelling
31	equal windows, lack of balconies
34	rhythm of windows
37	entry, porch, curtain wall
Factories, minimum level	
6	volume, scale
7	façades, dimensions, context (lorries)
12	ugly form, lorry, scale
13	immediately brings to mind the great industrial machine
17	little information, 'monotonous surface'
18	grandeur
19	closed space (prison)
21	form, scale
32	large mass
33	enclosed space
37	arrangement of masses, lorry, passage between two blocks
Factories, second level	
1	repetition of same elements, uniformity
12	façade
14	large wall
17	large proportions, repetition of monotonous façades
22	scale
6	monotony of façades, proportions
8	chimney, façade, roof
25	ugliness
28	chimney, doors, arcade, windows
29	big dimensions of the whole, connection between forms in evidence and probable functions, few openings, openings depending on the interior subdivision of spaces
31	large blocks, few windows
33	differentiated volumes, enclosure
34	rounded elements and elements of roof
Factories, third level	
3	modules, openings
4	importance of buildings
5	height of storeys, façade
7	forms of buildings, windows often less large
8	monstrous forms, scale of openings
9	sadness, ugliness, monumentality, monotony

Table A1. (continued)

Subject	Comment
Factories, third level	
14	dimension, play of 'full and empty'
16	working environment, service elements (doors, windows)
24	monotonous rhythms (façades), repetition (standardized windows)
26	scale, large openings, transparency
Factories, photographic level	
4	immense openings
10	windows, entry
11	play of volumes of different sizes, cold impression
12	entry, façade, form
15	windows, shape
19	mass, heaviness
21	façade, scale of openings
26	materials, volume, windows, sadness, façade
27	façade, form of windows, proportions
28	doors, bricks, windows, height
30	vast interior spaces
31	form, doors, environment
33	dimensions, simple volumes, transparency
35	monotony, cheapness
37	glazing, extent, structure
Tenement houses, minimum level	
6	volume, detail, greenery
8	scale, dimensions, proportions
9	representation of real needs and of their accommodation
12	form, symmetry, system, construction
14	'unité d'habitation'
17	typically modern housing
19	suspended box, free ground floor
24	grandeur, partial rhythm, cornice, suspension
26	scale, form
32	free ground floor
34	treatment of ground floor
	ground floor, slender pillars
Tenement houses, second level	
2	balconies
10	balconies
15	lines of balconies and windows, shape
19	continuity (loggias), horizontality
20	habitable, concentrated, regular
21	forms, façades, scales
24	dimensions of storeys, in general horizontally or vertically dominant
25	balconies
27	balconies, windows, proportions, form
28	balconies, windows
30	long balconies
31	horizontal lines, scale

Table A1. (continued)

Subject	Comment
Tenement houses, second level	
32	balconies
36	horizontal rhythm of the levels
37	large balconies
Tenement houses, third level	
2	loggia, balcony
4	particular signs of habitation
8	façade, windows, balconies, roof
13	image evocative of traditional urban tenement houses
14	scale, balconies, etc
15	shape of buildings, balconies and windows
16	ground floor, balconies, loggia, habitable environment
17	visibly tenement houses
21	façade, repetition of elements
22	scale, entries of building, opening
23	balconies
24	balconies, horizontality or verticality dominating
27	proportions and form
31	ground floor, scale, windows
35	typical residential building
37	shop, loggia
Tenement houses, photographic level	
1	balconies, entry, typological repetition
2	loggia, differentiated curtains, television antennas
4	different flats
6	façade, balconies
7	balconies, curtains in the windows
10	form, façade, curtains
11	volumes, balconies
12	loggia, curtain, proportions
13	image recalling known tenement houses
14	windows, balconies
17	visibly tenement houses
18	balconies, curtains
19	horizontality
20	situated in a town, animation, grandeur
22	functionality, dimensions
23	balconies, appearance of curtains
25	loggias
27	proportions
28	loggia, greenery at the entry, roof
29	relation of windows to wall surface, façade, repetition of loggias
31	cornices, balconies, curtains, ground floor
33	repetition, loggias, entries
34	rhythm, shop at ground floor
35	street, life, animation
37	shop, balcony

Table A1. (continued)

Subject	Comment
Churches, minimum level	
1	monumental form, inspires admiration, radiates calm
2	form, space
4	form
5	organization of volume, symbol
7	form, surrounds, landscape
8	form
13	dome
22	expression of will, dome (representation of heavens)
28	curves, pyramidal elements, heights
30	dome, rounded forms
32	dome, side chapels
33	structure: organization of volume
36	structuring and relationship between elements
37	sensitive traditional plan, radially positioned chapels, aisle, choir
Churches, second level	
4	particular signs (steeple)
5	form, dome
6	volume
15	volumes
18	dome
21	form
22	form
28	form, tower, steeple, windows
30	dome
33	unity of mass, expression, enclosure
35	dome, chapels, aisles
36	the coherence of the wall as opposed to the regular rhythm
37	traditional form, choir, aisle and radially positioned chapel
Churches, third level	
1	closed building: community, greenery, tower
2	steeple, abstract form
3	volume
10	form
11	different volumes, rectangle, triangles
14	steeple, form of the roof
16	element pointing high up
19	play of full volumes (church) and sculptural volumes (steeple)
23	special mythology, steeple
24	variety of volumes, sculptural verticality
29	form, freedom of the constituent elements, environment
32	steeple, sculptural aspect
35	church, steeple, modern architecture

Table A1. (continued)

Subject	Comment
Churches, photographic level	
4	particular signs
7	cross, bells, roof, opening in the roof
10	cross, bell
12	cross, form
15	volumes, façades without windows
17	visibly a church, cross
21	cross
27	cross, etc
28	steeple, cross, glazing
31	cross, bells
34	steeple
35	cross, roof, volume
36	the forms and the walls
Schools, minimum level	
1	students talking, tree, module of classrooms
4	form
8	volume, form, openings
9	severe lines, harmony in the mass: freedom of imagination
10	form, volumes, height
11	glazing (lighting), greater quantity of empty than of full space
13	transparency, opening, exterior module, no stairways
15	open ground floor, separated volumes
19	opening to the outside
20	clearness, impression of merriment and youth
22	scale, animation
23	volume, space, environment
28	windows, high entrance, greenery, passage between the two buildings
30	vertical pillars corresponding into division classrooms (classroom as module)
32	façade with large windows, two storeys, suspension
33	opening and proportion of volume (low and long), environment rhythm of façade
35	great glazing, orientation, two storeys
37	width of the span, grandeur of entrance
Schools, second level	
1	greenery, trees, each classroom has its window
2	openings of one classroom size
11	volumes with much glass for lighting
13	childlike scale
15	free ground floor, shape of windows
23	play of module within each storey
24	large windows, fairly low house, greenery
25	appearance
28	entries, façade
29	elements of the façade, their relationship with the depth of the building, the independence of one from the other, horizontal orientation of the building
30	scale approximating classroom size
	scale, large openings

Table A1. (continued)

Subject	Comment
Schools, third level	
1	repetition, same elements (classrooms), height of the building, tree
5	division into classrooms
11	glazed volume, different levels lending themselves to certain variations
13	openings
15	glazing and distinct volumes and ground floor
20	grouping, clear space, luminous
29	size of the whole, clear separation of the small units (classrooms)
32	entrance and dimensions of the window units in the façade
37	pattern, sun shades, access
Schools, photographic level	
2	dimension of interiors, no curtains, etc
3	volume, organization
6	volumes
11	volumes at different levels, many openings (light)
13	ground floor evoking play courses or covered assembly (?)
17	large classrooms
22	function, light and depth
23	important module
24	small height, glazing, greenery
32	large interior stairways, large windows
36	the structure of openings and the scale of openings
Houses, minimum level	
2	symbol: little house, gabled roof
3	volume
7	shape
8	particularity
11	composition of volumes and plans relatively similar
12	form
15	roof, shape, separated volumes
20	tranquility, sympathetic form
30	gabled roof, small scale
31	roof, dispersions between superstructure and buildings
	scale, roof, proportion
Houses, second level	
2	scale, screen chimney
3	roof
5	scale, greenery
6	volume, detail
7	shape, only one roof
8	scale
10	form, scale, porch
14	relatively small, penetration of volume
15	shape, entries and windows, form of volumes
18	form

Table A1. (continued)

Subject	Comment
Houses, second level (continued)	
20	small aspect
22	environment
25	dimensions and forms
27	elimination
29	relationship between dimensions of elements and dimensions of the whole, indication of functions of the elements
30	the large roof and the openings, scale
31	play of full and empty spaces, general aspect
32	rather small dimensions
33	freedom of volume
36	rhythm of volumes in relation to wall and details
Houses, third level	
1	greenery, variety of plan
2	scale, cloister, nature environment
3	dimension
6	volume, details
8	trees, scales, construction
9	variation or decoration, surface, greenery, large windows
10	scale, trees, form, veranda
14	small form of windows
17	villa in the green
22	scale, likeness
29	relationship of dimensions with the environment, scale of elements in relationship to the whole, functions of these elements
30	scale of the small house with terrace
33	freedom of volumes, environment
35	patio, greenery, large windows
37	big loggia, greenery, window
Houses, photographic level	
3	volume, size
5	scale, curtains
6	detail, façade
9	surfaces of stone, wood, glass and metal in contact with each other, variation between them, union with nature, natural forms
12	greenery, scale
14	small, environment
17	visibly a small dwelling
24	individuality, variety of volumes, greenery
29	relationship between house and nature, proportions of the elements, functions of the elements
30	villa, connection with nature, terrace
31	greenery, scale, form
33	freedom of volumes, environment
35	volume, loggia, greenery
37	large opening, terrace, greenery

Table A2. Study 1: factor analysis of similarity coefficients between eighteen façades, varimax rotated factor matrix.

Façade	Factor 1	Factor 2	Façade	Factor 1	Factor 2
1	0·85382	0·48263	10	0·93540	0·29591
2	−0·82088	−0·45807	11	0·18798	0·97679
3	0·34922	0·93069	12	−0·77536	−0·61284
4	−0·70222	−0·70097	13	−0·74340	−0·65875
5	0·86082	0·48492	14	−0·81077	−0·56749
6	0·98254	0·05007	15	0·59724	0·78540
7	0·51459	0·84927	16	−0·62346	−0·76366
8	−0·88159	−0·40417	17	0·93095	0·26627
9	0·17644	0·96349	18	0·61753	0·48709

Table A3. Study 1: factor analysis of similarity coefficients between twenty-two SD scales, varimax rotated factor matrix.

Façade	Factor 1	Factor 2	Factor 3	SD scales
1	0·55336	0·75326	0·11005	hot-cold
2	−0·14432	0·30446	−0·91009[a]	natural-technical
3	0·07151	0·84016[a]	−0·08019	meaningful-meaningless
4	0·89010[a]	0·34486	0·18636	inviting-repelling
5	−0·61501	−0·16585	−0·72455[b,c]	non-functional-functional
6	0·93223	0·24318	0·15667	pleasant-unpleasant
7	−0·18707	0·85967[a]	0·07682	loose-inflexible
8	−0·63063[c]	0·31217	0·42898	varied-dull
9	−0·92212[a]	0·24573	−0·06905	overdone-sober
10	−0·58027	0·73046	0·04335	artistic-unartistic
11	0·54679	−0·45429	−0·57323	exciting-calming
12	0·01144	0·65636[a]	0·26809	original-banal
13	0·72059	−0·58086	0·24466	well-thought-out-ill-formed
14	0·53904	0·33281	0·64933	alive-dead
15	0·08018	−0·15255	0·84843[a]	generous-stingy
16	−0·85613[a]	0·21441	0·35079	diverse-monotonous
17	0·35264	0·79222[a]	−0·40709	fertile-sterile
18	−0·88756[a]	0·33609	−0·23818	playful-serious
19	−0·11081	0·53605	−0·77170[b,c]	irregular-orderly
20	−0·03105	0·80176[a]	−0·31593	personal-impersonal
21	−0·16013	0·38895	0·76023[a]	expressive-inexpressive
22	0·89988[a]	0·31043	0·17783	friendly-unfriendly
	friendly pleasant	meaningful original personal loose	natural generous expressive	

[a]Very high loadings, ~90.
[b]High loadings, ~70+
[c]Loading not taken into consideration, although high.

Table A4. Mean values (upper lines) and standard deviations (lower lines) of German subjects 1 - 20 for the twenty-two SD scales for the eighteen façades; values for the new façades are set in bold type. The final two lines of the table are average values for the new and old façades.

F	SD scales																					
	1	2	3	4	5	6	7	8	9	10	11	12	13	14	15	16	17	18	19	20	21	22
1	**2.6**	**2.8**	**2.2**	**3.3**	**2.3**	**3.4**	**2.3**	**2.0**	**1.8**	**1.7**	**3.3**	**2.3**	**4.2**	**2.8**	**3.1**	**2.1**	**2.8**	**1.8**	**2.0**	**2.2**	**2.4**	**3.5**
	1.4	**1.1**	**1.3**	**1.4**	**1.7**	**1.3**	**1.4**	**1.2**	**1.2**	**1.0**	**1.1**	**1.2**	**1.5**	**1.4**	**1.1**	**0.9**	**1.2**	**1.2**	**1.4**	**1.0**	**1.4**	**1.5**
2	4.8	4.3	4.4	3.7	4.6	4.2	4.8	5.0	4.4	5.0	3.8	4.5	5.5	4.6	5.0	5.2	4.7	5.3	2.4	4.8	5.2	4.2
	1.3	1.2	1.1	1.2	1.4	1.4	1.4	1.5	1.0	0.8	1.0	1.2	1.4	1.4	0.8	1.2	0.9	0.8	1.2	1.3	0.9	1.3
3	**2.4**	**3.3**	**2.8**	**2.3**	**3.2**	**2.6**	**2.5**	**2.4**	**2.3**	**1.5**	**4.0**	**2.4**	**2.7**	**2.8**	**2.3**	**2.3**	**2.9**	**2.5**	**3.3**	**2.5**	**2.2**	**2.1**
	1.3	**1.3**	**1.6**	**1.0**	**1.9**	**1.1**	**1.4**	**1.4**	**1.2**	**0.8**	**1.1**	**1.3**	**1.3**	**1.5**	**1.0**	**1.5**	**1.3**	**1.4**	**1.8**	**1.3**	**1.2**	**1.4**
4	5.0	4.3	4.8	4.5	4.8	5.0	5.2	5.5	5.2	5.1	4.4	5.3	5.7	4.8	5.3	5.8	5.0	5.4	2.7	5.0	5.2	4.8
	1.2	1.4	1.1	1.3	1.3	1.2	1.5	1.1	1.0	1.0	1.0	1.2	1.2	1.6	1.2	1.0	1.0	0.9	1.2	1.2	1.2	1.6
5	**2.3**	**2.3**	**2.6**	**2.8**	**1.5**	**3.2**	**2.5**	**2.3**	**1.5**	**1.8**	**3.5**	**2.2**	**4.6**	**2.8**	**3.0**	**2.2**	**2.4**	**1.7**	**1.8**	**2.2**	**2.5**	**2.7**
	1.1	**1.4**	**1.1**	**1.3**	**0.6**	**1.2**	**1.6**	**1.3**	**0.8**	**1.0**	**0.9**	**1.0**	**1.4**	**1.3**	**0.9**	**1.3**	**0.9**	**0.8**	**1.2**	**1.3**	**1.2**	**1.3**
6	**3.5**	**2.3**	**2.7**	**4.3**	**1.8**	**4.3**	**2.5**	**2.5**	**1.7**	**2.5**	**3.2**	**3.2**	**5.0**	**3.8**	**3.5**	**2.3**	**3.0**	**1.8**	**1.5**	**2.3**	**3.2**	**4.3**
	1.3	**1.0**	**1.2**	**1.1**	**0.7**	**1.0**	**1.1**	**1.2**	**0.7**	**1.4**	**0.9**	**1.1**	**1.5**	**1.4**	**1.1**	**1.0**	**1.2**	**0.9**	**0.8**	**1.1**	**1.5**	**1.4**
7	**2.4**	**3.5**	**2.3**	**2.3**	**3.3**	**2.5**	**2.3**	**2.5**	**2.4**	**1.8**	**4.0**	**2.5**	**2.8**	**2.8**	**2.5**	**2.5**	**3.0**	**2.5**	**3.0**	**2.4**	**2.8**	**2.2**
	1.2	**1.1**	**1.1**	**1.3**	**1.8**	**1.2**	**1.3**	**1.4**	**1.4**	**1.1**	**0.5**	**1.1**	**1.5**	**1.8**	**1.1**	**1.4**	**1.3**	**1.1**	**1.5**	**1.1**	**1.5**	**1.5**
8	4.5	4.4	4.5	4.3	5.2	4.3	5.0	5.7	5.5	5.0	4.5	4.9	4.5	4.9	5.1	5.3	4.5	5.5	3.5	4.6	5.5	4.2
	1.5	1.0	1.7	1.6	1.4	1.5	1.6	1.0	1.1	0.9	1.4	1.4	1.4	1.8	1.7	1.5	1.2	0.8	1.5	1.3	1.2	1.7
9	3.3	3.9	3.3	3.1	3.7	3.3	3.3	3.3	2.6	2.7	3.8	3.3	3.8	3.4	3.3	3.0	3.5	3.3	3.0	3.4	3.3	3.3
	1.4	1.1	1.6	1.7	1.4	1.6	1.5	1.7	1.1	1.6	0.7	1.6	1.7	1.8	1.0	1.7	1.4	1.3	1.1	2.0	1.8	1.9
10	**3.0**	**2.3**	**2.7**	**3.8**	**2.2**	**3.9**	**3.3**	**3.0**	**2.4**	**2.3**	**3.5**	**2.8**	**4.2**	**3.1**	**3.0**	**2.8**	**3.0**	**1.9**	**2.0**	**2.7**	**2.6**	**3.7**
	1.5	**0.9**	**1.3**	**1.5**	**0.9**	**1.4**	**1.5**	**1.3**	**1.1**	**1.1**	**0.9**	**1.3**	**1.5**	**1.4**	**1.1**	**1.2**	**1.6**	**0.9**	**1.1**	**1.5**	**1.1**	**1.6**
11	3.2	3.8	3.0	2.8	4.5	3.0	3.5	3.1	3.5	2.9	4.0	3.2	3.7	3.0	3.1	3.3	3.3	3.3	3.0	2.8	2.8	2.5
	1.4	1.1	1.3	1.5	1.5	1.6	1.3	1.4	1.3	1.6	0.6	1.1	1.5	1.4	1.4	1.4	1.2	1.4	1.2	1.2	1.6	1.2
12	5.3	4.9	4.9	5.2	4.6	5.0	5.4	5.5	5.3	5.3	4.5	5.3	4.9	5.5	5.0	5.5	4.8	5.5	4.0	5.3	5.5	5.2
	0.8	1.0	1.2	1.1	1.4	1.2	1.1	0.9	1.0	1.3	1.1	1.2	1.3	1.4	1.2	1.3	1.3	0.9	1.2	1.0	1.1	1.3
13	5.0	4.5	5.0	4.8	5.8	4.8	5.3	5.6	6.1	5.0	5.5	5.3	5.2	5.4	5.9	5.8	4.9	5.7	4.3	5.0	5.5	5.0
	1.7	0.9	1.6	1.4	1.1	1.3	1.1	0.9	1.2	1.8	1.1	1.6	1.5	1.5	1.1	0.8	1.0	1.1	1.4	1.4	1.4	1.2
14	4.4	4.3	3.7	4.0	4.8	4.0	4.5	4.6	4.5	4.5	4.3	4.4	4.8	4.5	4.3	4.7	4.2	4.5	4.2	4.0	4.2	4.0
	1.4	1.1	1.5	1.8	1.4	1.7	1.1	1.3	1.4	1.4	0.9	0.9	1.6	2.0	1.1	1.7	1.3	1.1	1.6	1.3	1.7	1.6
15	**2.0**	**2.3**	**2.0**	**2.2**	**2.2**	**2.3**	**1.5**	**1.5**	**1.4**	**1.5**	**3.3**	**1.7**	**3.6**	**1.7**	**2.0**	**1.5**	**1.7**	**1.5**	**1.8**	**1.8**	**1.5**	**2.2**
	1.0	**1.3**	**0.9**	**1.3**	**1.4**	**1.2**	**0.8**	**0.7**	**0.6**	**0.7**	**0.9**	**1.0**	**1.8**	**0.8**	**1.1**	**0.6**	**0.9**	**0.8**	**0.9**	**1.0**	**0.7**	**1.2**
16	5.0	4.5	4.7	4.6	4.5	5.0	5.2	5.3	4.8	5.0	4.5	5.0	5.3	4.8	4.7	5.2	4.6	5.0	3.3	5.0	4.9	4.8
	1.1	1.5	1.6	1.4	1.5	1.2	1.5	1.0	1.3	1.4	0.9	1.7	1.2	1.6	1.2	1.3	1.1	0.9	1.3	1.3	1.5	1.7
17	**2.8**	**2.2**	**3.0**	**3.4**	**1.8**	**3.7**	**2.3**	**2.5**	**2.0**	**2.6**	**3.4**	**2.8**	**4.5**	**3.0**	**3.3**	**2.3**	**2.8**	**1.9**	**2.0**	**2.7**	**3.1**	**3.5**
	1.4	**1.3**	**1.7**	**1.4**	**0.8**	**1.4**	**1.6**	**1.4**	**1.3**	**1.4**	**0.8**	**1.5**	**1.9**	**1.6**	**1.1**	**1.3**	**1.3**	**1.0**	**0.9**	**1.5**	**1.7**	**1.6**
18	**3.8**	**3.0**	**3.0**	**3.3**	**3.7**	**3.4**	**3.8**	**3.7**	**2.8**	**2.9**	**4.2**	**3.8**	**3.4**	**3.8**	**3.5**	**3.8**	**3.9**	**3.6**	**4.0**	**3.5**	**3.8**	**3.5**
	1.6	**1.2**	**1.7**	**1.8**	**1.6**	**1.7**	**1.4**	**1.7**	**1.4**	**1.7**	**1.0**	**1.8**	**1.9**	**1.6**	**1.3**	**1.5**	**1.1**	**1.2**	**1.5**	**1.4**	**1.3**	**1.6**
	2.8	**2.7**	**2.6**	**3.1**	**2.4**	**3.3**	**2.6**	**2.5**	**2.0**	**2.1**	**3.6**	**2.6**	**3.9**	**3.0**	**2.9**	**2.4**	**2.8**	**2.1**	**2.4**	**2.5**	**2.7**	**3.1**
	4.5	4.3	4.3	4.1	4.7	4.3	4.7	4.8	4.7	4.5	4.4	4.6	4.8	4.5	4.6	4.9	4.4	4.8	3.4	4.4	4.7	4.2

Table A5. Mean values and standard deviations, German subjects 21–40. The final two lines of the table are average values for the new and old façades.

F	SD scales																					
	1	2	3	4	5	6	7	8	9	10	11	12	13	14	15	16	17	18	19	20	21	22
1	3.3	3.1	2.2	3.4	2.2	3.3	2.4	2.5	2.0	1.8	3.1	2.8	4.8	3.0	2.5	2.5	2.5	2.0	2.3	2.8	2.2	3.6
	1.7	1.6	0.9	1.9	1.0	1.8	1.2	1.4	0.9	1.0	1.1	1.0	1.6	1.5	0.9	1.1	1.1	0.9	1.3	1.8	1.1	1.6
2	4.3	4.3	4.3	3.8	4.7	4.0	4.3	4.2	4.4	4.3	4.2	4.4	4.7	4.5	4.8	4.3	4.0	4.2	3.2	4.0	4.3	3.5
	1.6	1.3	1.7	1.8	1.5	1.6	1.8	2.0	1.7	1.5	1.1	1.4	1.5	1.7	1.4	1.7	1.4	1.8	1.5	1.6	1.8	1.6
3	2.6	3.3	1.8	2.4	3.5	2.1	2.3	1.8	1.8	1.7	3.2	2.5	3.3	2.1	2.3	2.1	2.2	2.4	2.7	2.4	2.3	2.2
	1.7	1.3	0.8	1.5	1.8	1.1	1.3	1.0	0.7	0.8	1.0	1.5	1.9	1.2	1.1	1.0	1.1	1.0	1.5	1.5	1.5	1.2
4	4.5	4.6	4.8	4.5	5.0	4.3	5.3	5.4	5.5	4.8	4.7	4.7	4.7	5.0	5.5	5.4	4.8	5.7	3.2	4.6	4.8	4.4
	1.8	1.4	1.9	1.8	1.2	1.8	1.4	1.4	1.3	1.6	1.2	1.7	1.5	1.7	1.3	1.1	1.1	1.0	1.5	1.8	2.0	1.7
5	3.2	2.5	2.5	3.1	1.7	3.4	2.6	2.5	1.8	1.9	3.3	2.5	5.1	3.2	2.5	2.8	2.8	1.8	1.7	2.5	2.5	3.3
	1.7	1.4	1.3	1.6	0.8	1.6	1.8	1.7	0.7	1.1	1.3	1.2	1.8	1.8	1.0	1.6	1.5	0.9	0.7	1.7	1.3	1.7
6	3.0	1.8	2.7	4.0	1.7	3.6	2.3	2.5	1.4	1.7	3.0	2.7	5.2	2.8	2.9	1.8	2.3	1.6	1.5	2.3	2.7	3.5
	1.7	0.9	1.6	1.8	1.2	1.8	1.4	1.5	0.7	0.8	1.2	1.1	1.8	1.4	1.1	1.5	1.0	0.8	0.6	1.2	1.6	2.0
7	2.0	2.5	1.6	1.7	2.6	1.9	1.7	1.8	1.5	1.3	3.5	2.1	3.2	1.8	1.9	1.8	2.0	2.0	2.5	1.5	1.5	2.0
	1.1	1.1	0.7	0.9	1.8	1.2	0.7	1.4	0.8	0.6	0.9	1.2	2.0	0.6	0.9	1.4	1.0	1.1	1.6	0.7	0.7	1.1
8	3.8	4.2	3.8	3.5	4.5	3.5	3.8	4.5	5.2	4.0	4.2	4.0	4.5	4.2	4.9	4.8	4.0	4.8	3.5	4.0	4.3	3.6
	1.7	1.3	1.9	1.8	1.8	1.8	1.8	1.9	1.8	1.8	1.4	1.7	1.5	2.0	1.9	1.7	1.5	1.8	1.7	1.9	1.8	1.7
9	2.8	4.0	2.5	3.0	4.4	2.8	3.3	2.8	3.1	2.3	3.8	3.8	3.5	3.2	3.1	3.2	3.2	3.8	4.0	3.0	2.3	2.8
	2.0	1.3	1.6	1.9	1.6	1.8	1.8	1.9	1.7	1.3	1.3	1.4	1.4	1.6	1.4	1.9	1.4	1.5	1.9	1.6	1.3	1.5
10	3.1	2.5	2.6	3.5	1.5	3.7	2.9	3.2	1.7	2.6	3.0	3.0	5.3	3.5	3.0	2.4	2.9	1.8	1.7	2.8	2.6	3.3
	2.0	1.5	1.3	2.0	0.8	2.2	2.2	2.0	0.8	1.8	1.3	1.4	1.8	1.8	1.1	1.5	1.5	1.0	0.9	1.7	1.4	2.0
11	2.2	3.5	2.3	2.5	3.2	2.3	2.0	2.0	1.8	1.7	3.8	2.2	3.4	2.4	2.3	2.4	2.4	2.8	3.0	2.2	1.8	2.2
	1.0	0.8	0.9	1.0	1.5	1.1	1.1	0.9	0.6	0.7	0.9	0.9	1.6	1.0	0.9	0.9	0.8	0.8	1.6	1.0	0.8	1.4
12	4.1	4.4	4.3	4.1	4.8	4.3	4.8	4.8	4.8	4.0	3.8	4.0	4.4	4.2	4.3	4.8	4.3	4.7	3.7	4.0	4.1	4.0
	1.4	1.1	1.6	1.6	1.5	1.6	1.4	1.7	1.3	1.6	1.3	1.6	1.3	1.4	1.3	1.3	1.4	1.5	1.5	1.5	1.8	1.5
13	5.3	4.8	5.1	5.0	5.0	5.0	5.6	6.0	5.9	5.6	4.3	5.3	5.4	5.6	5.8	6.0	5.2	6.2	2.8	5.5	5.5	5.0
	1.1	1.0	1.3	1.4	1.8	1.5	1.2	0.9	0.9	1.4	1.3	1.3	0.8	1.0	0.8	0.9	0.8	0.8	1.5	1.4	1.4	1.5
14	5.0	4.5	4.6	4.8	4.9	5.0	4.9	5.2	5.3	5.0	3.9	4.8	5.2	5.0	5.2	5.5	4.9	5.8	3.4	5.2	5.3	4.8
	1.4	1.0	1.4	1.4	1.4	1.3	1.6	1.5	1.1	1.2	1.5	1.2	1.0	1.4	1.2	1.3	1.1	1.0	1.6	1.4	1.2	1.4
15	1.3	2.7	1.5	1.8	2.5	2.2	1.5	1.3	1.3	1.3	3.8	1.8	4.3	1.5	2.0	1.3	2.0	1.7	1.7	1.5	1.3	1.7
	0.6	1.6	0.8	0.8	1.6	1.1	0.9	0.7	0.6	0.6	0.6	1.1	1.9	0.8	1.1	0.7	1.1	0.7	0.7	0.7	0.6	0.9
16	5.4	4.3	5.2	5.3	4.8	5.3	5.6	6.0	5.2	5.4	3.6	4.9	5.6	5.3	5.3	5.8	5.2	5.6	3.0	5.5	5.7	5.2
	1.1	1.1	1.0	0.9	1.4	1.2	0.8	0.9	0.9	1.1	1.4	1.5	1.2	1.2	0.9	1.0	1.0	1.0	1.7	1.1	1.1	1.4
17	2.5	2.2	2.8	3.9	1.7	3.8	2.2	2.5	1.8	2.7	3.3	3.3	5.3	3.0	3.5	2.5	2.7	1.7	1.3	2.1	3.0	3.3
	1.6	1.5	1.4	1.4	1.4	1.6	1.2	1.5	0.9	1.5	1.1	1.1	2.0	1.9	1.4	1.1	1.4	0.7	0.7	1.1	1.5	1.6
18	3.0	2.0	2.2	3.0	3.3	3.2	2.3	2.8	2.0	1.9	3.5	2.7	3.3	3.0	2.7	2.8	2.8	2.2	2.8	2.5	2.7	3.3
	1.7	1.1	1.2	1.8	2.1	1.8	1.4	1.7	1.4	1.2	1.4	1.8	1.9	1.9	1.0	1.7	1.6	1.1	2.0	1.5	1.8	1.8
	2.7	2.5	2.2	3.0	2.3	3.0	2.2	2.3	1.7	1.9	3.3	2.6	4.4	2.7	2.6	2.2	2.5	1.9	2.0	2.3	2.3	2.9
	4.2	4.3	4.1	4.1	4.6	4.1	4.4	4.5	4.6	4.1	4.0	4.1	4.6	4.4	4.6	4.7	4.2	4.8	3.3	4.2	4.2	3.9

Table A6. Mean values and standard deviations, German subjects 41–60. The final two lines of the table are average values for the new and old façades.

F	SD scales																					
	1	2	3	4	5	6	7	8	9	10	11	12	13	14	15	16	17	18	19	20	21	22
1	3.7	2.8	2.9	4.2	2.5	3.9	3.2	2.7	2.0	1.8	3.5	3.2	5.0	3.5	3.0	2.5	2.8	2.0	2.0	4.2	3.2	4.3
	2.0	2.0	1.6	1.6	1.4	1.7	2.0	1.6	1.3	1.1	1.5	1.8	1.4	1.9	1.2	1.6	1.7	1.1	1.3	1.9	1.8	1.9
2	4.0	4.4	4.3	4.2	4.1	4.0	4.3	4.2	4.5	4.8	3.8	4.3	4.0	4.3	4.8	5.0	3.8	5.1	3.5	4.2	4.8	3.8
	1.9	1.1	1.6	1.9	2.1	1.8	1.7	1.9	1.8	1.8	1.4	1.8	1.9	1.8	1.6	1.6	1.7	1.5	2.0	1.7	1.6	1.7
3	2.5	3.8	2.3	2.7	3.8	3.0	2.3	2.5	2.7	1.8	4.3	3.0	3.0	2.6	2.9	2.8	2.7	3.3	4.3	2.6	2.3	2.0
	1.5	1.7	1.7	1.8	2.1	1.9	1.3	1.6	1.9	1.6	1.6	2.1	2.1	2.0	1.8	2.1	1.7	1.7	2.3	1.8	1.8	1.2
4	4.0	4.8	4.3	4.5	4.9	4.3	5.6	5.5	5.3	5.4	4.7	5.2	5.2	5.5	5.8	5.9	4.7	5.0	3.0	4.5	5.2	4.3
	2.1	1.5	1.4	1.9	2.0	2.0	1.5	1.2	1.5	1.3	1.7	1.5	1.8	1.3	1.1	0.9	1.3	1.9	1.6	2.0	1.6	2.1
5	3.0	3.3	2.9	3.4	2.7	3.8	3.0	3.0	2.3	2.1	3.5	3.2	4.5	3.3	2.8	3.0	3.0	2.0	2.0	3.3	3.2	3.4
	1.9	2.0	1.8	1.6	1.5	1.9	1.6	1.8	1.5	1.6	1.4	1.8	1.7	1.8	1.3	1.5	1.4	1.3	1.5	2.0	1.9	1.9
6	4.3	2.0	3.8	5.3	1.7	5.0	3.6	3.5	2.2	2.0	2.9	3.6	5.7	4.5	4.0	3.0	3.7	2.1	1.2	3.9	4.7	5.3
	1.8	1.4	1.9	1.5	1.0	1.5	2.3	1.5	1.6	1.3	1.3	2.0	1.6	1.8	1.8	1.6	1.6	1.8	0.5	1.9	1.9	1.4
7	2.5	3.0	2.6	3.1	2.5	3.3	2.7	2.3	2.2	1.7	3.5	1.7	3.8	2.3	2.2	1.8	2.5	1.7	2.8	2.7	2.3	2.7
	1.8	2.3	2.0	2.2	1.9	2.0	2.2	1.9	1.7	1.5	1.7	0.9	1.9	2.0	1.3	1.5	1.6	1.0	2.1	1.9	2.2	2.2
8	3.5	4.2	4.2	3.3	4.8	3.2	4.2	5.0	5.3	4.5	4.2	4.0	4.1	3.9	4.5	4.8	4.0	4.8	3.6	4.0	4.7	3.3
	2.2	1.6	2.0	2.1	2.0	2.0	2.4	1.7	1.5	1.8	1.5	2.1	2.1	2.2	1.5	1.8	1.7	1.7	2.3	1.8	1.8	2.0
9	2.9	4.5	2.6	2.5	4.0	2.3	2.2	2.6	3.0	3.0	3.7	3.2	3.3	2.5	3.1	2.9	3.0	3.2	2.9	2.6	2.5	2.5
	1.7	1.5	1.7	1.6	2.2	1.9	1.8	1.8	1.6	2.1	1.6	1.5	2.0	1.3	1.8	1.7	1.4	2.0	2.1	1.8	1.8	1.4
10	3.8	3.0	3.0	4.0	2.0	4.5	4.2	3.5	2.7	2.1	3.3	3.5	5.5	4.2	3.3	3.7	3.3	2.3	2.0	3.8	3.3	4.1
	2.0	2.0	1.6	1.8	1.6	1.8	2.1	1.9	1.6	1.3	1.6	1.7	1.0	1.8	1.4	1.9	1.7	1.3	1.1	2.0	2.0	1.7
11	2.7	3.7	2.6	2.5	4.0	2.8	2.7	3.0	2.9	2.5	3.8	3.3	3.5	3.0	3.5	2.9	2.8	3.7	3.7	3.0	2.7	3.2
	1.6	1.5	1.4	1.5	1.9	1.4	1.6	1.6	1.6	1.7	1.6	1.8	1.9	1.5	2.0	1.5	1.3	1.6	1.6	1.6	1.5	1.6
12	4.2	4.3	4.2	3.8	4.8	3.8	4.7	4.7	5.0	5.0	4.4	5.0	4.5	4.6	4.9	5.0	4.3	5.0	4.3	4.4	4.7	4.3
	1.8	1.7	1.4	1.9	1.5	1.8	1.8	2.0	1.7	1.5	1.5	1.7	1.8	1.8	1.5	1.8	1.5	1.6	1.8	1.6	1.8	1.7
13	4.8	4.7	5.3	5.2	5.0	4.8	5.1	5.8	5.3	5.8	5.5	5.7	5.0	5.5	5.9	5.9	4.8	6.0	4.1	4.8	5.5	4.7
	1.9	1.5	1.6	1.9	1.8	1.7	1.7	1.4	1.9	1.5	1.6	1.4	1.8	1.6	1.2	1.4	1.5	1.2	1.8	1.7	1.8	1.9
14	4.8	4.6	5.0	5.1	5.2	4.4	4.8	5.3	5.4	5.7	4.7	5.3	4.6	5.3	5.5	5.7	4.8	5.2	3.7	5.0	5.2	4.8
	1.9	1.7	1.5	1.6	1.4	1.7	1.5	1.7	1.6	1.5	1.3	1.5	1.5	1.4	1.5	1.4	1.4	1.8	1.8	1.5	1.5	1.7
15	1.9	3.3	2.1	2.3	2.7	2.3	1.8	1.8	2.3	1.8	3.9	2.0	3.0	1.9	2.2	1.7	2.3	2.0	2.6	2.2	1.8	2.4
	1.3	2.2	1.4	1.4	1.8	1.2	1.3	1.3	1.7	1.1	1.3	1.2	1.7	1.0	1.5	0.9	1.4	1.3	2.1	1.6	1.2	1.6
16	4.7	4.7	4.8	4.7	4.4	4.7	5.2	5.3	4.8	5.2	4.7	4.7	4.9	5.2	5.6	5.5	4.7	5.3	4.0	4.8	5.3	4.7
	1.6	1.1	1.6	1.7	1.9	1.7	1.3	1.6	1.3	1.7	1.4	1.6	1.7	1.6	1.2	0.9	1.0	1.5	1.8	1.8	1.1	1.6
17	3.9	2.5	3.7	4.3	1.8	4.8	3.5	3.3	2.3	2.4	3.7	3.7	5.4	3.5	4.3	3.0	3.0	2.2	2.0	3.3	4.3	4.3
	2.1	2.3	2.0	1.8	1.7	1.8	2.2	2.0	1.7	1.6	1.5	1.8	1.8	1.8	2.0	2.1	2.0	1.6	1.7	2.1	2.1	1.9
18	4.3	3.7	3.7	4.9	2.5	4.6	4.5	4.7	3.5	3.0	4.0	4.8	4.5	4.7	4.6	4.6	3.8	3.3	2.7	4.3	4.4	4.5
	2.0	1.9	1.7	1.7	1.6	2.0	1.8	1.8	1.3	1.9	1.4	1.8	2.0	2.0	1.2	1.7	1.7	1.9	2.1	1.8	2.1	1.9
	3.3	3.0	3.0	3.8	2.5	3.9	3.2	3.0	2.5	2.1	3.6	3.2	4.5	3.4	3.3	2.9	3.0	2.3	2.4	3.4	3.3	3.7
	4.0	4.4	4.1	4.0	4.6	3.8	4.3	4.6	4.6	4.7	4.4	4.5	4.3	4.4	4.8	4.8	4.1	4.8	3.6	4.1	4.5	4.0

Table A7. Mean values and standard deviations, Italian subjects 61–76. The final two lines of the table are average values for the new and old façades.

F	SD scales 1	2	3	4	5	6	7	8	9	10	11	12	13	14	15	16	17	18	19	20	21	22
1	2.9	2.9	3.3	3.1	3.0	3.5	3.1	2.5	2.9	2.5	3.2	3.0	3.5	3.4	3.6	2.8	3.6	2.6	2.4	3.5	3.7	3.6
	1.8	1.6	1.7	1.6	2.1	1.7	2.0	1.5	1.4	1.6	1.2	1.5	1.9	1.3	1.4	1.5	1.2	1.2	1.2	1.9	1.6	1.5
2	3.6	2.9	3.3	2.9	4.4	2.8	3.3	3.1	3.9	2.5	3.4	2.8	3.8	2.5	3.1	3.2	2.8	2.6	3.7	3.4	2.9	3.1
	1.7	1.3	1.6	1.7	1.6	1.9	1.9	1.9	2.0	1.7	1.3	1.9	2.0	1.8	1.5	1.5	1.3	1.4	2.0	1.6	1.7	1.9
3	2.6	2.6	3.1	2.9	4.1	2.6	2.8	2.8	3.0	2.3	4.1	3.2	2.5	2.7	3.4	2.8	2.9	2.8	3.8	3.4	3.6	3.3
	1.8	1.8	1.8	2.1	2.1	2.1	1.7	1.9	1.9	1.9	1.7	2.1	2.0	1.9	1.6	1.6	1.7	1.8	2.2	1.9	2.1	1.9
4	4.3	3.5	4.4	4.3	4.4	4.1	3.3	4.5	5.0	3.9	4.3	4.5	5.3	4.6	4.3	4.9	3.9	3.1	3.6	4.9	4.6	4.2
	1.7	1.6	1.4	1.9	2.0	1.9	2.0	2.1	1.7	2.2	1.4	1.8	1.8	1.8	1.3	2.0	1.4	1.5	1.9	1.9	1.7	1.8
5	3.8	2.7	3.3	3.8	2.9	3.9	3.4	3.1	3.0	2.9	4.0	3.2	3.3	3.5	3.9	3.2	3.7	3.0	2.6	3.1	3.1	3.7
	1.7	1.6	1.7	1.6	2.0	1.9	1.8	1.5	1.3	2.2	1.2	1.6	1.4	1.8	1.3	1.3	1.5	1.5	1.5	2.0	2.0	2.2
6	3.1	2.2	3.6	3.6	2.7	3.4	2.8	2.9	2.8	2.4	3.8	3.3	2.8	3.1	4.0	3.0	3.1	3.0	1.7	3.1	3.3	3.6
	2.3	1.8	2.1	2.0	2.1	2.2	1.8	1.7	1.7	2.2	1.2	2.0	1.6	1.9	1.5	2.1	1.8	2.0	1.6	1.9	2.0	1.9
7	2.3	2.4	3.1	2.3	4.1	2.2	1.9	2.0	3.6	2.1	3.7	2.1	2.9	2.4	2.5	2.4	2.8	2.1	3.1	2.4	2.8	2.6
	1.4	1.4	1.9	1.5	2.1	1.6	1.5	1.5	1.9	1.3	1.6	1.5	2.2	1.5	1.3	1.7	1.4	1.4	2.2	1.4	1.9	1.6
8	4.3	3.5	4.3	3.7	4.9	3.6	3.1	4.1	4.7	3.1	3.9	4.6	5.3	4.1	4.0	3.9	3.4	4.3	3.9	4.6	4.8	3.4
	1.9	1.9	2.0	1.8	1.9	1.8	1.5	2.0	2.0	1.8	1.2	1.6	1.9	2.0	1.4	1.6	1.6	1.8	2.1	1.7	1.7	1.7
9	3.9	3.1	3.8	3.7	3.6	3.2	3.4	3.3	3.3	3.1	4.1	3.0	3.7	3.4	3.2	2.9	3.3	2.9	3.0	3.8	4.2	3.8
	2.2	1.7	1.8	1.9	1.7	1.9	2.1	2.1	1.4	1.9	1.5	2.0	2.1	2.0	1.9	1.6	1.6	1.6	2.0	2.2	2.1	1.5
10	3.0	2.4	3.8	2.5	3.5	2.8	2.8	2.4	3.2	2.4	3.6	2.8	2.9	2.9	3.3	2.5	3.5	2.5	2.3	2.7	3.6	3.1
	1.8	1.5	1.8	1.5	2.1	1.9	1.9	1.4	1.6	1.5	1.3	1.8	1.7	1.6	1.1	1.3	1.2	1.3	1.3	1.5	1.9	1.9
11	3.3	3.1	3.3	3.1	5.0	3.6	3.0	3.3	3.8	2.9	4.0	3.6	4.1	3.5	3.3	3.4	3.3	2.8	4.4	4.5	3.8	3.5
	2.0	1.6	1.7	2.1	1.7	1.9	2.0	2.2	1.2	1.7	1.1	2.3	2.3	1.9	1.7	2.0	1.8	1.4	1.9	1.8	2.0	1.9
12	4.6	4.1	4.6	4.3	4.8	4.5	4.1	4.8	4.9	4.0	4.0	4.4	5.4	4.6	4.4	5.0	4.3	3.9	4.6	4.9	4.7	4.4
	1.8	1.5	1.4	2.1	1.9	2.0	1.6	1.9	1.7	1.9	1.2	1.8	1.5	1.8	1.4	1.6	1.6	1.7	1.5	1.6	1.5	2.0
13	3.8	3.6	4.1	3.4	4.9	3.3	3.2	4.9	5.8	3.0	4.8	4.4	5.7	4.4	3.8	4.8	4.1	3.6	4.5	4.4	4.2	3.3
	1.9	1.7	2.0	2.0	2.0	2.1	1.5	2.0	1.7	2.3	1.0	1.9	1.6	2.1	1.6	1.6	1.7	2.2	2.4	2.1	2.3	1.8
14	2.9	2.5	3.6	2.9	4.6	2.9	2.6	3.3	5.4	2.8	4.0	2.8	4.5	3.1	3.8	3.5	3.1	2.7	4.4	2.9	3.8	3.3
	1.8	1.2	1.7	1.4	1.6	1.3	1.7	1.7	1.5	1.8	1.4	1.5	2.0	1.8	1.2	2.0	1.7	1.6	2.3	1.9	2.0	1.3
15	2.3	1.8	2.8	2.9	3.6	2.7	2.4	2.9	3.0	2.5	3.0	2.1	2.5	2.8	2.6	2.1	2.4	2.5	2.5	2.6	2.8	3.1
	1.7	1.2	1.8	1.9	1.9	1.7	1.6	2.0	1.5	1.7	1.0	1.4	1.5	1.7	1.4	1.5	1.8	1.7	1.6	1.7	1.7	1.7
16	4.3	3.9	4.3	4.4	4.5	4.5	4.1	4.3	3.9	4.1	3.4	3.7	4.6	4.0	4.3	4.1	4.2	3.2	3.1	4.5	4.4	4.9
	2.1	1.5	1.7	2.0	2.2	2.2	1.9	2.2	2.0	1.5	1.3	1.9	2.3	1.9	1.5	2.2	1.2	1.6	1.7	1.6	1.6	1.9
17	2.6	1.8	3.1	2.8	2.9	3.1	2.4	3.1	3.0	2.6	4.5	2.9	2.8	3.4	3.6	2.5	3.3	2.9	2.2	2.5	3.3	3.3
	1.5	1.1	1.9	2.0	2.1	2.1	1.2	2.0	1.5	1.9	1.2	2.0	1.5	2.0	1.7	1.4	1.7	1.7	1.5	1.6	2.1	1.9
18	3.1	2.1	2.8	3.3	3.3	2.9	2.7	3.0	3.7	2.3	3.8	3.0	3.1	3.4	3.3	2.7	2.9	3.0	3.1	3.2	3.1	3.3
	1.5	1.1	1.5	1.7	1.9	1.7	1.6	1.6	2.0	1.3	1.4	1.9	1.3	2.0	1.4	1.7	1.2	1.8	2.0	1.9	1.6	2.0
	2.5	2.3	3.2	3.0	3.3	3.0	2.7	2.7	3.1	2.4	3.7	2.8	2.9	3.1	3.4	2.7	3.1	3.7	2.6	2.9	3.3	3.3
	3.9	3.2	4.0	3.6	4.1	3.6	3.3	3.9	4.5	3.3	4.0	3.8	4.7	3.8	3.8	4.0	3.6	3.2	3.9	4.2	4.2	3.8

Table A8. Mean values and standard deviations, subjects 1–76. The final two lines of the table are average values for the new and old façades.

F	SD scales 1	2	3	4	5	6	7	8	9	10	11	12	13	14	15	16	17	18	19	20	21	22
1	3.1	2.9	2.6	3.5	2.5	3.5	2.7	2.4	2.1	1.9	3.3	2.8	4.4	3.2	3.0	2.5	2.9	2.1	2.1	3.2	2.8	3.8
	1.8	1.6	1.4	1.7	1.6	1.6	1.7	1.4	1.2	1.2	1.2	1.4	1.7	1.5	1.2	1.3	1.4	1.1	1.3	1.8	1.6	1.6
2	4.2	4.0	4.1	3.7	4.5	3.8	4.2	4.2	4.3	4.3	3.8	4.1	4.5	4.0	4.5	4.5	3.9	4.4	3.1	4.1	4.4	3.7
	1.7	1.3	1.5	1.7	1.7	1.7	1.8	1.9	1.6	1.7	1.2	1.7	1.8	1.8	1.5	1.7	1.5	1.7	1.7	1.6	1.7	1.6
3	2.5	3.3	2.5	2.5	3.6	2.6	2.4	2.3	2.4	1.8	3.9	2.8	2.9	2.5	2.7	2.5	2.7	2.8	3.5	2.7	2.6	2.4
	1.5	1.5	1.5	1.6	2.0	1.6	1.4	1.6	1.5	1.3	1.4	1.8	1.8	1.7	1.4	1.6	1.5	1.5	2.0	1.6	1.7	1.5
4	4.5	4.3	4.6	4.4	4.8	4.4	4.9	5.3	5.3	4.9	4.5	4.9	5.2	5.0	5.3	5.5	4.6	4.9	3.1	4.7	4.9	4.4
	1.7	1.5	1.4	1.7	1.7	1.7	1.8	1.5	1.3	1.6	1.3	1.5	1.6	1.6	1.3	1.3	1.3	1.7	1.5	1.7	1.6	1.8
5	3.1	2.7	2.8	3.3	2.2	3.5	2.9	2.7	2.1	2.2	3.6	2.8	4.4	3.2	3.0	2.7	2.9	2.1	2.0	2.8	2.8	3.2
	1.7	1.6	1.5	1.5	1.4	1.6	1.7	1.6	1.2	1.5	1.2	1.5	1.7	1.7	1.2	1.5	1.4	1.2	1.3	1.8	1.6	1.8
6	3.5	2.0	3.2	4.3	1.9	4.1	2.8	2.9	2.0	2.1	3.2	3.2	4.7	3.6	3.6	2.5	3.0	2.1	1.5	2.9	3.4	4.2
	1.8	1.3	1.8	1.7	1.3	1.8	1.8	1.5	1.3	1.5	1.2	1.6	1.9	1.7	1.4	1.6	1.5	1.5	0.9	1.7	1.9	1.8
7	2.3	2.9	2.4	2.4	3.1	2.5	2.2	2.1	2.4	1.7	3.7	2.1	3.2	2.3	2.2	2.1	2.6	2.1	2.8	2.3	2.3	2.4
	1.4	1.6	1.6	1.6	2.0	1.6	1.5	1.6	1.6	1.2	1.3	1.2	1.9	1.6	1.2	1.5	1.4	1.2	1.8	1.4	1.7	1.6
8	4.0	4.1	4.2	3.7	4.9	3.7	4.1	4.9	5.2	4.2	4.2	4.4	4.6	4.3	4.7	4.8	4.0	4.9	3.6	4.3	4.8	3.6
	1.8	1.5	1.8	1.8	1.8	1.8	1.9	1.7	1.6	1.7	1.4	1.7	1.8	2.0	1.7	1.7	1.5	1.6	1.9	1.7	1.7	1.8
9	3.2	3.9	3.0	3.1	3.9	2.9	3.0	3.0	3.0	2.8	3.8	3.1	3.6	3.1	3.2	3.0	3.3	3.3	3.2	3.2	3.0	3.1
	1.8	1.5	1.7	1.8	1.8	1.8	1.8	1.9	1.4	1.7	1.3	1.6	1.8	1.7	1.5	1.7	1.4	1.6	1.8	1.9	1.9	1.6
10	3.3	2.6	3.0	3.5	2.2	3.8	3.3	3.0	2.4	2.4	3.4	3.1	4.6	3.5	3.1	2.8	3.2	2.1	2.0	3.0	3.0	3.6
	1.8	1.5	1.5	1.8	1.5	1.9	2.0	1.7	1.4	1.4	1.3	1.5	1.8	1.7	1.2	1.5	1.5	1.1	1.1	1.7	1.7	1.8
11	2.8	3.5	2.8	2.7	4.1	2.9	2.8	2.8	3.0	2.5	3.9	3.1	3.6	2.9	3.0	3.0	2.9	3.2	3.5	3.1	2.7	2.8
	1.5	1.3	1.4	1.5	1.8	1.6	1.6	1.6	1.4	1.5	1.1	1.6	1.8	1.5	1.6	1.5	1.3	1.4	1.7	1.6	1.6	1.6
12	4.5	4.5	4.5	4.3	4.7	4.4	4.8	5.0	5.0	4.6	4.2	4.7	4.8	4.7	4.7	5.1	4.4	4.8	4.1	4.6	4.8	4.5
	1.6	1.3	1.4	1.7	1.5	1.7	1.5	1.7	1.4	1.6	1.3	1.6	1.5	1.6	1.4	1.5	1.4	1.5	1.5	1.5	1.6	1.7
13	4.8	4.3	4.9	4.7	5.2	4.5	4.9	5.6	5.8	5.0	5.0	5.2	5.3	5.3	5.4	5.7	4.8	5.5	3.9	4.9	5.2	4.6
	1.7	1.6	1.7	1.8	1.7	1.7	1.6	1.4	1.5	2.0	1.3	1.6	1.4	1.6	1.4	1.3	1.3	1.7	1.9	1.6	1.8	1.7
14	4.3	4.1	4.3	4.3	4.9	4.1	4.3	4.7	5.1	4.6	4.3	4.4	4.8	4.6	4.7	4.9	4.3	4.7	3.9	4.4	4.7	4.3
	1.8	1.5	1.6	1.7	1.4	1.6	1.7	1.7	1.4	1.8	1.3	1.5	1.5	1.8	1.4	1.8	1.5	1.8	1.8	1.7	1.7	1.6
15	1.9	2.5	2.1	2.3	2.7	2.4	1.8	1.8	2.0	1.7	3.5	1.9	3.4	1.9	2.2	1.6	2.1	1.9	2.1	2.0	1.8	2.3
	1.2	1.7	1.3	1.4	1.7	1.3	1.2	1.4	1.3	1.1	1.0	1.2	1.8	1.2	1.3	1.0	1.3	1.2	1.5	1.3	1.2	1.4
16	4.9	4.4	4.7	4.8	4.6	4.9	5.1	5.3	4.7	4.9	4.1	4.6	5.1	4.9	5.0	5.2	4.7	4.9	3.3	5.0	5.1	4.9
	1.5	1.3	1.5	1.6	1.7	1.6	1.5	1.5	1.5	1.5	1.4	1.7	1.6	1.6	1.3	1.5	1.1	1.5	1.7	1.5	1.4	1.6
17	2.9	2.2	3.1	3.6	2.0	3.9	2.6	2.9	2.2	2.6	3.7	3.2	4.6	3.2	3.6	2.6	2.9	2.1	1.9	2.7	3.4	3.6
	1.8	1.6	1.8	1.7	1.6	1.8	1.7	1.7	1.4	1.5	1.2	1.6	2.0	1.8	1.6	1.5	1.6	1.3	1.3	1.7	1.9	1.7
18	3.6	2.7	2.9	3.6	3.2	3.6	3.3	3.6	2.9	2.5	3.9	3.6	3.6	3.7	3.5	3.5	3.4	3.0	3.1	3.4	3.5	3.7
	1.8	1.5	1.6	1.9	1.8	1.9	1.8	1.8	1.6	1.6	1.3	2.0	1.9	1.9	1.4	1.8	1.5	1.6	1.9	1.8	1.8	1.9
	2.9	2.6	2.7	3.2	2.6	3.3	2.7	2.6	2.3	2.1	3.6	2.8	4.0	3.0	3.0	2.5	2.9	2.6	2.3	2.8	2.8	3.2
	4.1	4.1	4.1	4.0	4.6	4.0	4.2	4.5	4.6	4.2	4.2	4.3	4.6	4.3	4.5	4.6	4.1	4.5	3.5	4.3	4.4	4.0

Table A9. Mean values and standard deviations, forty trade school students. The final two lines of the table are average values for the new and old façades.

F	SD scales 1	2	3	4	5	6	7	8	9	10	11	12	13	14	15	16	17	18	19	20	21	22
1	3.1	3.7	3.0	3.3	3.2	3.7	2.7	2.6	3.1	2.3	3.5	3.2	3.0	3.3	3.5	2.5	3.5	3.4	3.7	3.2	3.1	3.3
	1.6	1.2	1.2	1.6	1.8	1.3	1.4	1.4	1.4	1.4	1.3	1.2	1.4	1.5	1.2	1.5	1.0	1.3	1.9	1.3	1.5	1.7
2	3.5	4.1	4.0	3.6	4.3	3.5	3.6	4.1	4.2	5.0	4.2	4.5	4.2	3.8	4.0	4.1	3.8	4.2	3.7	4.0	4.3	3.4
	2.0	1.5	1.8	1.9	1.6	1.8	1.9	1.9	1.5	1.9	1.1	1.8	1.7	1.8	1.6	2.0	1.4	1.7	1.6	1.4	1.9	1.8
3	2.5	4.0	2.6	2.4	4.3	2.7	2.5	2.3	3.1	2.2	3.9	3.2	2.2	2.1	2.7	2.2	3.5	3.2	4.3	2.7	2.3	2.3
	1.4	1.5	1.3	1.2	1.7	1.4	1.4	1.5	1.3	1.4	1.0	1.6	1.3	1.4	1.6	1.2	1.4	1.4	1.8	1.4	1.5	1.3
4	3.9	4.1	4.8	4.3	3.9	4.2	4.0	4.6	4.3	5.5	4.4	4.9	4.8	4.3	4.8	5.0	3.8	4.0	3.2	4.3	5.1	4.4
	1.7	1.8	1.6	2.0	1.5	1.6	1.7	1.9	1.8	1.6	1.4	1.6	1.7	1.9	1.5	1.6	1.5	1.8	1.6	1.6	1.6	1.9
5	2.8	3.2	2.7	3.2	3.5	2.9	2.5	2.5	2.8	2.1	3.5	3.0	2.6	2.6	2.7	2.4	3.5	3.0	3.1	3.3	2.8	3.0
	1.9	1.7	1.4	1.7	1.9	1.7	1.7	1.4	1.5	1.4	1.3	1.6	1.6	1.8	1.8	1.5	1.5	1.5	1.7	1.5	1.7	1.7
6	3.5	3.3	3.7	3.9	2.7	4.0	3.2	3.2	3.0	2.5	3.3	3.6	3.4	3.3	3.2	3.0	3.2	2.8	2.5	3.4	3.2	3.9
	2.0	1.7	1.7	1.7	1.6	1.6	1.9	1.7	1.7	1.5	1.0	1.8	1.8	1.8	1.4	1.9	1.5	1.2	1.2	1.7	1.8	2.0
7	3.0	4.2	2.9	3.0	3.7	3.3	2.8	2.8	3.3	2.3	4.0	3.1	2.6	2.8	3.2	2.5	3.5	3.4	3.8	3.1	2.7	2.8
	1.5	1.2	1.4	1.5	1.6	1.6	1.4	1.7	1.4	1.3	0.8	1.4	1.5	1.6	1.4	1.6	1.3	1.1	1.7	1.5	1.4	1.6
8	3.8	4.3	4.5	4.2	4.3	3.9	4.0	4.5	4.4	5.3	4.3	4.7	4.5	4.3	4.6	4.5	4.0	4.5	3.8	3.8	4.7	3.9
	2.0	1.5	1.6	1.8	1.5	1.7	1.7	1.9	1.4	1.8	1.4	1.8	2.1	2.1	1.6	1.9	1.4	1.4	1.5	1.8	1.8	1.9
9	2.8	3.9	2.9	2.8	4.0	2.9	2.9	2.8	3.6	2.9	3.7	3.2	3.4	3.1	3.2	2.8	3.5	3.7	4.7	2.9	2.9	2.6
	1.5	1.4	1.4	1.6	1.5	1.5	1.7	1.4	1.7	1.6	1.2	1.4	1.6	1.6	1.4	1.6	1.4	1.5	1.6	1.4	1.6	1.5
10	3.5	3.5	3.3	3.5	3.0	3.5	3.3	3.1	3.1	2.8	3.6	3.4	3.2	3.4	3.4	2.6	3.3	3.3	2.9	3.3	3.0	3.3
	1.8	1.7	1.6	1.6	1.6	1.8	1.9	1.7	1.3	1.7	1.2	1.5	1.8	1.5	1.3	1.5	1.6	1.4	1.5	1.5	1.6	1.7
11	2.5	4.1	3.0	2.7	4.2	2.8	2.7	3.1	3.5	2.9	4.0	3.3	2.8	2.9	3.3	3.0	3.2	3.8	4.3	2.8	2.9	2.8
	1.6	1.3	1.4	1.5	1.4	1.3	1.5	1.8	1.4	1.7	1.1	1.8	1.6	1.9	1.5	1.8	1.5	1.5	1.8	1.4	1.7	1.8
12	3.3	4.0	3.7	3.2	4.1	3.4	3.4	3.6	3.8	4.3	4.3	3.5	3.6	3.7	4.0	4.1	3.5	4.1	4.2	3.5	3.6	3.1
	1.7	1.3	1.6	1.8	1.5	1.8	1.8	1.9	1.6	1.9	1.1	1.6	1.8	1.8	1.6	2.0	1.4	1.2	1.6	1.8	2.0	1.7
13	4.1	4.0	4.3	4.3	4.3	4.2	4.2	4.9	4.2	5.3	4.7	5.0	4.8	4.4	4.9	4.8	3.9	4.5	3.8	3.8	4.4	4.3
	1.9	1.7	1.8	1.8	1.5	1.7	1.6	1.8	1.5	1.9	1.7	1.7	1.6	2.0	1.4	1.4	1.4	1.6	1.7	1.8	1.8	2.0
14	3.2	4.1	3.5	3.3	4.3	3.3	3.6	3.6	4.0	4.2	4.2	3.8	3.7	3.0	3.8	3.7	3.5	3.8	4.2	3.3	3.6	3.4
	1.6	1.3	1.5	1.5	1.5	1.6	1.6	1.7	1.6	1.8	1.3	1.7	1.7	1.6	1.6	1.8	1.2	1.2	1.6	1.5	1.7	1.7
15	2.3	3.7	2.6	2.9	3.3	2.8	2.3	2.0	2.5	2.0	3.6	2.6	2.3	2.3	2.9	2.2	3.5	3.2	3.8	2.2	2.1	2.2
	1.5	1.6	1.5	1.5	1.7	1.4	1.4	1.1	1.3	1.4	1.3	1.4	1.3	1.5	1.7	1.4	1.8	1.6	1.9	1.4	1.4	1.5
16	3.7	4.3	3.7	3.5	4.7	3.7	3.5	4.0	4.3	4.3	4.0	3.7	3.6	3.6	3.9	4.1	3.5	4.6	4.1	3.6	4.1	3.8
	1.8	1.2	1.6	1.8	1.6	1.7	1.8	2.2	1.5	2.0	1.4	2.0	1.9	2.0	1.8	1.6	1.4	1.4	1.6	1.6	1.7	1.6
17	2.7	3.3	3.3	3.8	3.3	3.2	2.5	2.6	3.0	2.9	3.7	3.2	3.3	2.8	3.1	2.7	3.3	3.2	3.0	3.1	3.0	3.2
	1.7	1.6	1.7	2.1	1.6	1.5	1.5	1.6	1.5	1.6	1.1	1.6	2.0	1.8	1.5	1.8	1.5	1.4	1.7	1.5	1.6	1.7
18	3.4	3.5	2.8	3.2	3.8	3.2	3.0	2.9	3.3	2.6	3.8	3.2	3.2	3.3	3.5	3.2	3.4	3.1	3.8	3.4	2.9	3.1
	1.6	1.4	1.5	1.7	1.7	1.7	1.5	1.7	1.3	1.6	1.3	1.6	1.5	1.7	1.5	1.6	1.5	1.2	1.6	1.4	1.5	1.7
	3.0	3.6	3.0	3.2	3.4	3.3	2.8	2.7	3.0	2.4	3.7	3.2	2.9	2.9	3.1	2.6	3.4	3.2	3.4	3.1	2.8	3.0
	1.7	1.5	1.5	1.7	1.7	1.6	1.6	1.6	1.4	1.5	1.2	1.5	1.6	1.7	1.5	1.6	1.5	1.4	1.7	1.5	1.6	1.7
	3.4	4.1	3.8	3.5	4.2	3.6	3.5	3.9	4.0	4.4	4.2	4.1	3.9	3.7	4.0	4.0	3.6	4.2	4.0	3.5	4.0	3.5
	1.8	1.4	1.7	1.8	1.5	1.7	1.7	1.9	1.6	2.0	1.3	1.8	1.8	1.9	1.7	1.9	1.4	1.5	1.6	1.6	1.9	1.9

Table A10. Study 2: factor analysis of similarity coefficients between eighteen façades, varimax rotated factor matrix.

Façade	Factor 1	Factor 2
1	0·60798	0·74203
2	−0·74609	−0·54454
3	0·96027	0·19531
4	−0·93405	−0·28487
5	0·71429	0·64229
6	−0·03286	0·94412
7	0·86899	0·43661
8	−0·86933	−0·46775
9	0·94536	0·00745
10	0·28010	0·88027
11	0·96074	−0·03724
12	−0·23950	−0·75546
13	−0·89045	−0·40531
14	−0·08165	−0·81740
15	0·88102	0·42018
16	−0·54996	−0·69310
17	0·52103	0·71397
18	0·69871	0·54001

Transformation matrix

	Factor 1	Factor 2
Factor 1	0·80523	0·59296
Factor 2	0·59296	−0·80523

Table A11. Study 2: factor analysis of similarity coefficients between twenty-two SD scales, varimax rotated factor matrix.

Façade	Factor 1	Factor 2	Factor 3	
1	0·21124	0·93443[b]	−0·02453	hot-cold
2	−0·77657[a]	−0·27578	0·48106	natural-technical
3	0·55127	0·47396	−0·38032	meaningful-meaningless
4	−0·08965	−0·74938[a]	−0·54224	inviting-repelling
5	−0·51453	−0·39923	0·71256	nonfunctional-functional
6	−0·21631	0·85521	−0·33433	pleasant-unpleasant
7	0·50280	0·76505	0·13659	loose-inflexible
8	0·91462[b]	0·25019	−0·00681	varied-dull
9	0·22331	−0·04800	0·90942[b]	overdone-sober
10	0·91070[b]	−0·33801	0·00251	artistic-unartistic
11	−0·78731[a]	−0·51330	0·21286	exciting-calming
12	0·28232	−0·35606	−0·48781	original-banal
13	0·78024[b]	0·22416	−0·35748	well-thought-out-ill-formed
14	0·53228	0·88605[a]	−0·13571	alive-dead
15	0·35877	−0·41865	−0·30996	generous-stingy
16	0·93510	0·07087	0·04969	diverse-monotonous
17	−0·71004[a]	0·45199	0·30085	fertile-sterile
18	−0·26458	−0·34780	0·75379	playful-serious
19	−0·59414	−0·09485	−0·73575[a]	irregular-orderly
20	0·11367	0·87811[b]	0·00870	personal-impersonal
21	0·87706[b]	0·20228	−0·14758	expressive-inexpressive
22	0·26727	0·83623[b]	−0·34851	friendly-unfriendly

[a]High loadings, ~70+, but have not been taken into consideration.
[b]Very high loadings, ~90.

Table A12. Rank order of SD scales according to size of mean difference between 'old' and 'new' façades, study 1 and study 2.

Difference	SD scale	Difference	SD scale
Trade school students		German subjects 1-60	
2·03	aritistic-unartistic	2·69	playful-serious
1·40	diverse-monotonous	2·55	overdone-sober
1·22	varied-dull	2·41	artistic-unartistic
1·16	expressive-inexpressive	2·28	diverse-monotonous
1·08	well-thought-out-ill-formed	2·24	nonfunctional-functional
1·00	overdone-sober	2·06	varied-dull
0·96	playful-serious	1·80	loose-inflexible
0·91	generous-stingy	1·78	generous-stingy
0·89	original-banal	1·71	expressive-inexpressive
0·82	nonfunctional-functional	1·68	irregular-orderly
0·81	meaningful-meaningless	1·62	natural-technical
0·79	alive-dead	1·59	original-banal
0·78	loose-inflexible	1·56	meaningful-meaningless
0·57	irregular-orderly	1·54	personal-impersonal
0·54	exciting-calming	1·46	fertile-sterile
0·52	natural-technical	1·43	alive-dead
0·50	friendly-unfriendly	1·27	hot-cold
0·46	personal-impersonal	0·83	friendly-unfriendly
0·45	hot-cold	0·75	inviting-repelling
0·31	inviting-repelling	0·73	exciting-calming
0·30	pleasant-unpleasant	0·66	pleasant-unpleasant
0·23	fertile-sterile	0·32	well-thought-out-ill-formed
German subjects 1-40		Subjects 1-76	
2·80	playful-serious	2·32	overdone-sober
2·75	overdone-sober	2·24	playful-serious
2·45	diverse-monotonous	2·08	artistic-unartistic
2·33	artistic-unartistic	2·08	diverse-monotonous
2·30	varied-dull	2·02	nonfunctional-functional
2·28	nonfunctional-functional	1·88	varied-dull
2·14	loose-inflexible	1·56	loose-inflexible
1·94	expressive-inexpressive	1·54	expressive-inexpressive
1·94	personal-impersonal	1·49	generous-stingy
1·85	generous-stingy	1·49	personal-impersonal
1·77	meaningful-meaningless	1·48	natural-technical
1·74	natural-technical	1·44	original-banal
1·73	original-banal	1·39	meaningful-meaningless
1·64	alive-dead	1·28	alive-dead
1·64	fertile-sterile	1·25	fertile-sterile
1·60	hot-cold	1·22	hot-cold
1·14	irregular-orderly	1·19	irregular-orderly
1·10	friendly-unfriendly	0·76	friendly-unfriendly
1·04	inviting-repelling	0·72	inviting-repelling
1·04	pleasant-unpleasant	0·65	pleasant-unpleasant
0·73	exciting-calming	0·63	exciting-calming
0·56	well-thought-out-ill-formed	0·62	well-thought-out-ill-formed

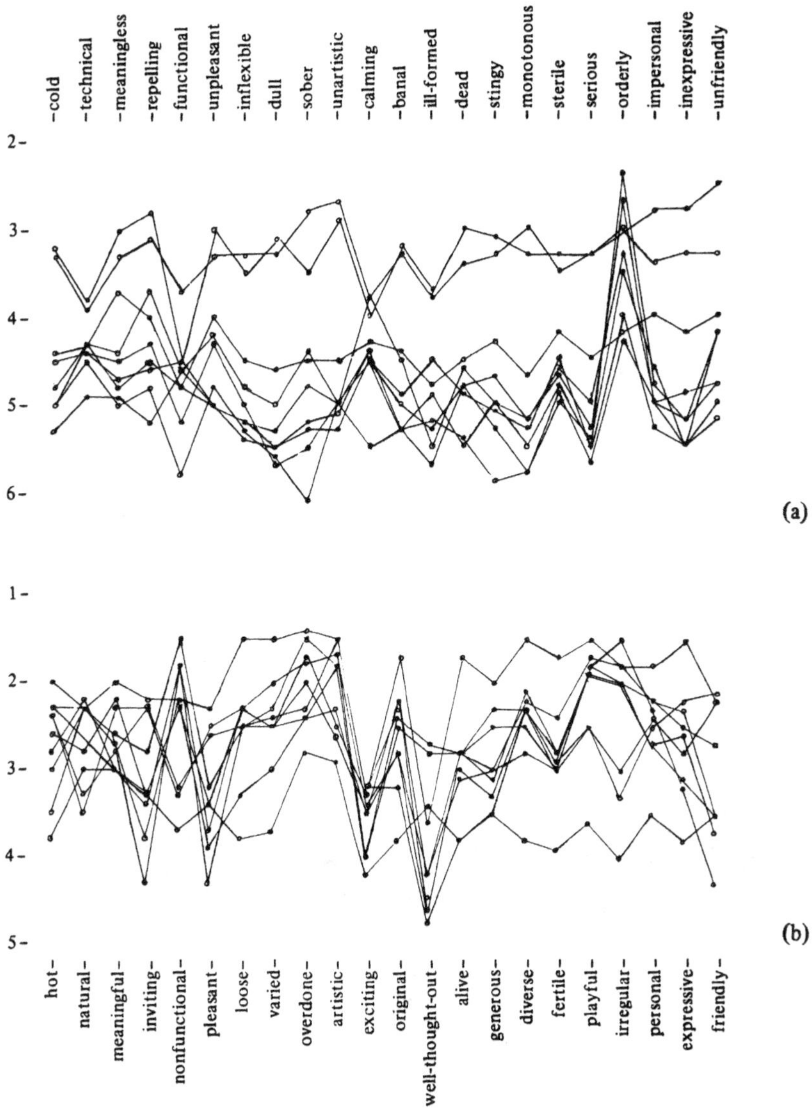

Figure A1. Façade profiles, German subjects 1-20: (a) 'old' façades; (b) 'new' façades.

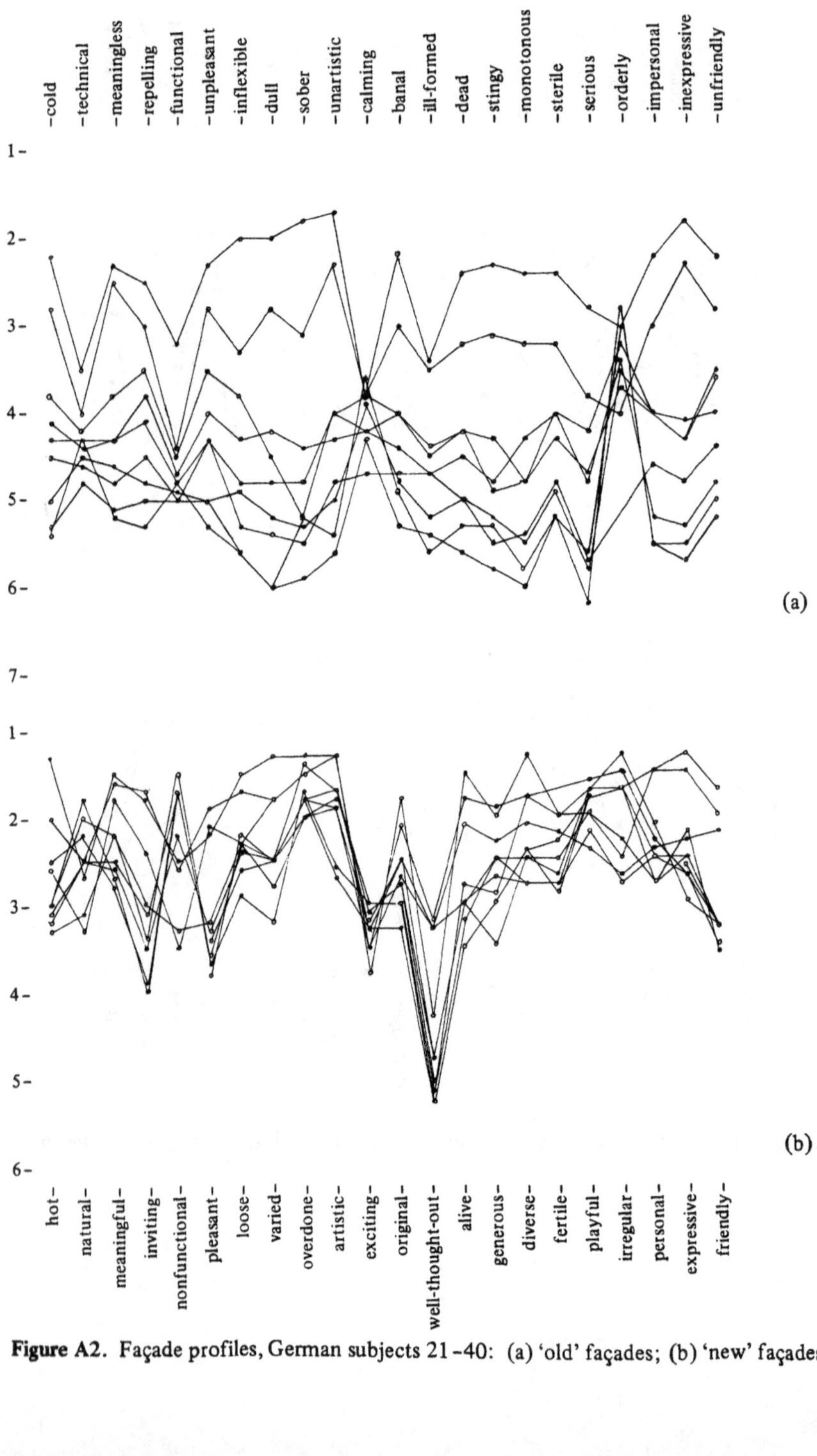

Figure A2. Façade profiles, German subjects 21–40: (a) 'old' façades; (b) 'new' façades.

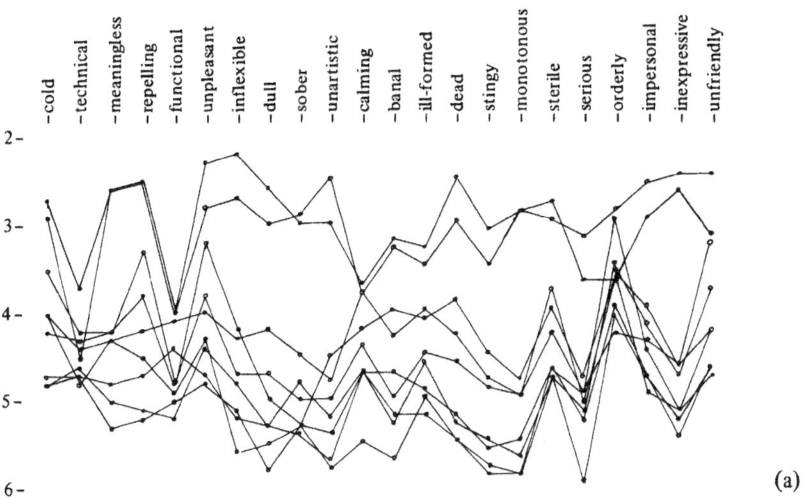

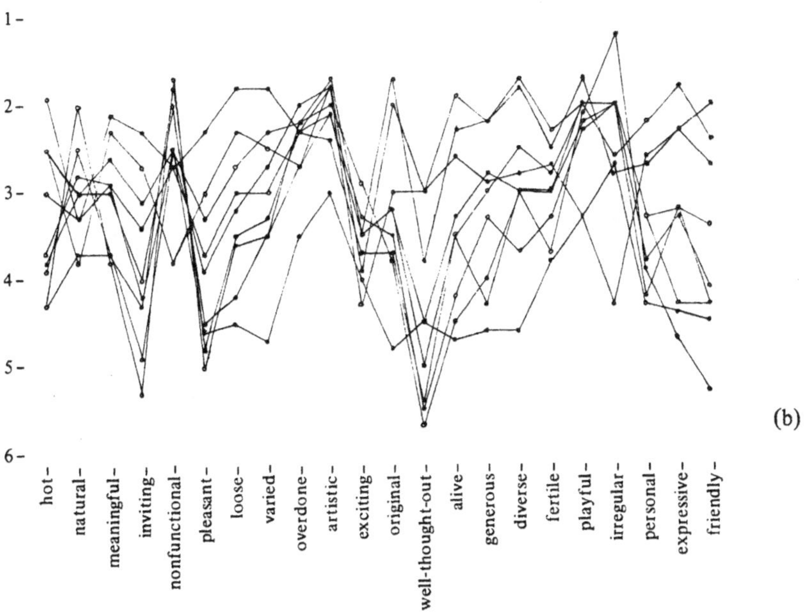

Figure A3. Façade profiles, German subjects 41-60: (a) 'old' façades; (b) 'new' façades.

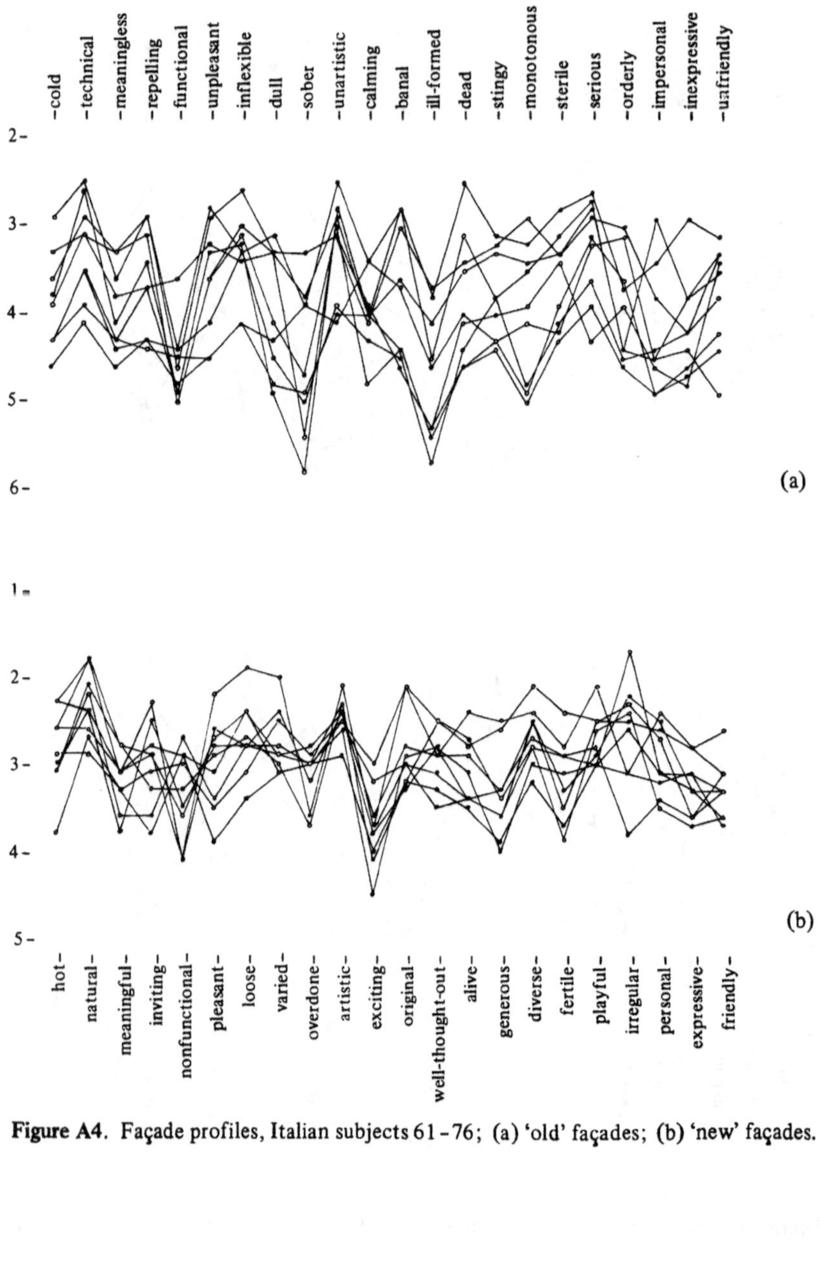

Figure A4. Façade profiles, Italian subjects 61-76; (a) 'old' façades; (b) 'new' façades.

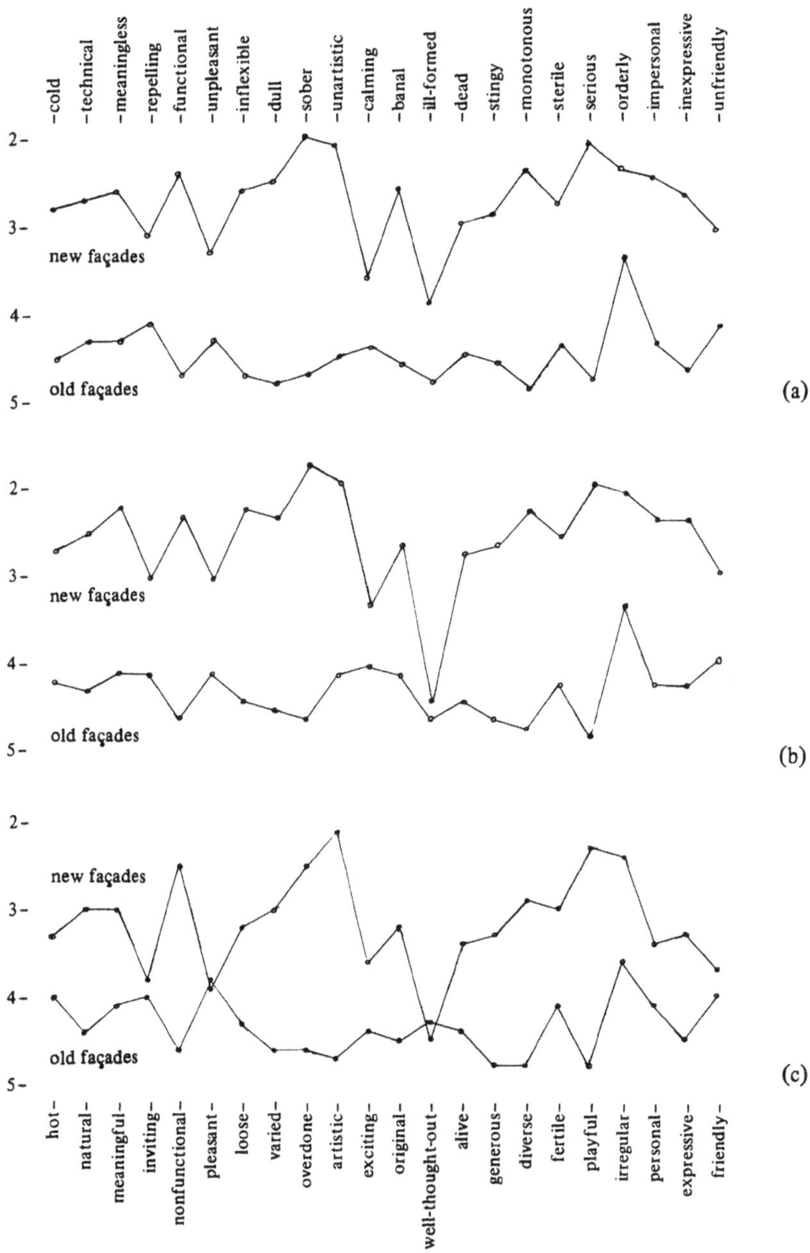

Figure A5. Average 'old' and 'new' façades: (a) German subjects 1-20; (b) German subjects 21-40; (c) German subjects 41-60; (d) Italian subjects 61-76; (e) German subjects 1-40; (f) German subjects 1-60; (g) German and Italian subjects 1-76.

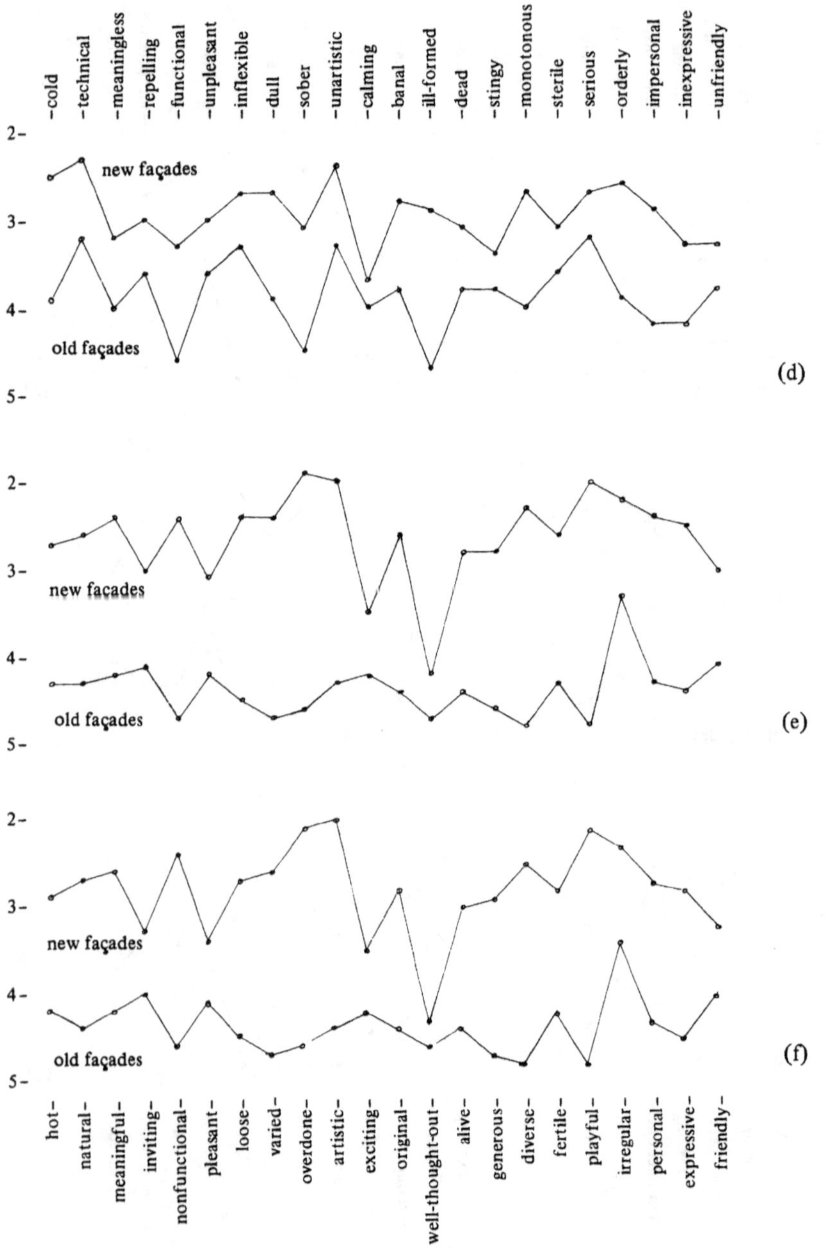

Figure A5. (continued).

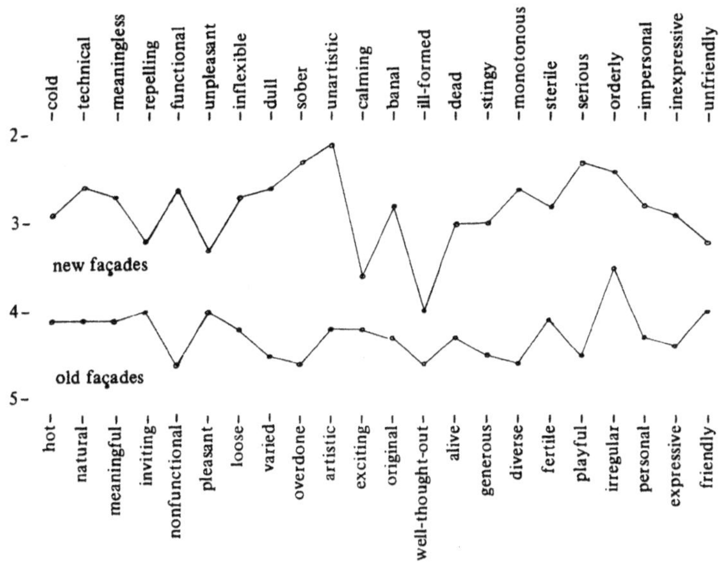

(g)

Figure A5. (continued).

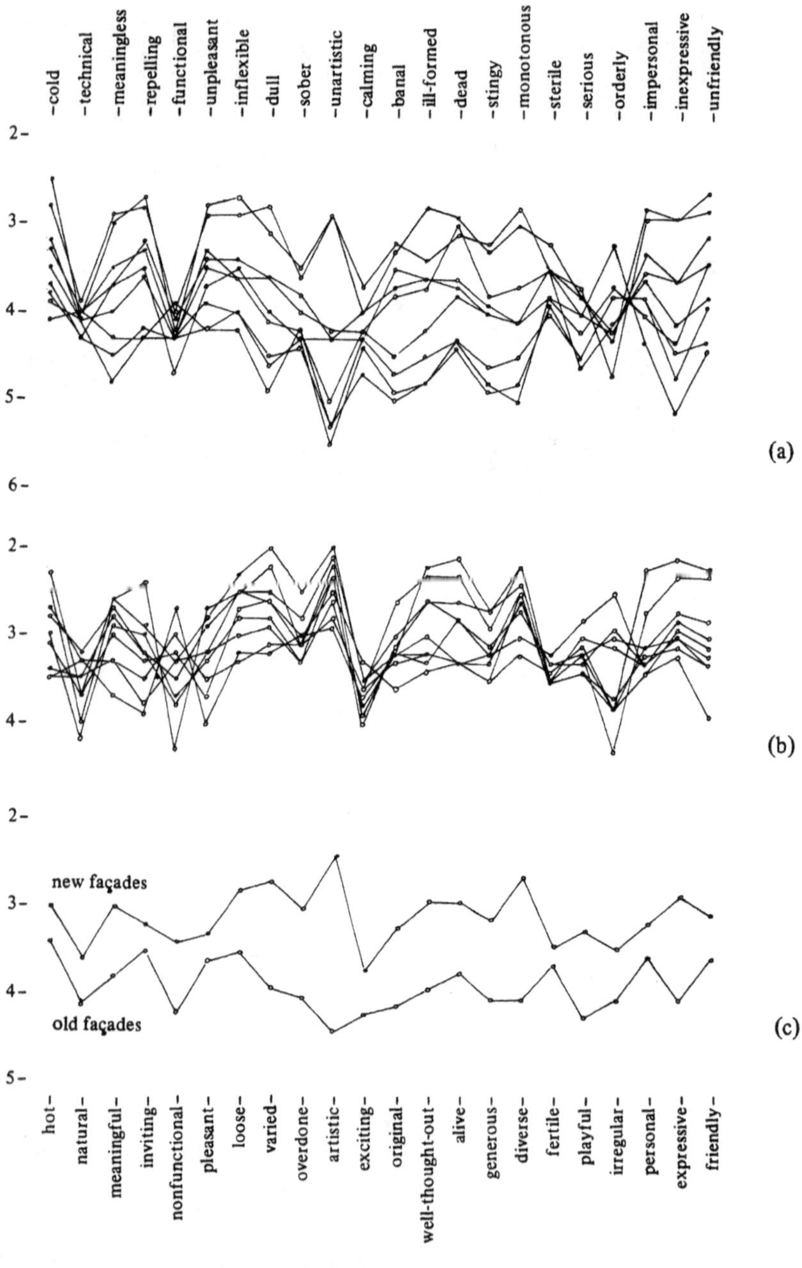

Figure A6. Façade profiles, forty trade school students: (a) 'old' façades; (b) 'new' façades; (c) average 'old' and 'new' façades.

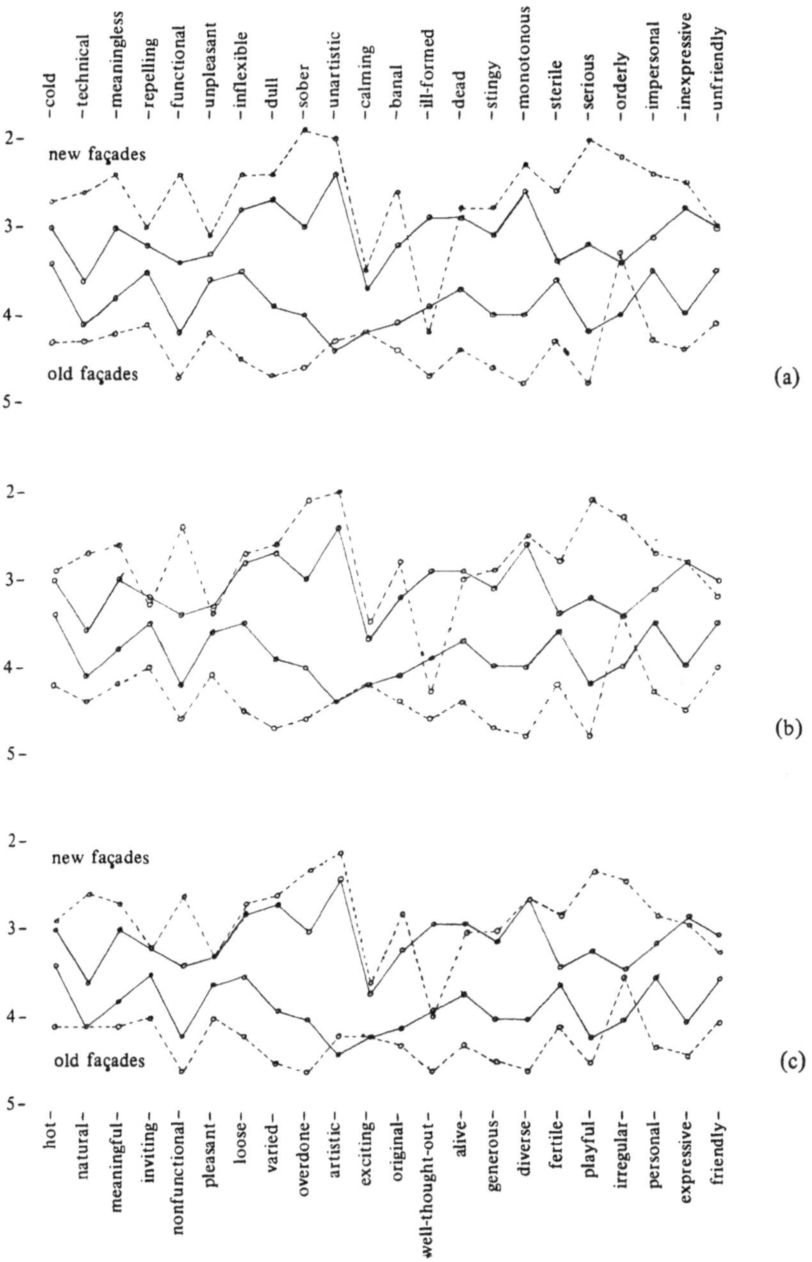

Figure A7. Comparison of average 'old' and 'new' façade profiles, study 1 versus study 2: (a) German subjects 1-40 (broken lines) versus forty trade school students (solid lines); (b) German subjects 1-60 versus trade school students; (c) German and Italian subjects 1-76 versus trade school students.

References

Agazzi G, Goetz M, Prati E, Ranc A, 1979 *Pratica e Rappresentazione dello Spazio Urbano* (Fondazione Ticino Nostro, Lugano)

Agrest D, Gandelsonas M, 1973 "Critical remarks on semiology of architecture" *Semiotica* 9(3) 252-271

Alberti L B, 1755/1965 *Ten Books on Architecture* translated by J Leoni, Ed. J Rykwert (Alec Tiranti, London)

Appleyard D, Lynch K, Myer J R, 1964 *The View from the Road* (MIT Press, Cambridge, Mass)

Arnheim R, 1971 *Entropy and Art* (University of California Press, Berkeley and Los Angeles)

Bachelard G, 1957 *La Poétique de l'Espace* (Presses Universitaires de France, Paris)

Barbey G, Gelber Ch, 1973 "The relationship between the built environment and human behavior: a survey and analysis of the existing literature" Institut de Recherche sur l'Environnement Construit, Ecole Polytechnique Fédérale, Lausanne

Barker R C, 1968 *Ecological Psychology: Concepts and Methods for Studying the Environment of Human Behavior* (Stanford University Press, Stanford, Calif.)

Barthes R, 1964 "Éléments de sémiologie" *Communications* **4** 91-135

Barthes R, 1965 *La Tour Eiffel* (Delpire, Paris)

Barthes R, 1967 "Semiologia e urbanistica" *Op. Cit. Selezione della Critica d'Arte Contemporanea* number 10 7-17

Barthes R, 1971 "Sémiologie et urbanisme" *Architecture Aujourd'hui* **153** 11-13

Baudrillard J, 1968 *Le Système des Objets* (Gallimard, Paris)

Baudrillard J, 1972 *Pour une Critique de l'Économie Politique du Signe* (Gallimard, Paris)

Becker H, Keim D, 1972 *Wahrnehmung in der städtischen Umwelt—möglicher Impuls für kollektives Handeln* (Kiepert, Berlin)

Bellman R E, Kalaba R E, Zadeh L A, 1964 "Abstraction and pattern classification" RM-4307 PR, Rand Corporation, Santa Monica, Calif.

Benevolo L, 1968 *Le Origini dell'Urbanistica Moderna* (Editori Laterza, Bari)

Bense M, 1956 *Aesthetica II (Ästhetische Information)* (Agis, Krefeld)

Bense M, 1968a "Einführung in die Informationsästhetik" in *Kunst und Kybernetik* (DuMont Schauberg, Köln) pp 28-41

Bense M, 1968b "Urbanismus und Semiotik" *Arch +* 3 23-25

Bense M, 1969a *Einführung in die informationstheoretische Ästhetik* (Rowohlt, Reinbek)

Bense M, 1969b "Urbanismus und Semiotik" in *Einführung in die informationstheoretische Ästhetik* (Rowohlt, Reinbek) pp 132-137

Bense M, 1971a "Urbanismus und Semiotik" in *Konzept 1, Architektur als Zeichensystem* Eds A Carlini, B Schneider (Studio Wasmuth, Tübingen)

Bense M, 1971b "Semiotik, Ästhetik, Urbanität" paper presented at a colloquium held during "Stuttgarter Tage 1971, Städtebau" Universität Stuttgart, September

Bense M, 1971c *Zeichen und Design. Semiotische Ästhetik* (Agis, Baden-Baden)

Bense M, 1973 "Architektursemiotik" in *Wörterbuch der Semiotik* Eds M Bense, E Walther (Kiepenheuer und Witsch, Köln)

Bense M, Walther E (Eds), 1973 *Wörterbuch der Semiotik* (Kiepenheuer und Witsch, Köln)

Bentmann R, Müller M, 1970 *Die Villa als Herrschaftsarchitektur* (Edition Suhrkamp, Frankfurt)

Berger A, Good L, 1963 "Architectural psychology in a psychiatric hospital" *Journal of the American Institute of Architects* (December) 76-80

Bergler R and others, 1975 *Das Eindrucksdifferential* (Hans Huber, Bern)

Berlyne D E, 1960 *Conflict, Arousal and Curiosity* (McGraw-Hill, New York)

Berndt H, 1968 *Das Gesellschaftsbild bei Stadtplanern* (Karl Krämer, Stuttgart)

Bernstein B, 1971 *Class, Codes and Control. Volume 1. Theoretical Studies Towards a Sociology of Language* (Paladin, St Albans, Herts)
Bertin J, 1967 *Sémiologie Graphique* (Mouton, Paris/Gauthier-Villars, Paris)
Bettini S, 1958 "Critica semantica e continuità storia dell'architettura europea" *Zodiac* 2 7-25
Birkhoff G, 1928 "Quelques éléments mathématiques de l'art" *Atti del Congresso Internazionale dei Matematici, Bologna* 315-333
Birkhoff G, 1932 "A mathematical theory of aesthetics" *The Rice Institute Pamphlet* V(19) 189-342
Birkhoff G D, 1950 *Collected Mathematical Papers* (The American Mathematical Society, New York)
Blomeyer G R, Helmholtz R M, 1976 "Semiotics in architecture" *Semiosis* 1 42-51
Bonta J P, 1972 "Towards a semiotic theory of graphic design languages" contribution to the International Semiotics Workshop, October (Semiotics Research Group, Ulm)
Bonta P, 1975 "Graficación y diseño. Lenguajes y sistemas gráficos en el proceso de diseño. Abstracción y concreción en el lenguaje gráfico" *Arquitecturas* January, 1-5; March, 28-30; May, 29-31
Bortz J, 1972 "Beiträge zur Anwendung der Psychologie auf den Städtebau. II. Erkundungsexperiment zur Beziehung zwischen Fassadengestaltung und ihrer Wirkung auf den Betrachter" *Zeitschrift für experimentelle und angewandte Psychologie* 19 226-281
Boudon P, 1969 "Sur un statut de l'objet: différer l'objet de l'objet" *Communications* 13 66-86
Boudon P, 1973 "Recherches sémiotiques sur le lieu" *Semiotica* 7 189-225
Boudon P, 1974 "Définition (sémiotique) des critères pour une théorie des lieux: architecture et methodologies" *MMI Bulletin des Seminaires Pédagogiques* (Institut de l'Environnement, Paris) 4 3-57
Boudon Ph, 1974 "Sémiologie des figures et syntaxe des formes" in *Figuration Graphique en Architecture* Eds Ph Boudon, J Guillerme, R Tabouret (AREA, Paris) fascicule 3a
Bourdieu P, 1968 "Éléments d'une théorie sociologique de la perception artistique" *Revue Internationale des Science Sociales* 20 640-664
Brandi C, 1960a *Le Due Vie* (Laterza, Bari)
Brandi C, 1960b *Segno e Immagine* (Il Saggiatore, Milano)
Brandi C, 1967 *Struttura e Architettura* (Einaudi, Torino)
Broadbent G, 1971 "Architektonische Semiotik" *Werk* 6 396-399
Broadbent G, 1972 "The analysis of signs in the designed environment" contribution to the International Semiotics Workshop, October (Semiotics Research Group, Ulm)
Broadbent G, 1973 "Communications" in *Design in Architecture* (John Wiley, Chichester, Sussex) chapter 11
Broadbent G, 1974 "Las estructuras profundas en la arquitectura" in *Arquitectura, Historia y Teoria de los Signos* Ed T Llorens (Publicationes del Colegio Oficial de Arquitectos de Cataluña y Baleares: La Gaya Ciencia, Barcelona)
Bunt R, Jencks C, Llorens T, 1974 "On the meaning of the built environment: a bibliography of sources in English language" *Versus* 8/9 101-128
Buttlar A von, Wetzig A, 1973 "Die Schönheit der Stadt—berechnet" *Süddeutsche Zeitung* number 103, 151-152
Buttlar A von, Selig H, Wetzig A, 1973 "The value of historic architecture" in *How to Play the Environment Game* Ed. T Crosby (Penguin Books, Harmondsworth, Middx) pp 74-81
Canter D (Ed.) 1970 *Architectural Psychology* (RIBA, London)

Canter D, Lee T (Eds), 1974 *Psychology and the Built Environment* (Architectural Press, London)
Canter D, Tagg S, 1972 "Can rooms be categorized?" contribution to the International Semiotics Workshop, October (Semiotics Research Group, Ulm)
Canter D, Tagg S, 1974 "Classificatión empírica de los aspectos del edificio y de sus atributos" in *Arquitectura, Historia y Teoria de los Signos* Ed. T Llorens (Publicationes del Colegio Oficial de Arquitectos de Cataluña y Baleares: La Gaya Ciencia, Barcelona)
Cappabianca A, 1967 "Architemi: appunti per una storia semantica dei materiali da costruzione" *Casabella* 316
Carroll J B, 1959 "Review of The Measurement of Meaning" *Language* **35** 58-77
Castels M, 1972 "Symbolique urbaine et mouvements sociaux" *Versus* **3/4** 81-92
Castex J, Depaule J C, Panerai P (Groupe Syntaxe), 1973 "Lectures d'architecture et la syntaxe de Frank Lloyd Wright" contribution to the Colloque de Sémiotique de l'Architecture, Centro Internazionale di Semiotica e Linguistica, Urbino, July (mimeo)
Castex J, Panerai P, 1974 "Structure de l'espace architectural" *Notes Méthodologiques en Architecture et en Urbanisme* **3/4** 39-66 (Centre MMI, Institut de l'Environnement, Paris)
Ceccato S, 1971 "Semiologie und Komplikationen" *Werk* **10** 686-688
Choay F, 1967 "Sémiologie et urbanisme" *Architecture d'Aujourd'hui* number 132 8-10
Choay F, 1970/1971 "Remarques à propos de sémiologie urbaine" *Architecture d'Aujourd'hui* number 153 (December/January) 9-10
Choay F, 1971 *Connexions* (L'Immobilière—Constructions de Paris, Paris)
Choay F, 1973a "Semiogeny and utopia as themes of the foundation of cities" contribution to the Colloque de Sémiotique de l'Architecture, Centro Internazionale di Semiotica e Linguistica, Urbino, July (mimeo)
Choay F, 1973b "Figures d'un discours méconnu" *Critique* **311** 293-317
Choay F, 1974a "Notes préliminaires à une sémiologie du discours sur la ville" *Notes Méthodologiques en Architecture et on Urbanisme* **3/4** 151-183 (Centre MMI, Institut de l'Environnement, Paris)
Choay F, 1974b "Sémiologie du discours des fondateurs de ville: les utopiques" *Cahier de l'École d'Architecture, Université de Genève* **1** 31-60
Chomsky N, 1965 *Aspects of the Theory of Syntax* (MIT Press, Cambridge, Mass)
Chomsky N, 1971 "Deep structure, surface structure and semantic interpretation" in *Semantics* Eds D D Steinberg, L A Jacobvitz (Cambridge University Press, London)
Chotlos J W, 1944 "Studies in language behavior: IV. A statistical and comparative analysis of individual written language samples" *Psychological Monographs* **56** 75-111
Choynowski M, 1969 "Dimensions of painting" *Exakte Ästhetik* **6** 89-90
Collins J B, Seaton R W, 1970 "Semantic dimensions as architectural discriminators" paper presented to the Western Psychological Association Annual Convention, San Francisco, Calif. April
Costa-Gomes-Ber K, Costa-Gomes F, Huynh Tu Kinh, 1973a "Étude historique d'un quartier urbain: le Rôtillon à Lausanne" *Habitation* **9** 27-82
Costa-Gomes-Ber K, Costa-Gomes F, Huynh Tu Kinh, 1973b " 'Microterritoires' dans des sites historiques et 'planification' urbaine" Diplôme, Département d'Architecture École Polytechnique Fédérale de Lausanne
Cronbach L J, Gleser G C, 1953 "Assessing similarity between profiles" *Physiological Bulletin* **50**(6) 456-473
Day H, 1965 "The motivation of aesthetic preference" paper read at the Second National Seminar on Design and Planning, Waterloo, Ontario, Canada (mimeo)

Dexter B, Lindsey S, 1976 "Architectural meaning: is it predictable" paper presented at EDRA 7 (mimeo)
Dexter B, Lindsey S, undated "Meaning in architecture: some theoretical issues" (mimeo)
Diebold R A, 1965 "A survey of psycholinguistic research, 1954-1964" in *Psycholinguistics* Eds C E Osgood, T A Sebeok (Indiana University Press, Bloomington)
Doblin J, 1965 "A proposed theory of design" in *Design and Planning* Ed. M Krampen (Hastings House, New York)
Dorfles G, 1958 *Le Oscillazioni del Gusto* (Lerici, Milano)
Dorfles G, 1959 "Valori semantici degli 'elementi di architettura' e dei 'caratteri distributivi'" *Domus* **360** 33-34
Dorfles G, 1962 "Valori communicativi e simbolici nell'architettura, nel disegno industriale e nella pubblicità" in *Simbolo, Communicazione, Consumo* (Einaudi, Torino) chapter 5
Dörner V, 1976 *Die Produktform als Mittel der Anmutungsgestaltung* (Peter Hanstein, Köln)
Downs R M, Stea D, 1974 *Image and Environment* (Edward Arnold, London)
Draghetti A, 1958/1959 *Il Segno Linguistico e il Problema dei Linguaggi Artistici* thesis, Facoltà di Lettere e Filosofia Università di Bologna
Draghetti A, 1961 "Architettura e linguaggio" *Il Verri* **6** 118-129
Dreyer C, 1974 "Einführung in die Architektursemiotik" Grundkurs Kommunikationstechnik SS 74, Informationseinheit 2, Fachbereich 1, Universität Stuttgart (mimeo)
Durand J N L, 1821; 1823; 1825 *Précis des Leçons d'Architecture Données à l'École Royale Polytechnique* chez l'auteur, à l'École Royale Polytechnique, Paris; volume I, volume II, volume III
Eco U, 1968 *La Struttura Assente* (Bompiani, Milano)
Eco U, 1971 "Per una analisi semantica dei segni architettonici" in *Le Forme del Contenuto* (Bompiani, Milano) pp 157-179
Eco U, 1976 *A Theory of Semiotics* (Indiana University Press, Bloomington)
Edwards E, 1964 *Information Transmission* (Chapman and Hall, London)
Ehrenfels C von, 1967 "Höhe und Reinheit der Gestalt" in *Gestalthaftes Sehen* Ed. F Weinhandl (Wissenschaftliche Buchgesellschaft, Darmstadt)
Eisenbeis M, 1974 "Théories de communication et création architecturale" Notes Méthodologiques en Architecture et en Urbanisme **3/4** 111-117 (Centre MMI, Institut de l'Environnement, Paris)
Eisenmann P, 1974 "Notas sobre arquitectura conceptual: estructura profunda dual" in *Arquitectura, Historia y Teoria de los Signos* Ed. T Llorens (Publicationes del Colegio Oficial de Arquitectos de Cataluña y Baleares: La Gaya Ciencia, Barcelona)
Ertel S, 1964 "Die emotionale Natur des 'semantischen' Raumes" *Psychologische Forschung* **28** 1-32
Ertel S, 1965a "Standardisierung eines Eindrucksdifferentials" *Zeitschrift für Experimentelle und Angewandte Psychologie* **12** 22-58
Ertel S, 1965b "Weitere Untersuchungen zur Standardisierung eines Eindrucksdifferentials" *Zeitschrift für experimentelle und angewandte Psychologie* **12** 177-208
Faccani R, Eco U (Eds), 1969 *I Sistemi di Segni e lo Strutturalismo Sovietico* (Bompiani, Milano)
Fauque R, 1973 "Pour une nouvelle approche sémiologique de la ville" *Espaces et Sociétés* **9** (Juillet) 15-27
Fechner G T, 1865 "Über die Fragen des goldenen Schnitts" *Archiv für die zeichnenden Künste* **11** 100-112
Fechner G T, 1876 *Vorschule der Aesthetik* (Breitkopf und Haertel, Leipzig)

Fisher H T, 1964 "A technique for processing complex statistical data into meaningful graphic form" in *Architecture and the Computer. Proceedings of the First Boston Architectural Center Conference* (The Boston Architectural Center, Boston, Mass) pp 13-18
Fowler H, 1965 *Curiosity and Exploratory Behavior* (Macmillan, New York)
Frank H, 1968 *Informationsästhetik. Grundlagenprobleme und erste Anwendung auf die Mime pure* (Schnelle, Quickborn)
Franke J, 1969 "Stadtbild. Zum Erleben der Wohnumgebung" *Stadtbauwelt* **24** 292-295
Franke J, Bortz J, 1972 "Beiträge zur Anwendung der Psychologie auf den Städtebau. I. Vorüberlegung und erste Erkundungsuntersuchung zur Beziehung zwischen Siedlungsgestaltung und Erleben der Wohnumgebung" *Zeitschrift für Experimentelle und Angewandte Psychologie* **19** 76-108
Fried M, 1963 "Grieving for a lost home" in *The Urban Condition* Ed. L G Duhl (Basic Books, New York)
Fusco R de, 1967 *Architettura come Mass-Medium* (Dedalo, Bari)
Fusco R de, Palmieri N, Pasca Raymondi G, 1966 "Note per una semiologia figurativa" *Op. Cit. Selezione della Critica d'Arte Contemporanea* number 7 49-66
Galbraith J K, 1958 *The Affluent Society* (Houghton Mifflin, Boston)
Gamberini J, 1953 *Per una Analisi Degli Elementi dell'Architettura* (Editrice Universitaria, Firenze)
Gamberini J, 1959 *Introduzione al Primo Corso di Elementi di Architettura. Gli Elementi dell'Architettura come 'parole' del Linguaggio Architettonico* (Coppini, Firenze)
Gamberini J, 1961 *Analisi Degli Elementi Costitutivi dell'Architettura* (Coppini, Firenze)
Gandelsonas M, 1972 "On reading architecture" *Progressive Architecture* **3** 68-87
Gardin J-C, 1958 "Four codes for the description of artifacts: an essay in archaeological technique and theory" *American Anthropologist* **60**(2) 335-357
Garnich R, 1968 *Konstruktion, Design, Ästhetik* (Rudolf Garnich, Esslingen)
Garroni E, 1964 *La Crisi Semantica delle Arti* (Officina, Roma)
Geddes P, 1949 *Cities in Evolution* (Williams and Norgate, London)
Gibson J J, 1954 "A theory of pictorial perception" *Audio-Visual Communication Review* **2** (winter) 3-23
Glassie H, 1973 "Structure and function, folklore and the artifact" *Semiotica* **7/8** 313-351
Goodman N, 1968 *Languages of Art. An Approach to a Theory of Symbols* (Bobbs-Merrill, Indianapolis, Ind.)
Graumann C F, 1975 "Interaktion und Kommunikation" in *Handbuch der Psychologie Volume 7. Sozialpsychologie. 2. Halbband* Ed. C F Graumann (Hogrefe, Göttingen) 1234-1242
Greimas A J, 1966 *Sémantique Structurale Recherche de Méthode* (Librairie Larousse, Paris)
Greimas A J, 1974 "Pour une sémiotique topologique" in *Notes Méthodologiques en Architecture et en Urbanisme* **3/4** 1-21 (Centre MMI, Institut de l'Environnement, Paris)
Guiraud P, 1973 *La Sémiologie* (Presses Universitaires de France, Paris)
Gunzenhäuser R, 1962 *Ästhetisches Maß und ästhetische Information* (Schnelle, Quickborn)
Guttman L, 1968 "A general non metric technique for finding the smallest coordinate space for a configuration of points" *Psychometrica* **33**(4) 469-506
Hall E T, 1966 *The Hidden Dimension* (Doubleday, New York)
Halprin L, 1965 "Motation" *Progressive Architecture* July 126-133

Hammad M and others (Groupe 107), 1973a *Sémiotique des Plans en Architecture* (Publication du Groupe 107, Paris)
Hammad M and others (Groupe 107), 1973b *Sémiotique d l'Espace* (Publication du Groupe 107, Paris)
Hammad M and others (Groupe '100 Tête'), 1973c "Pour une analyse sémiotique des plans d'architecture" contribution to the Colloque de Sémiotique de l'Architecture, Centro Internazionale di Semiotica e Linguistica, Urbino, July (mimeo)
Hammad M and others (Groupe 107), 1976 *Sémiotique des Plans en Architecture II* (Publication du Groupe 107, Paris)
Haug W F, 1972 *Kritik der Warenästhetik* (Edition Suhrkamp, Frankfurt)
Hauser A, 1951 *The Social History of Art* (Knopf, New York)
Hebb D O, 1961 *Organization of Behavior* (Science Editions, New York)
Heise D R, 1965 "Semantic differential profiles for 1000 most frequent English words" *Psychological Monographs: General and Applied* **79**(8) whole number 601
Heise D R, 1969 "Some methodological issues in semantic differential research" *Psychological Bulletin* **72** 406-422
Hempel C G, Oppenheim P, 1948 "Studies in the logic of explanation" *Philosophy of Science* **15** 135-175
Hershberger R G, 1972 "Toward a set of semantic scales to measure the meaning of architectural environments" *EDRA Three. Proceedings of the 3rd Annual Environmental Design Research Conference*, University of California, Los Angeles pages 6-4-1 - 6-4-10
Hillier B, Leaman A, 1973 "The architecture of architecture" in *Models and Systems in Architecture and Building* Ed. D Hawkes (Construction Press, London)
Hillier B, Leaman A, Stansall P, Bedford M, 1976 "Space syntax" *Environment and Planning B* **3** 147-185
Hjelmslev L, 1968 *Prolégomènes a une Théorie du Langage* (Les Éditions de Minuit, Paris)
Hofstaetter P R, 1955 "Über Ähnlichkeit" *Psyche* **9** 54-79
Holschneider J, 1969 *Schlüsselbegriffe der Architektur und Stadtbaukunst* (Schnelle, Quickborn)
Honikman B (Ed.), 1971 *Proceedings of the Architectural Psychology Conference at Kingston Polytechnic* 1-4 September, 1970 (Kingston Polytechnic and RIBA publications, London)
Hudson W, 1960 "Pictorial depth perception in sub-cultural groups in Africa" *Journal of Social Psychology* **52** 183-208
Ivanov V V, Toporov V N, Zaliznjak A A, 1969 "Possibilità di uno studio tipologico-strutturale di alcuni sistemi semiotici modellizzanti" in *I Sistemi dei Segni e lo Strutturalismo Sovietico* Eds R Faccani, U Eco (Bompiani, Milano)
Jakobson R, 1963 *Essais de Linguistique Générale* (Les Éditions de Minuit, Paris)
Jenkins J J, Russel W A, Suci G J, 1958 "An atlas of semantic profiles for 360 words" *American Journal of Psychology* **71** 688-699
Jencks C, 1969 "History as myth" in *Meaning in Architecture* Eds C Jencks, G Baird (Cresset Press, London)
Jencks C, 1971 *Architecture 2000, Prediction and Methods* (Praeger, New York)
Jencks C, 1972 "Rhetoric and architecture" *Architectural Association Quarterly* **4**(3) 4-17
Jencks C, 1974 "Retórica y arquitectura" in *Arquitectura, Historia y Teoria de los Signos* Ed. T Llorens (Publicationes del Colegio Oficial de Arquitectos de Cataluña y Baleares: La Gaya Ciencia, Barcelona)
Joedicke J, Dirlewanger H, Geisler E, Magnago V, 1975 "Zwischenbericht. Grundlagen der Gestaltung weitgespannter Flächentragwerke" Sonderforschungsbereich 64 an der Universität Stuttgart
Johnson S C, 1967 "Hierarchical clustering schemes" *Psychometrica* **32**(3) 241-254

Kiefer G R, 1970 *Zur Semiotisierung der Umwelt. Eine exemplarische Erörterung der sekundären Architektur* dissertation, Universität Stuttgart

Kiemle M, 1967 *Ästhetische Probleme der Architektur unter dem Aspekt der Informationsästhetik* (Schnelle, Quickborn)

Klaus G, 1963 *Semiotik und Erkentnistheorie* (VEB Deutscher Verlag der Wissenschaften, Berlin)

Koenig G K, 1964 *Analisi del Linguaggio Architettonico* (Libreria Editrice Fiorentina, Firenze)

Koenig G K, 1970 *Architettura e Communicazione: Preceduta da Elementi di Analisi del Linguaggio Architettonico* (Libreria Editrice Fiorentina, Firenze)

Komorita S S, Bass A R, 1967 "Attitude differentiation and evaluative scales of the semantic differential" *Journal of Personality and Social Psychology* **6** 241-244

Krampen M, 1969 "225 scales for SD research in architecture" Lehrstuhl für Zeichnen und Modellieren, Universität Stuttgart (mimeo)

Krampen M, 1971 "Das Messen von Bedeutung in Architektur, Stadtplanung und Design. Teil 1: Das Polaritätsprofil als Meßinstrument" *Werk* **1** 57-60

Krampen M, 1973a "Dalla semiologia della communicazione alle logica degli instrument" *Op. Cit. Selezione della Critica d'Arte Contemporanea* number 28 24-44

Krampen M, 1973b "Iconic signs, supersigns and models" *Versus* **4** 101-108

Krampen M, 1974a "Una posible analogia entre (psico-) lingüistica y arquitectura: la 'type-token ratio" in *Arquitectura, Historia y Teoria de los Signos* Ed. T Llorens (La Gaya Ciencia, Barcelona)

Krampen M, 1974b "A possible analogy between (psycho-) linguistic and architectural measurement—the type-token ratio (TTR)" in *Psychology and the Built Environment* Eds D Canter, T Lee (Architectural Press, London)

Krampen M, Breitenbücher P, Frehse S, 1971 "Das Messen von Bedeutung in Architektur und Stadtplanung. Teil 2: Bedeutungsmäßige Differenzierung zweier Gebäude" *Werk* **2** 137-140

Kumata H, 1957 *A Factor Analytic Investigation of Semantic Structures Across Two Selected Cultures* PhD dissertation, University of Illinois, Urbana

Kumata H, Schramm W, 1956 "A pilot study of cross-cultural methodology" *Public Opinion Quarterly* **20** 229-238

Kürth H, Kutschmar A, 1969 *Baustilfibel* (Volk und Wissen, Berlin)

Lagrange M-S, 1973 *Analyse Sémiologique et Histoire de l'Art. T.A. Documents 3* (Editions Klincksieck, Paris)

Laumann E O, House J S, 1970 "Living room styles and social attributes. The patterning of material artifacts in a modern urban community" *Sociology and Social Research* **54**(3) 321-342

Ledrut R, 1973a "Parole et silence de la ville" *Espaces et Sociétés* July 3-14

Ledrut R, 1973b *Les Images de la Ville* (Éditions Anthropos, Paris)

Lefebvre H, 1968 *La Vie Quotidienne dans le Monde Moderne* (Gallimard, Paris)

Lessing O (Ed.), 1882 *Ausgeführte Bauornamente Berlins. Sammlung hervorragender Ornamentausführungen der bedeutendsten Architekten, Bildhauer und Maler Berlins* (100 Tafeln). (Ernst Wasmuth, Berlin)

Lessing O (Ed.), 1884 *Ausgeführte Bauornamente der Neuzeit. Sammlung hervorragender Ornamentsausführungen der bedeutendsten Architekten und Bildhauer in Deutschland und Österreich* (100 Tafeln) (Ernst Wasmuth, Berlin)

Lévi-Strauss C, 1958 *Anthropologie Structurale* (Librairie Plon, Paris)

Lindekens R, 1971 *Éléments pour une Sémiotique de la Photographie* (Didier, Paris)

Lingoes J C, 1970 "A general nonparametrical model for representing objects in a joint metric space" in *Archélogie et Calculateurs. Problèmes Sémiologiques et Mathématiques: Colloques Internationaux du Centre National de la Recherche Scientifique* **534** 277-298

Lingoes J C, 1973 *The Guttman-Lingoes Nonmetric Program Series* (Mathesis Press, Ann Arbor, Mich.)
Llorens T, 1974 "El studio del significado de la arquitectura: problemas y criterios epistemológicos" in *Arquitectura, Historia y Teoria de los Signos* Ed. T Llorens (Publicationes del Colegio Oficial de Arquitectos de Cataluña y Baleares: La Gaya Ciencia, Barcelona)
Lotman J M, 1969 "Il problema di una tipologia della cultura" in *I Sistemi di Segni e lo Strutturalismo Sovietico* Eds R Faccani, U Eco (Bompiani, Milano) pp 309-318
Lotman J M, Uspenskij B A, 1975 *Tipologia della Cultura* (Bompiani, Milano)
Lynch K, 1960 *The Image of the City* (MIT Press, Cambridge, Mass)
Maldonado T, 1973 "Objekte, Waren, Bedürfnisse. Zur gegenwärtigen Diskussion um eine marxistische Anthropologie" in *Semiologie und Theorie der Architektur* Ed. M Krampen, Bericht über ein Seminar an der École d'Architecture, Université de Genève (mimeo)
Marcus S, 1973 "Genetic grammar and architecture" contribution to the Colloque de Sémiotique de l'Architecture, Centro Internazionale di Semiotica e Linguistica, Urbino, July (mimeo)
Marcus S, 1974 "Linguistic structures and generative devices in molecular genetics" *Cahiers de Linguistique Théorique et Appliquée* **11**(1) 77-104
Marin L, 1974 "Espace dans le texte et espace du texte" in *Notes Méthodologiques en Architecture et en Urbanisme* 3/4 119-141 (Centre MMI, Institut de l'Environnement, Paris)
Markuzon V, 1972 "An attempt to redefine the 'language of architecture' in terms of semantics" *Architectural Association Quarterly* **4**(4) 41-48
Marx K, Engels F, 1970 *Werke* Volumen 3, 4, 13, 23, 26 (Dietz, Berlin)
Menna F, 1967 "Design, communicazione estetica e mass-media" *Edilizia Moderna* **85** 32-37
Menna F, 1971 "Design, ästhetische Kommunikation und Massenmedien" in *Konzept 1, Architektur als Zeichensystem* Eds A Carlini, B Schneider (Studio Wasmuth, Tübingen)
Miller G A, 1967 "Psycholinguistic approaches to the study of communication" in *Journeys in Science* Ed. D L Arm (The University of New Mexico Press, Albuquerque) pp 22-73
Miller G A, 1969 "A psychological method to investigate verbal concepts" *Journal of Mathematical Psychology* **6** 169-191
Moles A A, 1958 *Théorie de l'Information et Perception Esthétique* (Flammarion, Paris)
Moles A A, 1971 *Informationstheorie und ästhetische Wahrnehmung* (DuMont Schauberg, Köln)
Moles A A, 1972 *Théorie des Objets* (Éditions Universitaires, Paris)
Monke F, 1968 *Grundrißentwicklung und Aussehen des Berliner Miethauses von 1850 bis 1914 dargestellt an Beispielen aus dem Stadtteil Moabit. I and II* doctoral dissertation, Fakultät für Architektur, Technische Universität, Berlin
Morris Ch, 1955 *Signs, Language and Behavior* (George Braziller, New York)
Mukařovský J, 1970 *Kapitel aus der Ästhetik* (Edition Suhrkamp, Frankfurt)
Nattiez J J, 1973 "La linguistica: una nuova via per l'analisi musicale" *Versus* **5** 2-18
Neisser U, 1967 *Cognitive Psychology* (Meredith, New York)
Nunally J, 1962 "The analysis of profile data" *Psychological Bulletin* **59**(4) 311-319
Orlik P, 1965 "Eine Modellstudie zur Psychophysik des Polaritätsprofils" *Zeitschrift für experimentelle und angewandte Psychologie* **12** 614-647
Orlik P, 1967 "Eine Technik zur erwartungsgetreuen Skalierung psychologischer Merkmalsräume aufgrund des Polaritätsprofils *Zeitschrift für experimentelle und angewandte Psychologie* **14** 616-650

Osgood C E, 1958 *Method and Theory in Experimental Psychology* (Oxford University Press, New York)
Osgood C E, 1959 "Semantic space revisited: a reply to Uriel Weinreich's review of *The Measurement of Meaning*" *Word* **15**(1) 192-200
Osgood C E, 1960 "The cross-cultural generality of visual-verbal synesthetic tendencies" *Behavioral Science* **5**(2) 146-169
Osgood C E, 1971 "Explorations in semantic space: a personal diary" *Journal of Social Issues* **27**(4) 5-64
Osgood C E, Suci G J, 1952 "A measure of relation determined by both mean difference and profile information" *Psychological Bulletin* **49** 251-262
Osgood C E, Suci G J, Tannenbaum P H, 1957 *The Measurement of Meaning* (University of Illinois Press, Urbana)
Ostrowetsky S, 1972 "De l'urbain à l'urbain" *Cahiers Internationaux de Sociologie* **52** 93-110
Ostrowetsky S, 1974 "Une typologique du lieu" in *Notes Méthodologiques en Architecture et en Urbanisme* **3/4** 185-199 (Centre MMI, Institut de l'Environnement, Paris)
Ostrowetsky S, Bordreuil S, Milanoff C, Ronchi Y, 1973 "Reproduction of regional styles in architecture" contribution to the Colloque de Sémiotique de l'Architecture, Centro Internazionale di Semiotica e Linguistica, Urbino, July (mimeo)
Panofsky E, 1957 *Meaning in the Visual Arts* (Doubleday, Garden City, NY)
Parr A E, 1966 "Über die Konsequenzen der Monotonie. Ein psychologischer Exkurs" *Bauwelt* **20** 571-578 (English: "City and psyche" *Yale Review* **25**(1) 71-85, 1965)
Peirce C S, 1960 *Collected Papers* volumes I-VI, Eds C Hartshorne, P Weiss; volumes VII-VIII Ed. A W Burkes (Harvard University Press, Cambridge, Mass)
Prieto L J, 1966 *Messages et Signaux* (Presses Universitaires de France, Paris)
Prieto L J, 1971 "Notes pour une sémiologie de la communication artistique" *Werk* **4** 248-251
Prieto L J, 1973 "Signe et instrument" in *Littérature, Histoire, Linguistique. Receuil d'Études Offert à Bernard Gagnebin* (Éditions l'Age d'Homme, Lausanne)
Prieto L J, 1975a *Études de Linguistique et de Sémiologie Générales* (Librairie Droz, Genève)
Prieto L J, 1975b *Pertinence et Pratique* (Les Éditions de Minuit, Paris)
Prieto L J, 1979 "Sull' integrazione, Postscriptum" in *Pratica e Rappresentazione dello Spazio Urbano* G Agazzi, M Goetz, E Prati, A Ranc (Fondazione Ticino Nostro, Lugano)
Prochansky H M, Ittelson W H, Rivlin L G, 1970 *Environmental Psychology* (Holt, Rinehart and Winston, New York)
Rapoport A, Hawkes R, 1970 "The perception of urban complexity" *Journal of the American Institute of Planners* **36**(2) 106-111
Rapoport A, Kantor R, 1967 "Complexity and ambiguity in environmental design" *Journal of the American Institute of Planners* **33**(4) 210-221
Rauen B, 1971 "Problematik der Informationsästhetik" *Sprache im technischen Zeitalter* **37** 80-85
Raymond H, 1974 "Habitat, modeles culturels et architecture" *Architecture d'Aujourd'hui* (July/August) 174
Reinecke H P, 1966a "Fundamentals for the acoustic investigations of noise in offices" *Kommunikation* **2** 29-38
Reinecke H P, 1966b "Acoustic and psychometric investigations concerning the qualities of noise in offices" *Kommunikation* **2** 44-45
Resnikow L O, 1968 *Erkennungstheoretische Fragen der Semiotik* (VEB Deutscher Verlag der Wissenschaften, Berlin)

Rossi-Landi F, 1968 *Il Linguaggio come Lavoro e come Mercato* (Bompiani, Milano)
Rossi-Landi F, 1972 "Omologia della riproduzione sociale" *Ideologia* **16/17** 43-103
Rossi-Landi F, 1975 *Linguistics and Economics* (Mouton, The Hague)
Rubinstein S L, 1972 *Sein und Bewußtsein* (Akademie, Berlin)
Sanoff H, 1970 "House form and preference" in *EDRA 2. Proceedings of the 2nd Annual Design Research Association Conference Pittsburgh, Pa,* Eds J Archea, C Eastman, pp 334-339
Sanoff H, 1973 "Youth's perception and categorizations of residential cues" in *EDRA 4. Environmental Design Research* Ed. W F Preiser (Dowden, Hitchinson and Ross, Stroudsburg, Pa) volume 1, pp 84-97
Saussure F de, 1971 *Cours de Linguistique Générale* (Payot, Paris)
Saussure F de, 1976 *Cours de Linguistique Générale, Edition Critique Préparé par Tullio de Mauro* (Payot, Paris)
Scalvini M L, 1968 "Per una teoria dell'architettura" *Op. Cit. Selezione della Critica d'Arte Contemporanea* number 13, 30-44
Scalvini M L, 1971 "Über das Signifikat in der Architektur" *Werk* **6** 390-393
Scalvini M L, 1972 "Semiotic analysis: are there common basic rules for different design disciplines?" contribution to the International Semiotics Workshop, October (Semiotics Research Group, Ulm)
Scalvini M L, 1975 *L'Architettura come Semiotica Connotativa* (Bompiani, Milano)
Schoof H, 1965 *Idealstädte und Stadtmodelle als theoretische Planungskonzepte. Beitrag zur räumlich funktionalen Organisation der Stadt* doctoral dissertation, Technische Hochschule Karlsruhe
Sebeok T A, 1972 " 'Semiotica' e affini" *Versus* **3** 1-11
Sebeok T A, 1977 "Jakob von Uexküll. Neglected figures in the history of semiotic inquiry I" lecture delivered on 27 August at the III Wiener Symposium über Semiotik (mimeo)
Segall M H, Campbell D T, Herskovits M J, 1966 *The Influence of Culture on Visual Perception* (Bobbs-Merrill, Indianapolis, Ind.)
Segaud M, 1973 "Anthropologie de l'espace, catalogue ou projet?" *Espaces et Sociétés* July 29-38
Shafer E L, Richards T A, 1974 "A comparison of viewer reactions to outdoor scenes and photographs of those scenes" in *Psychology and the Built Environment* Eds D Canter, T Lee (Architectural Press, London)
Shannon C E, Weaver W, 1949 *The Mathematical Theory of Communication* (University of Illinois Press, Urbana)
Shepard R N, 1962a "The analysis of proximities. Multidimensional scaling with an unknown distance function. I" *Psychometrica* **27**(2) 125-140
Shepard R N, 1962b "The analysis of proximities. Multidimensional scaling with an unknown distance function. II" *Psychometrica* **27**(3) 219-246
Shepard R N, 1966 "Report from Bell Laboratories. 'Subjective space'—a new method of uncovering meaning" *IEEE Spectrum* February 8
Siegel S, 1956 *Nonparametric Statistics for the Behavioral Sciences* (McGraw-Hill, New York)
Sieverts T, 1968 "Information einer Geschäftsstraße—Umwelt als Informationsfeld" *Stadtbauwelt* **20** 1496-1498 (also: *Werk* 1969 **10**)
Simmat W E, 1969 "Das 'semantic differential' als Instrument der Kunstanalyse" *Exakte Ästhetik* **6** 69-88
Skinner B F, 1957 *Verbal Behavior* (Appleton-Century-Crofts, New York)
Smith R G, 1961 "A semantic differential for theatre concepts" *Speech Monographs* **28** 1-8
Snider J G, Osgood C E (Eds), 1969 *Semantic Differential Techniques: A Sourcebook* (Aldine, Chicago)

Sommer R, 1965 "The significance of space" *Journal of the American Institute of Architects* May, 63-65
Summerson J, 1963 *The Classical Language of Architecture* (MIT Press, Cambridge, Mass)
Tagg S K, 1971 "Smallest space analysis. A technique for finding an optimum layout from an association matrix" Department of Psychology, University of Surrey, Guildford (mimeo)
Tagg S K, 1974 "The subjective meaning of rooms: some analyses and investigations" in *Psychology and the Built Environment* Eds D Canter, T Lee (Architectural Press, London) pp 65-70
Thiel P, 1961a "A sequence: experience notation for architectural and urban spaces" *Town Planning Review* **32**(2) April 33-52
Thiel P, 1961b "To the Kamakura Station" *Landscape* **11** (Autumn) 6-9
Thiel P, 1962 "An old garden, a new tool, and our future cities" *Landscape Architecture* July 2-8
Thiel P, Nitschke G, 1968a "Anatomie des gelebten Raumes. Entwicklung einer Raumnotierungstechnik" *Bauen und Wohnen* **9** 313-320; **10** 1-6
Thiel P, Nitschke G, 1968b "Entwicklung einer modernen Darstellungsmethode bewegungs-, zeit- und stimmungsstrukturierten Umwelterlebnisses" *Bauen und Wohnen* **12** 1-12, 15
Triandis H C, Osgood C E, 1958 "A comparative factorial analysis of semantic structures in monolingual Greek and American college students" *Journal of Abnormal and Social Psychology* **57** 187-196
Uexküll J von, 1935 "Der Kampf um den Himmel" *Neue Rundschau* 367-379
Uexküll J von, 1940 "Bedeutungslehre" *Bios (Leipzig)* volume 10
Uexküll J von, 1957 "A stroll through the worlds of animals and men: a picture book of invisible worlds" in *Instinctive Behavior: The Development of a Modern Concept* Ed. C H Schiller (International Universities Press, New York)
Uexküll J von, Kriszat G, 1934/1970 *Streifzüge durch die Umwelten von Tieren und Menschen (und) Bedeutungslehre* (S Fischer, Frankfurt)
Uexküll T von, 1973 "Information als Mitteilung und Formung" *Praxis der Psychotherapie* **18** 137-150
Ullmann S, 1962 *Semantics. An Introduction to the Science of Meaning* (Basil Blackwell, Oxford)
Veblen T, 1963 *The Theory of the Leisure Class* (Mentor Books, New York)
Vitruvius, 1960 *The Ten Books on Architecture* translated by M H Morgan (Dover, New York)
Volli P, 1974a "Pour une utilisation sémiotique de 'ensemble nuançe" summary of a paper delivered at the first congress of the International Association for Semiotic Studies (IASS), Milan, 2-6 June (mimeo)
Volli U, 1974b "Sémiologie architecturale et structure iconique" *Cahier de l'École d'Architecture Université de Genève* **1** 61-67
Volpe G della, 1964 *Critica del Gusto* (Feltrinelli, Milano)
Wallis M, 1973 "Semantic and symbolic elements in architecture: iconology as a first step toward an architectural semiotic" *Semiotica* **7/8** 220-238
Walther E, 1962 "Die Begründung der Zeichentheorie bei Charles Sanders Peirce" *Grundlagenstudien aus Kybernetik und Geisteswissenschaft* **3** 33-44
Walther E, 1969 "Abriß der Semiotik" *Arch+* **2** H8
Walther E, 1974 *Allgemeine Zeichenlehre. Einführung in die Grundlagen der Semiotik* (Deutsche Verlagsanstalt, Stuttgart)
Weinreich U, 1958 "Travels through semantic space" *Word* **14**(2-3) 346-366
Weinreich U, 1959 "A rejoinder" *Word* **15**(1) 200-201

Wells R, 1957 "A mathematical approach to meaning" *Cahiers Francaise de Saussure* **15** 117-136

Wells R, 1958 paper presented at the eighth International Congress of Linguistics, 1957, in *Proceedings of the 8th International Congress of Linguistics* (Oslo University Press, Oslo)

Whorf D L, 1953 "Linguistic factors in the terminology of Hopi architecture" *International Journal of American Linguistics* **19** 141-145

Wiener N, 1954 *The Human Use of Human Beings* (Doubleday, Garden City, NY)

Winkel G H, 1972 "Some human dimensions of urban design" paper presented at the international conference on Aufgaben und Methoden der Stadtgestaltung, University of Stuttgart (mimeo)

Wright L, 1973 "How many bits?" *Architectural Review* number 914, 251-252

Zadeh L A, 1965 "Fuzzy Sets" *Information and Control* **8** 338-353

Zeyer und Drechsler (Bildhauer), ca 1905 *Dekorative Bildhauerarbeiten. Vorbilder für Verzierungen von Fassaden, Innenräumen und kunstgewerblichen Gegenständen. Nach Entwürfen hervorragender Architekten und nach eigenen Kompositionen ausgeführt* (erste und zweite Serie) (Bruno Hessling, Berlin)

Z'Graggen F J, 1974 "Vers une conception scientifique du logement. Production systematique de plans-masses à l'aide de l'ordinateur" diplôme, Département d'Architecture, École Polytechnique Fédérale de Lausanne, Lausanne

Zimmermann D, 1966 *ZAFO. Eine allgemeine Formenordnung für Werkstücke—Gestaltung, Handhabung und Rationalisierungserfolg* (Güntor Grossmann, Stuttgart)

Name index

Subject index

www.ingramcontent.com/pod-product-compliance
Lightning Source LLC
LaVergne TN
LVHW011108300926
849160LV00008B/315